Optimization by
Variational Methods

OPTIMIZATION BY
VARIATIONAL METHODS

MORTON M. DENN

Associate Professor of Chemical Engineering
University of Delaware

McGRAW-HILL BOOK COMPANY

New York   St. Louis   San Francisco   London
Sydney   Toronto   Mexico   Panama

OPTIMIZATION BY VARIATIONAL METHODS

Library of Congress Catalog Card Number 73-75167
16395

123456789 M A M M 7654321069

To my wife and children
for the understanding and support
that permits books to be written

Preface

The development of a systematic theory of optimization since the mid-1950s has not been followed by widespread application to the design and control problems of the process industries. This surprising and disappointing fact is largely a consequence of the absence of effective communication between theorists and process engineers, for the latter typically do not have sufficient mathematical training to look past the sophistication with which optimization theory is usually presented and recognize the practical significance of the results. This book is an attempt to present a logical development of optimization theory at an elementary mathematical level, with applications to simple but typical process design and control situations.

The book follows rather closely a course in optimization which I have taught at the University of Delaware since 1965 to classes made up of graduate students and extension students from local industry, together with some seniors. Approximately half of the students each year have been chemical engineers, with the remainder made up of other types of engineers, statisticians, and computer scientists. The only formal mathematical prerequisites are a familiarity with calculus through Taylor series and a first course in ordinary differential equations, together with the maturity and problem awareness expected of students at this level. In two sections of Chapter 3 and in Chapter 11 a familiarity with partial differential equations is helpful but not essential. With this background it is possible to proceed in one semester from the basic elements of optimization to an introduction to that current research which is of direct process significance. The book deviates from the course in containing more material of an advanced nature than can be reasonably covered in one semester, and in that sense it may be looked upon as comprising in part a research monograph.

Chapter 1 presents the essential features of the variational approach within the limited context of problems in which only one or a finite number of discrete variables is required for optimization. Some of this material is familiar, and a number of interesting control and design problems can be so formulated. Chapter 2 is concerned with the parallel variational development of methods of numerical computation for such problems. The approach of Chapter 1 is then resumed in Chapters 3 to 5, where the scope of physical systems analyzed is gradually expanded to include processes described by several differential equations with magnitude limits on the decisions which can be made for optimization. The optimal control of a flow reactor is a typical situation.

In Chapters 6 and 7 much of the preceding work is reexamined and unified in the context of the construction of Green's functions for linear systems. It is only here that the Pontryagin minimum principle, which dominates modern optimization literature, is first introduced in its complete form and carefully related to the more elementary and classical material which is sufficient for most applications.

Chapter 8 relates the results of optimization theory to problems of design of practical feedback control systems for lumped multivariable processes.

Chapter 9 is an extension of the variational development of principles of numerical computation first considered in Chapter 2 to the more complex situations now being studied.

Chapters 10 and 11 are concerned with both the theoretical development and numerical computation for two extensions of process significance. The former deals with complex structures involving recycle and bypass streams and with periodic operation, while the latter treats distributed-parameter systems. Optimization in periodically operated and distributed-parameter systems represents major pertinent efforts of current research.

Chapter 12 is an introduction to dynamic programming and Hamilton-Jacobi theory, with particular attention to the essential equivalence in most situations between this alternate approach and the variational approach to optimization taken throughout the remainder of the book. Chapter 12 can be studied at any time after the first half of Chapters 6 and 7, as can any of the last five chapters, except that Chapter 9 must precede Chapters 10 and 11.

Problems appear at the end of each chapter. Some supplement the theoretical developments, while others present further applications.

In part for reasons of space and in part to maintain a consistent mathematical level I have omitted any discussion of such advanced topics as the existence of optimal solutions, the Kuhn-Tucker theorem, the control-theory topics of observability and controllability, and optimization under uncertainty. I have deliberately refrained from using matrix notation in the developments as a result of my experience in teaching this material; for I have found that the very conciseness afforded by matrix notation masks the significance of the manipulations being performed for many students, even those with an adequate background in linear algebra. For this reason the analysis in several chapters is limited to two-variable processes, where every term can be conveniently written out.

In preparing this book I have incurred many debts to colleagues and students. None will be discharged by simple acknowledgement, but some must be explicitly mentioned. My teacher, Rutherford Aris, first introduced me to problems in optimization and collaborated in the development of Green's functions as the unifying approach to variational problems. J. R. Ferron, R. D. Gray, Jr., G. E. O'Connor, A. K. Wagle, and particularly J. M. Douglas have been most helpful in furthering my understanding and have permitted me to use the results of our joint efforts. The calculations in Chapter 2 and many of those in Chapter 9 were carried out by D. H. McCoy, those in Section 10.8 by G. E. O'Connor, and Figures 6.1 to 6.4 were kindly supplied by A. W. Pollock. My handwritten manuscript was expertly typed by Mrs. Frances Phillips. For permission to use copyrighted material I am grateful to my several coauthors and to the following authors and publishers:

The American Chemical Society for Figures 4.2, 5.7, 5.8, 9.1, 9.2, 9.7 to 9.13, 10.5 to 10.15, 11.8 to 11.12, which appeared in *Industrial and Engineering Chemistry Monthly* and *Fundamentals Quarterly*.

Taylor and Francis, Ltd., for Figures 11.1 to 11.7 and Sections 11.2 to 11.7, a paraphrase of material which appeared in *International Journal of Control*.

R. Aris, J. M. Douglas, E. S. Lee, and Pergamon Press for Figures 5.9, 5.15, 9.3, and Table 9.8, which appeared in *Chemical Engineering Science*.

D. D. Perlmutter and the American Institute of Chemical Engineers for
    Figures 8.1 and 8.2, which appeared in *AIChE Journal*.

Several of my colleagues at the University of Delaware have shaped
my thinking over the years about both optimization and pedagogy, and
it is my hope that their contributions to this book will be obvious at least
to them.   J. M. Douglas and D. D. Perlmutter have kindly read the
entire manuscript and made numerous helpful suggestions for improvement.   For the decision not to follow many other suggestions from
students and colleagues and for the overall style, accuracy, and selection
of material, I must, of course, bear the final responsibility.

Morton M. Denn

Contents

Optimization by
Variational Methods

Introduction

OPTIMIZATION AND ENGINEERING PRACTICE

The optimal design and control of systems and industrial processes has long been of concern to the applied scientist and engineer, and, indeed, might be taken as a definition of the function and goal of engineering. The practical attainment of an optimum design is generally a consequence of a combination of mathematical analysis, empirical information, and the subjective experience of the scientist and engineer.   In the chapters to follow we shall examine in detail the principles which underlie the formulation and resolution of the practical problems of the analysis and specification of optimal process units and control systems.   Some of these results lend themselves to immediate application, while others provide helpful insight into the considerations which must enter into the specification and operation of a working system.

The formulation of a process or control system design is a trial-and-error procedure, in which estimates are made first and then information is sought from the system to determine improvements.   When a sufficient mathematical characterization of the system is available, the effect of changes about a preliminary design may be obtained analytically, for the perturbation techniques so common in modern applied mathematics and engineering analysis have their foundation in linear analysis, despite the nonlinearity of the system being analyzed. Whether mathematical or experimental or a judicious combination of both, perturbation analysis lies at the heart of modern engineering practice.

Mathematical optimization techniques have as their goal the development of rigorous procedures for the attainment of an optimum in a system which can be characterized mathematically. The mathematical characterization may be partial or complete, approximate or exact, empirical or theoretical. Similarly, the resulting optimum may be a final implementable design or a guide to practical design and a criterion by which practical designs are to be judged. In either case, the optimization techniques should serve as an important part of the total effort in the design of the units, structure, and control of a practical system.

Several approaches can be taken to the development of mathematical methods of optimization, all of which lead to essentially equivalent results. We shall adopt here the variational method, for since it is grounded in the analysis of small perturbations, it is the procedure which lies closest to the usual engineering experience. The general approach is one of assuming that a preliminary specification has been made and then enquiring about the effect of small changes. If the specification is in fact the optimum, any change must result in poorer performance and the precise mathematical statement of this fact leads to *necessary conditions*, or equations which define the optimum. Similarly, the analysis of the effect of small perturbations about a nonoptimal specification leads to computational procedures which produce a better specification. Thus, unlike most approaches to optimization, the variational method leads rather simply to both necessary conditions and computational algorithms by an identical approach and, furthermore, provides a logical framework for studying optimization in new classes of systems.

BIBLIOGRAPHICAL NOTES

An outstanding treatment of the logic of engineering design may be found in

D. F. Rudd and C. C. Watson: "Strategy of Process Engineering," John Wiley & Sons, Inc., New York, 1968

Mathematical simulation and the formulation of system models is discussed in

A. E. Rogers and T. W. Connolly: "Analog Computation in Engineering Design," McGraw-Hill Book Company, New York, 1960

R. G. E. Franks: "Mathematical Modeling in Chemical Engineering," John Wiley & Sons, Inc., New York, 1967

Perturbation methods for nonlinear systems are treated in such books as

W. F. Ames: "Nonlinear Ordinary Differential Equations in Transport Processes," Academic Press, Inc., New York, 1968
————: "Nonlinear Partial Differential Equations in Engineering," Academic Press, Inc., New York, 1965
R. E. Bellman: "Perturbation Techniques in Mathematics, Physics, and Engineering," Holt, Rinehart and Winston, Inc., New York, 1964
W. J. Cunningham: "Introduction to Nonlinear Analysis," McGraw-Hill Book Company, New York, 1958
N. Minorsky: "Nonlinear Oscillations," D. Van Nostrand Company, Inc., Princeton, N.J., 1962

Perhaps the most pertinent perturbation method from a system analysis viewpoint, the Newton-Raphson method, which we shall consider in detail in Chaps. 2 and 9, is discussed in

R. E. Bellman and R. E. Kalaba: "Quasilinearization and Nonlinear Boundary Value Problems," American Elsevier Publishing Company, New York, 1965

1
Optimization with Differential Calculus

1.1 INTRODUCTION

A large number of interesting optimization problems can be formulated in such a way that they can be solved by application of differential calculus, and for this reason alone it is well to begin a book on optimization with an examination of the usefulness of this familiar tool. We have a further motivation, however, in that all variational methods may be considered to be straightforward extensions of the methods of differential calculus. Thus, this first chapter provides the groundwork and basic principles for the entire book.

1.2 THE SIMPLEST PROBLEM

The simplest optimization problem which can be treated by calculus is the following: $\mathcal{E}(x_1, x_2, \ldots, x_n)$ is a function of the n variables $x_1, x_2, \ldots, x_n$. Find the particular values $\bar{x}_1, \bar{x}_2, \ldots, \bar{x}_n$ which cause the function $\mathcal{E}$ to take on its minimum value.

We shall solve this problem in several ways. Let us note first that the minimum has the property that

$$\mathcal{E}(x_1, x_2, \ldots, x_n) - \mathcal{E}(\bar{x}_1, \bar{x}_2, \ldots, \bar{x}_n) \geq 0 \tag{1}$$

Suppose that we let $x_1 = \bar{x}_1 + \delta x_1$, where δx_1 is a small number in absolute value, while $x_2 = \bar{x}_2$, $x_3 = \bar{x}_3$, $\ldots$, $x_n = \bar{x}_n$. If we divide Eq. (1) by δx_1, we obtain, depending upon the algebraic sign of δx_1,

$$\frac{\mathcal{E}(\bar{x}_1 + \delta x_1, \bar{x}_2, \ldots, \bar{x}_n) - \mathcal{E}(\bar{x}_1, \bar{x}_2, \ldots, \bar{x}_n)}{\delta x_1} \geq 0 \qquad \delta x_1 > 0 \tag{2a}$$

or

$$\frac{\mathcal{E}(\bar{x}_1 + \delta x_1, \bar{x}_2, \ldots, \bar{x}_n) - \mathcal{E}(\bar{x}_1, \bar{x}_2, \ldots, \bar{x}_n)}{\delta x_1} \leq 0 \qquad \delta x_1 < 0 \tag{2b}$$

The limit of the left-hand side as $\delta x_1 \to 0$ is simply $\partial \mathcal{E} / \partial x_1$, evaluated at $\bar{x}_1, \bar{x}_2, \ldots, \bar{x}_n$. From the inequality (2a) this partial derivative is nonnegative, while from (2b) it is nonpositive, and both inequalities are satisfied only if $\partial \mathcal{E} / \partial x_1$ vanishes. In an identical way we find for all x_k, $k = 1, 2, \ldots, n$,

$$\frac{\partial \mathcal{E}}{\partial x_k} = 0 \tag{3}$$

at the minimizing values $\bar{x}_1, \bar{x}_2, \ldots, \bar{x}_n$. We thus have n algebraic equations to solve for the n unknowns, $\bar{x}_1, \bar{x}_2, \ldots, \bar{x}_n$.

It is instructive to examine the problem somewhat more carefully to search for potential difficulties. We have, for example, made a rather strong assumption in passing from inequalities (2a) and (2b) to Eq. (3), namely, that the partial derivative in Eq. (3) exists at the values x_1, $x_2, \ldots, x_n$. Consider, for example, the function

$$\mathcal{E}(x_1, x_2, \ldots, x_n) = |x_1| + |x_2| + \cdots + |x_n| \tag{4}$$

which has a minimum at $x_1 = x_2 = \cdots = x_n = 0$. Inequalities (1) and (2) are satisfied, but the partial derivatives in Eq. (3) are not defined at $x_k = 0$. If we assume that one-sided derivatives of the function $\mathcal{E}$ exist everywhere, we must modify condition (3) to

$$\lim_{x_k \to \bar{x}_k^-} \frac{\partial \mathcal{E}}{\partial x_k} \leq 0 \tag{5a}$$

$$\lim_{x_k \to \bar{x}_k^+} \frac{\partial \mathcal{E}}{\partial x_k} \geq 0 \tag{5b}$$

with Eq. (3) implied if the derivative is continuous.

In problems which describe real situations, we shall often find that physical or economic interpretations restrict the range of variables we may consider. The function

$$\mathcal{E} = \tfrac{1}{2}(x + 1)^2 \tag{6}$$

for example, has a minimum at $x = -1$, where Eq. (3) is satisfied, but if the variable x is restricted to nonnegative values (an absolute temperature, or a cost, for example), the minimum occurs at the *boundary* $x = 0$. In writing the inequalities (2a) and (2b) we have assumed that we were free to make δx_1 positive or negative as we wished. But if $\bar{x}_1$ lies at a lower bound of its allowable region, we cannot decrease x_1 any more, so that inequality (2b) is inadmissable, and similarly, if $\bar{x}_1$ lies at an upper bound, inequality (2a) is inadmissable. We conclude, then, that $\bar{x}_k$ either satisfies Eq. (3) [or inequalities (5a) and (5b)] or lies at a lower bound where $\partial\mathcal{E}/\partial x_k \geq 0$, or at an upper bound where $\partial\mathcal{E}/\partial x_k \leq 0$; that is,

$$\frac{\partial\mathcal{E}}{\partial x_k} \begin{cases} \leq 0 & x_k \text{ at upper bound} \\ = 0 & x_k \text{ between bounds [or (5a) and (5b)]} \\ \geq 0 & x_k \text{ at lower bound} \end{cases} \tag{7}$$

Thus, to minimize Eq. (6) subject to $0 \leq x < \infty$ we find $\partial\mathcal{E}/\partial x = 1 > 0$ at $x = 0$.

Several more restrictions need to be noted. First, if we were seeking to *maximize* $\mathcal{E}$, we could equally well minimize $-\mathcal{E}$. But from Eq. (3),

$$\frac{\partial(-\mathcal{E})}{\partial x_k} = -\frac{\partial\mathcal{E}}{\partial x_k} = 0 = \frac{\partial\mathcal{E}}{\partial x_k}$$

which is identical to the condition for a minimum, so that if $\bar{x}_k$ is an interior point where the derivative is continuous, we have not yet found a way to distinguish between a maximizing and minimizing point. Indeed, a point at which Eq. (3) is satisfied need be neither a maximum nor a minimum; consider $\mathcal{E} = (x_1)^2 - (x_2)^2$. Furthermore, while inequalities (1) and (2) are true for all $x_1, x_2, \ldots, x_n$ from the meaning of a minimum, and hence define a *global* minimum, when we allow δx_1 to become small and obtain Eqs. (3), (5), and (7), we obtain conditions which are true for any *local* minimum and which may have many solutions, only one of which leads to the true minimum value of $\mathcal{E}$.

These last points may be seen by considering the function

$$\mathcal{E}(x) = \tfrac{1}{3}x^3 - x + \tfrac{2}{3} \tag{8}$$

shown in Fig. 1.1. If we calculate $d\mathcal{E}/dx = 0$, we obtain

$$\frac{d\mathcal{E}}{dx} = x^2 - 1 = 0 \tag{9}$$

or

$$\bar{x} = \pm 1 \tag{10}$$

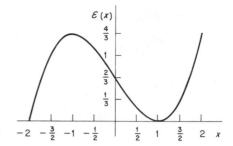

Fig. 1.1 A function with a local minimum and maximum: $\mathcal{E}(x) = \frac{1}{3}x^2 - x + \frac{2}{3}$.

and we see by inspection of the figure that $\mathcal{E}$ is a minimum at $x = +1$ and a maximum at $x = -1$, with $\mathcal{E}(1) = 0$. But for $x < -2$ we have $\mathcal{E}(x) < 0$, so that unless we restrict x to $x \geq -2$, the point $\bar{x} = 1$ is only a local minimum. If, for example, we allow $-3 \leq x < \infty$, then at the boundary $x = -3$ we have

$$\frac{d\mathcal{E}}{dx}\bigg|_{x=-3} = (-3)^2 - 1 = 8 > 0 \tag{11}$$

and so the point $\bar{x} = -3$ also satisfies Eq. (7) and is in fact the global minimum, since

$$\mathcal{E}(-3) = -\tfrac{16}{3} < \mathcal{E}(+1) = 0 \tag{12}$$

1.3 A VARIATIONAL DERIVATION

In order to motivate the development in later chapters, as well as to obtain a means of distinguishing between maxima and minima, we shall consider an alternative treatment of the problem of the previous section. For simplicity we restrict attention to the case $n = 2$, but it will be obvious that the method is general. We seek, then, the minimum of the function $\mathcal{E}(x_1, x_2)$, and for further simplicity we shall assume that the function has continuous first partial derivatives and that the minimum occurs at interior values of x_1, x_2.

If $\mathcal{E}$ has bounded second partial derivatives, we may use a two-dimensional Taylor series to express $\mathcal{E}(\bar{x}_1 + \delta x_1, \bar{x}_2 + \delta x_2)$ in terms of $\mathcal{E}(\bar{x}_1, \bar{x}_2)$, where δx_1, δx_2 are any small changes (variations) in x_1, x_2 which are consistant with the problem formulation:

$$\mathcal{E}(\bar{x}_1 + \delta x_1, \bar{x}_2 + \delta x_2) - \mathcal{E}(\bar{x}_1, \bar{x}_2) = \frac{\partial \mathcal{E}}{\partial x_1} \delta x_1 + \frac{\partial \mathcal{E}}{\partial x_2} \delta x_2$$
$$+ o(\max |\delta x_1|, |\delta x_2|) \tag{1}$$

where the partial derivatives are evaluated at $x_1 = \bar{x}_1$, $x_2 = \bar{x}_2$ and the notation $o(\epsilon)$ refers to a term which goes to zero faster than ϵ; that is,

$$\lim_{\epsilon \to 0} \frac{o(\epsilon)}{\epsilon} = 0 \tag{2}$$

In Eq. (1),

$$o(\max |\delta x_1|, |\delta x_2|) = \frac{1}{2} \left[\frac{\partial^2 \mathcal{E}}{\partial x_1^2} (\delta x_1)^2 + 2 \frac{\partial^2 \mathcal{E}}{\partial x_1 \partial x_2} \delta x_1 \, \delta x_2 \right.$$
$$\left. + \frac{\partial^2 \mathcal{E}}{\partial x_2^2} (\delta x_2)^2 \right] \tag{3}$$

where the second partial derivatives are evaluated at some point

$$[\rho \bar{x}_1 + (1 - \rho)(\bar{x}_1 + \delta x_1), \rho \bar{x}_2 + (1 - \rho)(\bar{x}_2 + \delta x_2)] \qquad 0 \leq \rho \leq 1$$

If $\bar{x}_1$, $\bar{x}_2$ are the minimizing values then, from Eq. (1) and Eq. (1) of the preceding section,

$$\frac{\partial \mathcal{E}}{\partial x_1} \delta x_1 + \frac{\partial \mathcal{E}}{\partial x_2} \delta x_2 + o(\max |\delta x_1|, |\delta x_2|) \geq 0 \tag{4}$$

Now δx_1 and δx_2 are arbitrary, and Eq. (4) must hold for any variations we choose. A particular pair which suits our needs is

$$\delta x_1 = -\epsilon \frac{\partial \mathcal{E}}{\partial x_1} \qquad \delta x_2 = -\epsilon \frac{\partial \mathcal{E}}{\partial x_2} \tag{5}$$

where ϵ is a small positive constant and the partial derivatives are again evaluated at $\bar{x}_1$, $\bar{x}_2$. If these partial derivatives both vanish, inequality (4) is satisfied trivially; if not, we may write

$$-\epsilon \left[\left(\frac{\partial \mathcal{E}}{\partial x_1} \right)^2 + \left(\frac{\partial \mathcal{E}}{\partial x_2} \right)^2 \right] + o(\epsilon) \geq 0 \tag{6}$$

or, dividing by ϵ and taking the limit as $\epsilon \to 0$, with Eq. (2),

$$\left(\frac{\partial \mathcal{E}}{\partial x_1} \right)^2 + \left(\frac{\partial \mathcal{E}}{\partial x_2} \right)^2 \leq 0 \tag{7}$$

But a sum of squares can be nonpositive if and only if each term in the sum vanishes identically; thus, if $\bar{x}_1$ and $\bar{x}_2$ are the minimizing values, it is necessary that

$$\frac{\partial \mathcal{E}}{\partial x_1} = \frac{\partial \mathcal{E}}{\partial x_2} = 0 \tag{8}$$

We may now obtain a further condition which will allow us to distinguish between maxima and minima by assuming that $\mathcal{E}$ is three times differentiable and carrying out the Taylor series an additional term:

$$\mathcal{E}(\bar{x}_1 + \delta x_1, \bar{x}_2 + \delta x_2) - \mathcal{E}(\bar{x}_1, \bar{x}_2) = \frac{\partial \mathcal{E}}{\partial x_1} \delta x_1 + \frac{\partial \mathcal{E}}{\partial x_2} \delta x_2$$

$$+ \frac{1}{2} \left[\frac{\partial^2 \mathcal{E}}{\partial x_1{}^2} (\delta x_1)^2 + 2 \frac{\partial^2 \mathcal{E}}{\partial x_1 \partial x_2} \delta x_1 \delta x_2 + \frac{\partial^2 \mathcal{E}}{\partial x_2{}^2} (\delta x_2)^2 \right]$$

$$+ o(\max |\delta x_1|^2, |\delta x_2|^2) \geq 0 \quad (9)$$

The terms with arrows through them vanish by virtue of Eq. (8). If we now set

$$\delta x_1 = \epsilon \alpha_1 \qquad \delta x_2 = \epsilon \alpha_2 \qquad (10)$$

where ϵ is an infinitesimal positive number and α_1, α_2 are arbitrary finite numbers of any algebraic sign, Eq. (9) may be written, after dividing by ϵ^2,

$$\frac{1}{2} \sum_{i=1}^{2} \sum_{j=1}^{2} \alpha_i \alpha_j \frac{\partial^2 \mathcal{E}}{\partial x_i \partial x_j} + \frac{o(\epsilon^2)}{\epsilon^2} > 0 \quad (11)$$

and letting $\epsilon \to 0$, we find that for *arbitrary* α_i, α_j it is necessary that

$$\sum_{i=1}^{2} \sum_{j=1}^{2} \alpha_i \alpha_j \frac{\partial^2 \mathcal{E}}{\partial x_i \partial x_j} \geq 0 \quad (12)$$

The collection of terms

$$\begin{bmatrix} \dfrac{\partial^2 \mathcal{E}}{\partial x_1 \partial x_1} & \dfrac{\partial^2 \mathcal{E}}{\partial x_1 \partial x_2} \\ \dfrac{\partial^2 \mathcal{E}}{\partial x_1 \partial x_2} & \dfrac{\partial^2 \mathcal{E}}{\partial x_2 \partial x_2} \end{bmatrix}$$

is called the *hessian matrix* of $\mathcal{E}$, and a matrix satisfying inequality (12) for arbitrary α_1, α_2 is said to be *positive semidefinite*. The matrix is said to be *positive definite* if equality holds only for α_1, α_2 both equal to zero.

The positive semidefiniteness of the hessian may be expressed in a more convenient form by choosing particular values of α_1 and α_2. For example, if $\alpha_1 = 1$, $\alpha_2 = 0$, Eq. (12) becomes

$$\frac{\partial^2 \mathcal{E}}{\partial x_1 \partial x_1} \geq 0 \quad (13)$$

and similarly for $\partial^2 \mathcal{E} / (\partial x_2 \partial x_2)$. On the other hand, if $\alpha_1 = -\partial^2 \mathcal{E} / (\partial x_1 \partial x_2)$, $\alpha_2 = \partial^2 \mathcal{E} / (\partial x_1 \partial x_1)$, then Eq. (12) becomes the more familiar result

$$\frac{\partial^2 \mathcal{E}}{\partial x_1{}^2} \frac{\partial^2 \mathcal{E}}{\partial x_2{}^2} - \left(\frac{\partial^2 \mathcal{E}}{\partial x_1 \partial x_2} \right)^2 \geq 0 \quad (14)$$

Thus we have proved that, subject to the differentiability assumptions which we have made, *if the minimum of* $\mathcal{E}(x_1,x_2)$ *occurs at the interior point* $\bar{x}_1$, $\bar{x}_2$, *it is necessary that the first and second partial derivatives of* $\mathcal{E}$ *satisfy conditions* (8), (13), *and* (14). It is inequality (13) which distinguishes a minimum from a maximum and (14) which distinguishes both from a *saddle*, at which neither a minimum nor maximum occurs despite the vanishing of the first partial derivatives.

Let us note finally that if Eq. (8) is satisfied and the hessian is positive definite (not semidefinite) at $\bar{x}_1$, $\bar{x}_2$, then for sufficiently small $|\delta x_1|$, $|\delta x_2|$ the inequality in Eq. (9) will always be satisfied. Thus, the vanishing of the first partial derivatives and the positive definiteness of the hessian are sufficient to ensure a (local) minimum at $\bar{x}_1$, $\bar{x}_2$.

1.4 AN OPTIMAL-CONTROL PROBLEM: FORMULATION

The conditions derived in the previous two sections may be applied to an elementary problem in optimal control. Let us suppose that we have a dynamic system which is described by the differential equation

$$\dot{y} = \frac{dy}{dt} = F(y,u) \tag{1}$$

where u is a control variable and y is the state of the system (y might be the temperature in a mixing tank, for example, and u the flow rate of coolant in a heat exchanger). The system is designed to operate at the steady-state value y_{ss}, and the steady-state control setting u_{ss} is found by solving the algebraic equation

$$F(y_{ss},u_{ss}) = 0 \tag{2}$$

If the system is disturbed slightly from the steady state at time $t = 0$, we wish to guide it back in an optimal fashion. There are many criteria which we might use to define the optimal fashion, but let us suppose that it is imperative to keep the system "close" to y_{ss}. If we determine the deviation from steady state $y(t) - y_{ss}$ at each time, then a quantity which will be useful for defining both positive and negative deviations is the square, $(y - y_{ss})^2$, and a reasonable measure of the total deviation will be $\int_0^\theta (y - y_{ss})^2\, dt$, where θ is the total control time. On the other hand, let us suppose that we wish to hold back the control effort, which we may measure by $\int_0^\theta (u - u_{ss})^2\, dt$. If ρ^2 represents the relative values of control effort and deviation from steady state, then our control problem is as follows.

Given an initial deviation y_0 and a system defined by Eq. (1), find the control settings $u(t)$ which minimize

$$\mathcal{E} = \frac{1}{2} \int_0^\theta [(y - y_{ss})^2 + \rho^2 (u - u_{ss})^2] \, dt \tag{3}$$

For purposes of implementation it is usually helpful to have the control function u not in terms of t but of the state y. [This is called *feedback control*.]

In later chapters we shall develop the necessary mathematics for the solution of the control problem just stated. At this point we must make a further, but quite common, simplification. [We assume that the deviations in y and u are sufficiently small to permit expansion of the function $F(y,u)$ in a Taylor series about y_{ss}, u_{ss} retaining only linear terms.] Thus we write

$$\dot{y} \approx F(y_{ss}, u_{ss}) + \frac{\partial F}{\partial y}\bigg|_{\substack{y = y_{ss} \\ u = u_{ss}}} (y - y_{ss}) + \frac{\partial F}{\partial u}\bigg|_{\substack{y = y_{ss} \\ u = u_{ss}}} (u - u_{ss}) \tag{4}$$

Letting

$$x = y - y_{ss} \qquad w = \frac{\partial F}{\partial u}(u - u_{ss})$$

and noting that $\dot{x} = \dot{y}$ and $F(y_{ss}, u_{ss}) = 0$, we have, finally,

$$\dot{x} = Ax + w \qquad x(0) = x_0 \tag{5}$$

$$\min \mathcal{E} = \frac{1}{2} \int_0^\theta (x^2 + c^2 w^2) \, dt \tag{6}$$

It is sometimes convenient to consider control policies which are constant for some interval; that is,

$$w(t) = w_n = \text{const} \qquad (n - 1)\Delta \leq t < n\Delta \tag{7}$$

Equation (5) then has the solution

$$x(n\Delta) = x[(n - 1)\Delta]e^{A\Delta} - \frac{w_n(1 - e^{A\Delta})}{A} \tag{8}$$

and Eq. (6) may be approximated by

$$\min \mathcal{E} = \frac{1}{2}\Delta \sum_{n=1}^{N} [x^2(n\Delta) + c^2 w_n^2] \tag{9}$$

Letting

$$x_n = x(n\Delta) \qquad u_n = -\frac{w_n(1 - e^{A\Delta})}{A}$$

$$C^2 = \frac{c^2 A^2}{(1 - e^{A\Delta})^2} \qquad a = e^{A\Delta} \tag{10}$$

we obtain the equivalent control problem

$$x_n = ax_{n-1} + u_n \qquad x_0 \text{ given} \tag{11}$$

$$\min \mathcal{E} = \frac{1}{2} \sum_{n=1}^{N} (x_n{}^2 + C^2 u_n{}^2) \tag{12}$$

where, since Δ is a constant, we have not included it in the summation for $\mathcal{E}$.

1.5 OPTIMAL PROPORTIONAL CONTROL

The most elementary type of control action possible for the system defined by Eqs. (5) and (6) of the preceding section is *proportional* control, in which the amount of control effort is proportional to the deviation, a large deviation leading to large control, a small deviation to little. In that case we set

$$w(t) = Mx(t) \tag{1}$$
$$\dot{x} = (A + M)x \qquad x(0) = x_0 \tag{2}$$

and we seek the optimal control setting M:

$$\min \mathcal{E}(M) = \tfrac{1}{2}(1 + c^2 M^2) \int_0^\theta x^2 \, dt \tag{3}$$

The solution to Eq. (2) is

$$x(t) = x_0 e^{(A+M)t} \tag{4}$$

and Eq. (3) becomes

$$\min \mathcal{E}(M) = \frac{x_0{}^2(1 + c^2 M^2)}{2} \int_0^\theta e^{2(A+M)t} \, dt \tag{5}$$

or

$$\min \mathcal{E}(M) = \frac{x_0{}^2}{4} \frac{(1 + M^2 c^2)}{A + M} (e^{2(A+M)\theta} - 1) \tag{6}$$

Since x cannot be allowed to grow significantly larger because of the control action, it is clear that we require for stability that

$$A + M \leq 0 \tag{7}$$

This is the concept of *negative feedback*. The optimal control setting is then seen from Eq. (6) to depend on A, c, and θ but not on x_0, which is simply a multiplicative constant. Since we shall usually be concerned with controlling for long times compared to the system *time constant* $|A + M|^{-1}$, we may let θ approach infinity and consider only the problem

$$\min \mathcal{E}(M) = -\frac{1 + M^2 c^2}{A + M} \tag{8}$$

We obtain the minimum by setting the derivative equal to zero

$$\frac{d\varepsilon}{dM} = -\frac{2Mc^2}{A+M} + \frac{1+M^2c^2}{(A+M)^2} = 0 \tag{9}$$

or

$$M = -A \pm \sqrt{\frac{1+A^2c^2}{c^2}} \tag{10}$$

Condition (7) requires the negative sign, but it will also follow from considering the second derivative.

For a minimum we require $d^2\varepsilon/dM^2 \geq 0$. This reduces to

$$-2\frac{1+A^2c^2}{(A+M)^3} \geq 0 \tag{11}$$

which yields a minimum for the stable case $A + M \leq 0$, or

$$w = -\left(A + \sqrt{\frac{1+A^2c^2}{c^2}}\right) x \tag{12}$$

1.6 DISCRETE OPTIMAL PROPORTIONAL CONTROL

We shall now consider the same problem for the discrete analog described by Eqs. (11) and (12) of Sec. 1.4,

$$x_n = ax_{n-1} + u_n \tag{1}$$

$$\min \varepsilon = \frac{1}{2} \sum_{n=1}^{N} (x_n^2 + C^2 u_n^2) \tag{2}$$

We again seek the best control setting which is proportional to the state at the beginning of the control interval

$$u_n = mx_{n-1} \tag{3}$$

so that Eq. (1) becomes

$$x_n = (a + m)x_{n-1} \tag{4}$$

which has a solution

$$x_n = x_0(a + m)^n \tag{5}$$

It is clear then that a requirement for stability is

$$|a + m| \leq 1 \tag{6}$$

For simplicity in the later mathematics it is convenient to substitute Eq. (4) into (3) and write

$$u_n = \frac{m}{a+m} x_n \equiv Mx_n \tag{7}$$

and

$$x_n = \frac{a}{1 - M} x_{n-1} = \left(\frac{a}{1 - M}\right)^n x_0 \tag{8}$$

From Eq. (2) we then seek the value of M which will minimize

$$\mathcal{E}(M) = \frac{1}{2} x_0^2 (1 + M^2 C^2) \sum_{n=1}^{N} \left(\frac{a}{1 - M}\right)^{2n} \tag{9}$$

or, using the equation for the sum of a geometric series,

$$\mathcal{E}(M) = \frac{1}{2} x_0^2 \frac{a^2 (1 + M^2 C^2)}{(1 - M)^2 - a^2} \left[1 - \left(\frac{a}{1 - M}\right)^{2N}\right] \tag{10}$$

As in the previous section, we shall assume that the operation is sufficiently long to allow $N \to \infty$, which reduces the problem to

$$\min \mathcal{E}(M) = \frac{1}{2} \frac{1 + M^2 C^2}{(1 - M)^2 - a^2} \tag{11}$$

Setting $d\mathcal{E}/dM$ to zero leads to the equation

$$M^2 + \left(a^2 - 1 + \frac{1}{C^2}\right) M - \frac{1}{C^2} = 0 \tag{12}$$

or

$$M = \frac{1}{2C^2} [C^2 (1 - a^2) - 1 \pm \{[C^2 (1 - a^2) - 1]^2 + 4C^2\}^{1/2}] \tag{13}$$

The positivity of the second derivative implies

$$M \leq \frac{C^2 (1 - a^2) - 1}{2C^2} \tag{14}$$

so that the negative sign should be used in Eq. (13) and M is negative. The stability condition, Eq. (6), is equivalent to

$$1 - M \geq |a| \tag{15}$$

From Eq. (14),

$$1 - M \geq \frac{C^2 + 1}{2C^2} + \frac{1}{2} a^2 \geq \frac{1}{2}(1 + a^2) \geq |a| \tag{16}$$

where the last inequality follows from

$$a^2 - 2|a| + 1 = (|a| - 1)^2 \geq 0 \tag{17}$$

Thus, as in the previous section, the minimizing control is stable.

In Sec. 1.4 we obtained the discrete equations by setting

$$a = e^{A\Delta} \qquad u_n = -w_n \frac{1 - e^{A\Delta}}{A}$$
$$C^2 = \frac{c^2 A^2}{1 - e^{A\Delta}} \tag{18}$$

If we let Δ become infinitesimal, we may write

$$a \approx 1 + A\Delta \qquad u_n \approx w_n \Delta \qquad C^2 \approx \frac{c^2}{\Delta^2} \tag{19}$$

Equation (12) then becomes

$$M^2 + 2AM\Delta - \frac{\Delta^2}{c^2} \approx 0 \tag{20}$$

or

$$M = -\left(A + \sqrt{\frac{1 + A^2 c^2}{c^2}}\right)\Delta \tag{21}$$

and

$$w_n = -\left(A + \sqrt{\frac{1 + A^2 c^2}{c^2}}\right)x_n \tag{22}$$

which is identical to Eq. (12) of Sec. 1.5.

1.7 DISCRETE OPTIMAL CONTROL

We now consider the same discrete system

$$x_n = ax_{n-1} + u_n \tag{1}$$

but we shall not assume that the form of the controller is specified. Thus, rather than seeking a single proportional control setting we wish to find the *sequence* of settings $u_1, u_2, \ldots, u_N$ that will minimize

$$\mathcal{E} = \frac{1}{2}\sum_{n=1}^{N}(x_n^2 + C^2 u_n^2) \tag{2}$$

If we know all the values of x_n, $n = 1, 2, \ldots, N$, we can calculate u_n from Eq. (1), and so we may equivalently seek

$$\min \mathcal{E}(x_1, x_2, \ldots, x_N) = \frac{1}{2}\sum_{n=1}^{N}[x_n^2 + C^2(x_n - ax_{n-1})^2] \tag{3}$$

We shall carry out the minimization by setting the partial derivatives of $\mathcal{E}$ with respect to the x_n equal to zero.

Differentiating first with respect to x_N, we obtain

$$\frac{\partial \mathcal{E}}{\partial x_N} = x_N + C^2(x_N - ax_{N-1}) = x_N + C^2 u_N = 0 \tag{4}$$

or

$$u_N = -\frac{1}{C^2} x_N \tag{5}$$

Next,

$$\frac{\partial \mathcal{E}}{\partial x_{N-1}} = x_{N-1} + C^2(x_{N-1} - ax_{N-2}) - aC^2(x_N - ax_{N-1})$$
$$= x_{N-1} + C^2 u_{N-1} - aC^2 u_N = 0 \tag{6}$$

and combining Eqs. (1), (5), and (6),

$$\frac{\partial \mathcal{E}}{\partial x_{N-1}} = x_{N-1} + C^2 u_{N-1} + \frac{a^2 C^2}{1 + C^2} x_{N-1} = 0 \tag{7}$$

or

$$u_{N-1} = -\frac{1}{C^2} \frac{1 + C^2 + a^2 C^2}{1 + C^2} x_{N-1} \tag{8}$$

In general,

$$x_n + C^2(x_n - ax_{n-1}) - aC^2(x_{n+1} - ax_n) = 0$$
$$n = 1, 2, \ldots, N - 1 \tag{9}$$

and the procedure of Eqs. (4) to (8) may be applied in order from $n = N$ to $n = 1$ to find the optimal control settings $u_1, u_2, \ldots, u_N$.

We note from Eq. (9) that the optimal control will always be a proportional controller of the form

$$u_n = M_n x_n = \frac{aM_n}{1 - M_n} x_{n-1} \tag{10}$$

with, from Eq. (5),

$$M_N = -\frac{1}{C^2} \tag{11}$$

If we substitute into Eq. (9), we obtain

$$x_n + C^2 M_n x_n - a^2 C^2 \frac{M_{n+1}}{1 - M_{n+1}} x_n = 0 \tag{12}$$

or, if $x_n \neq 0$,

$$M_n M_{n+1} + \frac{1 + a^2 C^2}{C^2} M_{n+1} - M_n - \frac{1}{C^2} = 0 \tag{13}$$

with Eq. (11) equivalent to the condition

$$M_{N+1} = 0 \tag{14}$$

Equation (13) is a *Riccati* difference equation, and the unique solution can be obtained by applying Eq. (13) recursively from $n = N$ to $n = 1$. An analytical solution can be found by the standard technique of introducing a new variable

$$\frac{1}{M_n - (1/2C^2)[1 + a^2C^2 - C^2 \pm \sqrt{(1 + a^2C^2 - C^2)^2 + 4C^2}]}$$

where either the positive or negative sign may be used, and solving the resulting linear difference equation (see Appendix 1.1 at the end of the chapter). The solution satisfying Eq. (14) may be put in the form

$$M_n = 1 - 2a^2C^2 \frac{[1 + C^2 + a^2C^2 - \sqrt{(1 + a^2C^2 - C^2)^2 + 4C^2}]^{n-1}}{[1 + C^2 + a^2C^2 - \sqrt{(1 + a^2C^2 - C^2)^2 + 4C^2}]^n}$$
$$\frac{+ \alpha[1 + C^2 + a^2C^2 + \sqrt{(1 + a^2C^2 - C^2)^2 + 4C^2}]^{n-1}}{+ \alpha[1 + C^2 + a^2C^2 + \sqrt{(1 + a^2C^2 - C^2)^2 + 4C^2}]^n} \tag{15}$$

where

$$\alpha = \left[\frac{1 + C^2 + a^2C^2 - \sqrt{(1 + a^2C^2 - C^2)^2 + 4C^2}}{1 + C^2 + a^2C^2 + \sqrt{(1 + a^2C^2 - C^2)^2 + 4C^2}} \right]^N$$
$$\frac{1 + C^2 - a^2C^2 - \sqrt{(1 + a^2C^2 - C^2)^2 + 4C^2}}{1 + C^2 - aC^2 + \sqrt{(1 + a^2C^2 - C^2)^2 + 4C^2}} \tag{16}$$

If we take the control time to be infinite by letting $N \to \infty$, we shall obtain $\lim x_N \to 0$, so that Eq. (4) will be automatically satisfied without the need of the boundary condition (11). We find that

$$\lim_{N \to \infty} \alpha = 0$$

and M_n becomes a constant,

$$M_n = -\frac{1}{2} \left[a^2 - 1 + \frac{1}{C^2} + \sqrt{\left(a^2 - 1 + \frac{1}{C^2} \right)^2 + \frac{4}{C^2}} \right] \tag{17}$$

which is the optimal proportional control found in Sec. 1.6.

The result of this section, that *the optimal control for the linear system described by Eq. (1) and quadratic objective criterion by Eq. (2) is a proportional feedback control with the control setting the solution of a Riccati equation and a constant for an infinite control time,* is one of the important results of modern control theory, and it generalizes to larger systems. We shall return to this and related control problems in Sec. 1.10 and in later chapters.

1.8 LAGRANGE MULTIPLIERS

In the previous section we were able to substitute the equation describing the process, Eq. (1), directly into the objective function, Eq. (2), and thus put the problem in the framework of Sec. 1.1, where we needed only to set partial derivatives of the objective to zero. It will not always be possible or desirable to solve the system equations for certain of the variables and then substitute, and an alternative method is needed. One such method is the introduction of *Lagrange multipliers*.

For notational convenience we shall again restrict our attention to a system with two variables, x_1 and x_2. We seek to minimize $\mathcal{E}(x_1,x_2)$, but we assume that x_1 and x_2 are related by an equation

$$g(x_1,x_2) = 0 \tag{1}$$

It is convenient to consider the problem geometrically. We can plot the curve described by Eq. (1) in the x_1x_2 plane, and for every constant we can plot the curve defined by

$$\mathcal{E}(x_1,x_2) = \text{const} \tag{2}$$

A typical result is shown in Fig. 1.2.

The solution is clearly the point at which the line $g = 0$ intersects the curve of constant $\mathcal{E}$ with the least value, and if both curves are continuously differentiable at the point, then they will be tangent to each other. Thus, they will possess a common tangent and a common normal. Since the direction cosines for the normal to the curve $g = 0$ are proportional to $\partial g/\partial x_i$ and those of $\mathcal{E} = \text{const}$ to $\partial \mathcal{E}/\partial x_i$ and for a common normal the direction cosines must coincide, it follows then that at the optimum it is necessary that $\partial g/\partial x_i$ be proportional to $\partial \mathcal{E}/\partial x_i$; that is, a

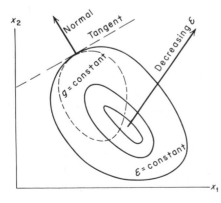

Fig. 1.2 Contours and a constraint curve.

necessary condition for optimality is

$$\frac{\partial \mathcal{E}}{\partial x_1} + \lambda \frac{\partial g}{\partial x_1} = 0 \qquad (3a)$$

$$\frac{\partial \mathcal{E}}{\partial x_2} + \lambda \frac{\partial g}{\partial x_2} = 0 \qquad (3b)$$

We shall now obtain this result somewhat more carefully and more generally. As before, we suppose that we somehow know the optimal values $\bar{x}_1$, $\bar{x}_2$, and we expand $\mathcal{E}$ in a Taylor series to obtain

$$\mathcal{E}(\bar{x}_1 + \delta x_1,\ \bar{x}_2 + \delta x_2) - \mathcal{E}(\bar{x}_1, \bar{x}_2) = \frac{\partial \mathcal{E}}{\partial x_1} \delta x_1 + \frac{\partial \mathcal{E}}{\partial x_2} \delta x_2$$
$$+\ o(\max |\delta x_1|, |\delta x_2|) \geq 0 \quad (4)$$

Now, however, we cannot choose δx_1 and δx_2 in any way that we wish, for having chosen one we see that the constraint, Eq. (1), fixes the other. We can, therefore, freely choose one variation, δx_1 or δx_2, but not both.

Despite any variations in x_1 and x_2, we must satisfy the constraint. Thus, using a Taylor series expansion,

$$g(\bar{x}_1 + \delta x_1,\ \bar{x}_2 + \delta x_2) - g(\bar{x}_1, \bar{x}_2) = \frac{\partial g}{\partial x_1} \delta x_1 + \frac{\partial g}{\partial x_2} \delta x_2$$
$$+\ o(\max |\delta x_1|, |\delta x_2|) = 0 \quad (5)$$

If we multiply Eq. (5) by an *arbitrary constant* λ (the Lagrange multiplier) and add the result to Eq. (4), we obtain

$$\left(\frac{\partial \mathcal{E}}{\partial x_1} + \lambda \frac{\partial g}{\partial x_1} \right) \delta x_1 + \left(\frac{\partial \mathcal{E}}{\partial x_2} + \lambda \frac{\partial g}{\partial x_2} \right) \delta x_2$$
$$+\ o(\max |\delta x_1|, |\delta x_2|) \geq 0 \quad (6)$$

We are now free to choose any λ we please, and it is convenient to define λ so that the variation we are not free to specify vanishes. Thus, we *choose* λ such that at $\bar{x}_1$, $\bar{x}_2$

$$\frac{\partial \mathcal{E}}{\partial x_2} + \lambda \frac{\partial g}{\partial x_2} = 0 \qquad (7)$$

which is Eq. (3b). We now have

$$\left(\frac{\partial \mathcal{E}}{\partial x_1} + \lambda \frac{\partial g}{\partial x_1} \right) \delta x_1 + o(\max |\delta x_1|, |\delta x_2|) \geq 0 \qquad (8)$$

and by choosing the special variation

$$\delta x_1 = -\ \epsilon \left(\frac{\partial \mathcal{E}}{\partial x_1} + \lambda \frac{\partial g}{\partial x_1} \right) \qquad (9)$$

it follows, as in Sec. 1.3, that at the minimum of $\mathcal{E}$ it is necessary that

$$\frac{\partial \mathcal{E}}{\partial x_1} + \lambda \frac{\partial g}{\partial x_1} = 0 \tag{10}$$

which is Eq. (3a). Equations (1), (7), and (10) provide three equations for determining the *three* unknowns $\bar{x}_1$, $\bar{x}_2$, and λ. We note that by including the constraint in this manner we have introduced an additional variable, the multiplier λ, and an additional equation.

It is convenient to introduce the *lagrangian* $\mathcal{L}$, defined

$$\mathcal{L}(x_1, x_2, \lambda) = \mathcal{E}(x_1, x_2) + \lambda g(x_1, x_2) \tag{11}$$

Equations (7) and (10) may then be written as

$$\frac{\partial \mathcal{L}}{\partial x_1} = \frac{\partial \mathcal{L}}{\partial x_2} = 0 \tag{12}$$

while Eq. (1) is simply

$$\frac{\partial \mathcal{L}}{\partial \lambda} = 0 \tag{13}$$

Thus, we reformulate the necessary condition in the *Lagrange multiplier rule*, as follows:

> *The function $\mathcal{E}$ takes on its minimum subject to the constraint equation (1) at a stationary point of the lagrangian $\mathcal{L}$.*

(A stationary point of a function is one at which all first partial derivatives vanish.) For the general case of a function of n variables $\mathcal{E}(x_1, x_2, \ldots, x_n)$ and m constraints

$$g_i(x_1, x_2, \ldots, x_n) = 0 \qquad i = 1, 2, \ldots, m < n \tag{14}$$

the lagrangian† is written

$$\mathcal{L}(x_1, x_2, \ldots, x_n, \lambda_1, \lambda_2, \ldots, \lambda_m) = \mathcal{E}(x_1, x_2, \ldots, x_n)$$
$$+ \sum_{i=1}^{m} \lambda_i g_i(x_1, x_2, \ldots, x_n) \tag{15}$$

---

† We have not treated the most general situation, in which the lagrangian must be written

$$\mathcal{L} = \lambda_0 \mathcal{E} + \sum_{i=1}^{m} \lambda_i g_i$$

So-called irregular cases do exist in which $\lambda_0 = 0$, but for all regular situations λ_0 may be taken as unity without loss of generality.

and the necessary condition for a minimum is

$$\frac{\partial \mathcal{L}}{\partial x_i} = 0 \qquad i = 1, 2, \ldots, n \tag{16a}$$

$$\frac{\partial \mathcal{L}}{\partial \lambda_i} = 0 \qquad i = 1, 2, \ldots, m \tag{16b}$$

It is tempting to try to improve upon the multiplier rule by retaining second-order terms in the Taylor series expansions of $\mathcal{E}$ and g and attempting to show that the minimum of $\mathcal{E}$ occurs at the minimum of $\mathcal{L}$. This is often the case, as when g is linear and $\mathcal{E}$ is convex (the hessian of $\mathcal{E}$ is positive definite), but it is not true in general, for the independent variation cannot be so easily isolated in the second-order terms of the expansion for $\mathcal{L}$.   It is easily verified, for example, that the function

$$\mathcal{E} = x_1(1 + x_2) \tag{17}$$

has both a local minimum and a local maximum at the constraint

$$x_1 + (x_2)^2 = 0 \tag{18}$$

while the stationary points of the lagrangian

$$\mathcal{L} = x_1(1 + x_2) + \lambda x_1 + \lambda(x_2)^2 \tag{19}$$

are neither maxima nor minima but only saddle points.

The methods of this section can be extended to include inequality constraints of the form

$$g(x_1, x_2) \geq 0 \tag{20}$$

but that would take us into the area of nonlinear programming and away from our goal of developing variational methods of optimization, and we must simply refer the reader at this point to the specialized texts.

1.9 A GEOMETRICAL EXAMPLE

As a first example in the use of the Lagrange multiplier let us consider a problem in geometry which we shall find useful in a later discussion of computational techniques.   We consider the linear function

$$\mathcal{E} = ax_1 + bx_2 \tag{1}$$

and the quadratic constraint

$$g = \alpha(x_1)^2 + 2\beta x_1 x_2 + \gamma(x_2)^2 - \Delta^2 = 0 \tag{2}$$

where

$$\alpha > 0 \qquad \alpha\gamma - \beta^2 > 0 \tag{3}$$

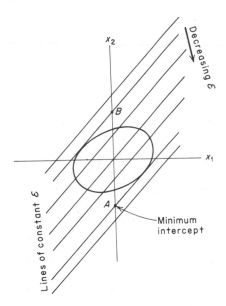

Fig. 1.3 A linear objective with elliptic constraint.

The curve $g = 0$ forms an ellipse, while for each value of ε Eq. (1) defines a straight line. We seek to minimize ε subject to constraint (2); i.e., as shown in Fig. 1.3, we seek the intersection of the straight line and ellipse leading to the minimum intercept on the x_2 axis.

As indicated by the multiplier rule, we form the lagrangian

$$\mathcal{L} = ax_1 + bx_2 + \lambda\alpha(x_1)^2 + 2\lambda\beta x_1 x_2 + \lambda\gamma(x_2)^2 - \lambda\Delta^2 \tag{4}$$

We then find the stationary points

$$\frac{\partial\mathcal{L}}{\partial x_1} = a + 2\lambda\alpha x_1 + 2\lambda\beta x_2 = 0 \tag{5a}$$

$$\frac{\partial\mathcal{L}}{\partial x_2} = b + 2\lambda\beta x_1 + 2\lambda\gamma x_2 = 0 \tag{5b}$$

while $\partial\mathcal{L}/\partial\lambda = 0$ simply yields Eq. (2). Equations (5) are easily solved for x_1 and x_2 as follows:

$$x_1 = -\frac{a\gamma - b\beta}{2\lambda(\alpha\gamma - \beta^2)} \tag{6a}$$

$$x_2 = -\frac{b\alpha - a\beta}{2\lambda(\alpha\gamma - \beta^2)} \tag{6b}$$

λ can then be obtained by substitution into Eq. (2), and the final result is

$$x_1 = \pm\Delta\frac{a\gamma - b\beta}{\sqrt{(\alpha\gamma - \beta^2)(\gamma a^2 + \alpha b^2 - 2ab\beta)}} \tag{7a}$$

$$x_2 = \pm\Delta\frac{b\alpha - a\beta}{\sqrt{(\alpha\gamma - \beta^2)(\gamma a^2 + \alpha b^2 - 2ab\beta)}} \tag{7b}$$

The ambiguity ($\pm$) sign in Eqs. (7) results from taking a square root, the negative sign corresponding to the minimum intercept (point A in Fig. 1.3) and the positive sign to the maximum (point B).

1.10 DISCRETE PROPORTIONAL CONTROL WITH LAGRANGE MULTIPLIERS

We shall now return to the problem of Sec. 1.7, where we wish to minimize

$$\varepsilon = \frac{1}{2} \sum_{n=1}^{N} (x_n{}^2 + C^2 u_n{}^2) \tag{1}$$

and we write the system equation

$$x_n - a x_{n-1} - u_n \equiv g_n(x_n, x_{n-1}, u_n) = 0 \tag{2}$$

Since we wish to find the minimizing $2N$ variables $x_1, x_2, \ldots, x_N, u_1, u_2, \ldots, u_N$ subject to the N constraints (2), we form the lagrangian

$$\mathcal{L} = \frac{1}{2} \sum_{n=1}^{N} (x_n{}^2 + C^2 u_n{}^2) + \sum_{n=1}^{N} \lambda_n(x_n - a x_{n-1} - u_n) \tag{3}$$

Taking partial derivatives of the lagrangian with respect to x_1, $x_2, \ldots, x_N, u_1, u_2, \ldots, u_N$, we find

$$\frac{\partial \mathcal{L}}{\partial u_n} = C^2 u_n - \lambda_n = 0 \qquad n = 1, 2, \ldots, N \tag{4a}$$

$$\frac{\partial \mathcal{L}}{\partial x_N} = x_N + \lambda_N = 0 \tag{4b}$$

$$\frac{\partial \mathcal{L}}{\partial x_n} = x_n + \lambda_n - a\lambda_{n+1} = 0 \qquad n = 1, 2, \ldots, N - 1 \tag{4c}$$

Equation (4b) may be included with (4c) by defining

$$\lambda_{N+1} = 0 \tag{5}$$

We thus have simultaneous difference equations to solve

$$x_n - a x_{n-1} - \frac{\lambda_n}{C^2} = 0 \qquad x_0 \text{ given} \tag{6a}$$

$$x_n + \lambda_n - a\lambda_{n+1} = 0 \qquad \lambda_{N+1} = 0 \tag{6b}$$

with the optimal controls u_n then obtained from Eq. (4a).

Equations (6a) and (6b) represent a structure which we shall see repeated constantly in optimization problems. Equation (6a) is a difference equation for x_n, and (6b) is a difference equation for λ_n, but they are coupled. Furthermore, the boundary condition x_0 for Eq. (6a) is given for $n = 0$, while for Eq. (6b) the condition is given at $n = N + 1$. Thus, our problem requires us to solve coupled difference equations with split boundary conditions.

For this simple problem we can obtain a solution quite easily.   If, for example, we assume a value for λ_0, Eqs. (6) can be solved simultaneously for λ_1, x_1, then λ_2, x_2, etc.   The calculated value for λ_{N+1} will probably differ from zero, and we shall then have to vary the assumed value λ_0 until agreement is reached.   Alternatively, we might assume a value for x_N and work backward, seeking a value of x_N which yields the required x_0. Note that in this latter case we satisfy all the conditions for optimality for whatever x_0 results, and as we vary x_N, we produce a whole family of optimal solutions for a variety of initial values.

We have already noted, however, that the particular problem we are considering has a closed-form solution, and our past experience, as well as the structure of Eq. (6), suggests that we seek a solution of the form

$$\lambda_n = C^2 M_n x_n \tag{7a}$$

or

$$u_n = M_n x_n \tag{7b}$$

We then obtain

$$x_{n+1} - ax_n - M_{n+1}x_{n+1} = 0 \tag{8a}$$
$$x_n + C^2 M_n x_n - aC^2 M_{n+1}x_{n+1} = 0 \tag{8b}$$

and eliminating x_n and x_{n+1}, we again get the Riccati equation

$$M_n M_{n+1} + \frac{1 + a^2 C^2}{C^2} M_{n+1} - M_n - \frac{1}{C^2} = 0 \tag{9}$$

with the boundary condition

$$M_{N+1} = 0 \tag{10}$$

for a finite control period.   Note that as $N \to \infty$, we cannot satisfy Eq. (10), since the solution of the Riccati equation becomes a constant, but in that case $x_N \to 0$ and the boundary condition $\lambda_{N+1} = 0$ is still satisfied.

1.11 OPTIMAL DESIGN OF MULTISTAGE SYSTEMS

A large number of processes may be modeled, either as a true representation of the physical situation or simply for mathematical convenience, by the sequential *black-box* structure shown in Fig. 1.4.   At each stage the system undergoes some transformation, brought about in part by the decision u_n to be made at that stage.   The transformation is described by a set of algebraic equations at each stage, with the number of equa-

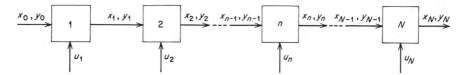

Fig. 1.4 Schematic of a multistage system.

tions equal to the minimum number of variables needed in addition to the decision variable to describe the state of the process.

For example, let us suppose that two variables, x and y, describe the state. The variable x might represent an amount of some material, in which case one of the equations at each stage would be an expression of the conservation of mass. Should y be a temperature, say, the other equation would be an expression of the law of conservation of energy. We shall use the subscript n to denote the value of x and y in the process stream leaving stage n. Thus, our system is described by the $2N$ algebraic equations

$$\phi_n(x_n,x_{n-1},y_n,y_{n-1},u_n) = 0 \qquad n = 1, 2, \ldots, N \tag{1a}$$
$$\psi_n(x_n,x_{n-1},y_n,y_{n-1},u_n) = 0 \qquad n = 1, 2, \ldots, N \tag{1b}$$

where presumably x_0 and y_0 are given. We shall suppose that the state of the stream leaving the last stage is of interest, and we wish to choose $u_1, u_2, \ldots, u_N$ in order to minimize some function $\mathcal{E}(x_N,y_N)$ of that stream.

This problem can be formulated with the use of $2N$ Lagrange multipliers, which we shall denote by $\lambda_1, \lambda_2, \ldots, \lambda_N, \Lambda_1, \Lambda_2, \ldots, \Lambda_N$. The lagrangian is then

$$\mathcal{L} = \mathcal{E}(x_N,y_N) + \sum_{n=1}^{N} \lambda_n\phi_n(x_n,x_{n-1},y_n,y_{n-1},u_n)$$
$$+ \sum_{n=1}^{N} \Lambda_n\psi_n(x_n,x_{n-1},y_n,y_{n-1},u_n) \tag{2}$$

and the optimum is found by setting the partial derivatives of $\mathcal{L}$ with respect to x_n, y_n, u_n, λ_n, and Λ_n to zero, $n = 1, 2, \ldots, N$. At stage N we have

$$\frac{\partial\mathcal{E}}{\partial x_N} + \lambda_N\frac{\partial\phi_N}{\partial x_N} + \Lambda_N\frac{\partial\psi_N}{\partial x_N} = 0 \tag{3a}$$

$$\frac{\partial\mathcal{E}}{\partial y_N} + \lambda_N\frac{\partial\phi_N}{\partial y_N} + \Lambda_N\frac{\partial\psi_N}{\partial y_N} = 0 \tag{3b}$$

$$\lambda_N\frac{\partial\phi_N}{\partial u_N} + \Lambda_N\frac{\partial\psi_N}{\partial u_N} = 0 \tag{3c}$$

or, eliminating λ_N and Λ_N from (3c) with (3a) and (3b),

$$\frac{\partial \mathcal{E}}{\partial y_N}\left(\frac{\partial \phi_N}{\partial u_N}\frac{\partial \psi_N}{\partial x_N} - \frac{\partial \psi_N}{\partial u_N}\frac{\partial \phi_N}{\partial x_N}\right) - \frac{\partial \mathcal{E}}{\partial x_N}\left(\frac{\partial \phi_N}{\partial u_N}\frac{\partial \psi_N}{\partial y_N} - \frac{\partial \psi_N}{\partial u_N}\frac{\partial \phi_N}{\partial y_N}\right) = 0 \quad (4)$$

For all other n, $n = 1, 2, \ldots, N - 1$, we obtain

$$\lambda_n \frac{\partial \phi_n}{\partial x_n} + \lambda_{n+1}\frac{\partial \phi_{n+1}}{\partial x_n} + \Lambda_n \frac{\partial \psi_n}{\partial x_n} + \Lambda_{n+1}\frac{\partial \psi_{n+1}}{\partial x_n} = 0 \qquad (5a)$$

$$\lambda_n \frac{\partial \phi_n}{\partial y_n} + \lambda_{n+1}\frac{\partial \phi_{n+1}}{\partial y_n} + \Lambda_n \frac{\partial \psi_n}{\partial y_n} + \Lambda_{n+1}\frac{\partial \psi_{n+1}}{\partial y_n} = 0 \qquad (5b)$$

$$\lambda_n \frac{\partial \phi_n}{\partial u_n} + \Lambda_n \frac{\partial \psi_n}{\partial u_n} = 0 \qquad (5c)$$

Again we find that we have a set of coupled difference equations, Eqs. (1a), (1b) and (5a), (5b), with initial conditions x_0, y_0, and final conditions λ_N, Λ_N given by Eqs. (3a) and (3b). These must be solved simultaneously with Eqs. (3c) and (5c). Any solution method which assumes values of λ_0, Λ_0 or x_N, y_N and then varies these values until agreement with the two neglected boundary conditions is obtained now requires a search in *two* dimensions. Thus, if ν values of x_N and of y_N must be considered, the four difference equations must be solved ν^2 times. For this particular problem the dimensionality of this search may be reduced, as shown in the following paragraph, but in general we shall be faced with the difficulty of developing efficient methods of solving such boundary-value problems, and a large part of our effort in later chapters will be devoted to this question.

Rewriting Eq. (5c) for stage $n + 1$,

$$\lambda_{n+1}\frac{\partial \phi_{n+1}}{\partial u_{n+1}} + \Lambda_{n+1}\frac{\partial \psi_{n+1}}{\partial u_{n+1}} = 0 \qquad (5d)$$

we see that Eqs. (5a) to (5d) make up four linear and homogeneous equations for the four variables λ_n, λ_{n+1}, Λ_n, Λ_{n+1}, and by a well-known result of linear algebra we shall have a nontrivial solution if and only if the determinant of coefficients in Eqs. (5a) to (5d) vanishes; that is,

$$\begin{vmatrix} \dfrac{\partial \phi_n}{\partial x_n} & \dfrac{\partial \phi_{n+1}}{\partial x_n} & \dfrac{\partial \psi_n}{\partial x_n} & \dfrac{\partial \psi_{n+1}}{\partial x_n} \\[2mm] \dfrac{\partial \phi_n}{\partial y_n} & \dfrac{\partial \phi_{n+1}}{\partial y_n} & \dfrac{\partial \psi_n}{\partial y_n} & \dfrac{\partial \psi_{n+1}}{\partial y_n} \\[2mm] \dfrac{\partial \phi_n}{\partial u_n} & 0 & \dfrac{\partial \psi_n}{\partial u_n} & 0 \\[2mm] 0 & \dfrac{\partial \phi_{n+1}}{\partial u_{n+1}} & 0 & \dfrac{\partial \psi_{n+1}}{\partial u_{n+1}} \end{vmatrix} = 0 \qquad (6)$$

or

$$\left(\frac{\partial \psi_n}{\partial x_n}\frac{\partial \phi_n}{\partial u_n} - \frac{\partial \phi_n}{\partial x_n}\frac{\partial \psi_n}{\partial u_n}\right)\left(\frac{\partial \psi_{n+1}}{\partial y_n}\frac{\partial \phi_{n+1}}{\partial u_{n+1}} - \frac{\partial \phi_{n+1}}{\partial y_n}\frac{\partial \psi_{n+1}}{\partial u_{n+1}}\right)$$
$$-\left(\frac{\partial \psi_n}{\partial y_n}\frac{\partial \phi_n}{\partial u_n} - \frac{\partial \phi_n}{\partial y_n}\frac{\partial \psi_n}{\partial u_n}\right)\left(\frac{\partial \psi_{n+1}}{\partial x_n}\frac{\partial \phi_{n+1}}{\partial u_{n+1}} - \frac{\partial \phi_{n+1}}{\partial x_n}\frac{\partial \psi_{n+1}}{\partial u_{n+1}}\right) = 0 \quad (7)$$

[Equation (7) has been called a *generalized Euler equation.*]

Equations (1a), (1b), (4), and (7) provide $3N$ design equations for the optimal x_n, y_n, and decision variables. One particular method of solution, which requires little effort, is to assume u_1 and obtain x_1 and y_1 from Eqs. (1a) and (1b). Equations (1a), (1b), and (7) may then be solved simultaneously for x_2, y_2, u_2 and sequentially in this manner until obtaining x_N, y_N, and u_N. A check is then made to see whether Eq. (4) is satisfied, and if not, a new value of u_1 is chosen and the process repeated. A plot of the error in Eq. (4) versus choice of u_1 will allow rapid convergence.

1.12 OPTIMAL TEMPERATURES FOR CONSECUTIVE REACTIONS

As a specific example of the general development of the preceding section, and in order to illustrate the type of simplification which can often be expected in practice, we shall consider the following problem.

A chemical reaction

$$X \rightarrow Y \rightarrow \text{products}$$

is to be carried out in a series of well-stirred vessels which operate in the steady state. The concentrations of materials X and Y will be denoted by lowercase letters, so that x_n and y_n represent the concentrations in the flow leaving the nth reaction vessel, and because the tank is well stirred, the concentration at all points in the vessel is identical to the concentration in the exit stream.

If the volumetric flow rate is q and the volume of the nth vessel V_n, then the rate at which material X enters the vessel is qx_{n-1}, the rate at which it leaves is qx_n, and the rate at which it decomposes is $V_n r_1(x_n,u_n)$, where r_1 is an experimentally or theoretically determined rate of reaction and u_n is the temperature. The expression of conservation of mass is then

$$qx_{n-1} = qx_n + V_n r_1(x_n) \tag{1}$$

In a similar way, for y_n,

$$qy_{n-1} = qy_n + V_n r_2(x_n,y_n) \tag{2}$$

where r_2 is the net rate of decomposition of y_n and includes the rate of formation, which is proportional to the rate of decomposition of x_n. In particular, we may write

$$r_1(x_n,u_n) = k_1(u_n)F(x_n) \tag{3a}$$
$$r_2(x_n,y_n,u_n) = -\nu k_1(u_n)F(x_n) + k_2(u_n)G(y_n) \tag{3b}$$

where k_1 and k_2 will generally have the *Arrhenius form*

$$k_i(u_n) = k_{i0} \exp\left(\frac{-E_i'}{u_n}\right) \tag{4}$$

Defining the *residence time* θ_n by

$$\theta_n = \frac{V_n}{q} \tag{5}$$

we may write the transformation equations corresponding to Eqs. (1a) and (1b) of Sec. 1.11 as

$$\phi_n = x_{n-1} - x_n - \theta_n k_1(u_n)F(x_n) \tag{6a}$$
$$\psi_n = y_{n-1} - y_n + \nu\theta_n k_1(u_n)F(x_n) - \theta_n k_2(u_n)G(y_n) \tag{6b}$$

We shall assume that Y is the desired product, while X is a valuable raw material, whose relative value measured in units of Y is ρ. The value of the product stream is then $\rho x_N + y_N$, which we wish to maximize by choice of the temperatures u_n in each stage. Since our formulation is in terms of minima, we wish to minimize the negative of the values, or

$$\mathcal{E}(x_N,y_N) = -\rho x_N - y_N \tag{7}$$

If Eqs. (6a) and (6b) are substituted into the generalized Euler equation of the preceding section and k_1 and k_2 are assumed to be of Arrhenius form [Eq. (4)], then after some slight grouping of terms we obtain

$$\frac{1 + \theta_n k_1(u_n)F'(x_n)}{1 + \theta_n k_2(u_n)G'(y_n)} \frac{E_2'k_2(u_n)G(y_n)}{E_1'k_1(u_n)F(x_n)} + \frac{\nu\theta_n k_2(u_n)G'(y_n)}{1 + \theta_n k_2(u_n)G'(y_n)}$$
$$= \frac{E_2'k_2(u_{n+1})G(y_{n+1})}{E_1'k_1(u_{n+1})F(x_{n+1})} \tag{8}$$

where the prime on a function denotes differentiation. Equation (8) can be solved for u_{n+1} in terms of x_{n+1}, y_{n+1}, x_n, y_n, and u_n:

$$\exp\left(-\frac{1}{u_{n+1}}\right) = \left\{\frac{k_{10}}{k_{20}}\left[\frac{k_2(u_n)}{k_1(u_n)}\frac{1 + \theta_n k_1(u_n)F'(x_n)}{1 + \theta_n k_2(u_n)G'(y_n)}\frac{G(y_n)F(x_{n+1})}{G(y_{n+1})F(x_n)}\right.\right.$$
$$\left.\left. + \nu\theta_n\frac{E_1'}{E_2'}\frac{k_2(u_n)G'(y_n)F(x_{n+1})}{[1 + \theta_n k_2(u_n)G'(y_n)]G(y_{n+1})}\right]\right\}^{1/(E_2'-E_1')}$$
$$\equiv [B(x_n,x_{n+1},y_n,y_{n+1},u_n)]^{1/(E_2'-E_1')} \tag{9}$$

and upon substitution Eqs. (6a) and (6b) become

$$x_{n-1} - x_n - \theta_n k_{10}[B(x_{n-1},x_n,y_{n-1},y_n,u_{n-1})]^{E_1'/(E_2'-E_1')}F(x_n) = 0$$

$$(10a)$$

$$y_{n-1} - y_n + \nu(x_{n-1} - x_n) \\ - \theta_n k_{20}[B(x_{n-1},x_n,y_{n-1},y_n,u_{n-1})]^{E_2'/(E_2'-E_1')}G(y_n) = 0 \quad (10b)$$

The boundary condition, Eq. (4) of Sec. 1.11, becomes

$$k_1(u_N)F(x_N)(\rho - \nu) + \frac{E_2'}{E_1'} k_2(u_N)G(y_N)[1 + \theta_N k_1(u_N)F'(x_N)] \\ + \rho\theta_N k_1(u_N)k_2(u_N)F(x_N)G'(y_N) = 0 \quad (11)$$

The computational procedure to obtain an optimal design is then to assume u_1 and solve Eqs. (6a) and (6b) for x_1 and y_1, after which Eqs. (10a) and (10b) may be solved successively for $n = 2, 3, \ldots, N$. Equation (11) is then checked at $n = N$ and the process repeated for a new value of u_1. Thus only $2N$ equations need be solved for the optimal temperatures and concentrations.

This is perhaps an opportune point at which to interject a note of caution. We have assumed in the derivation that any value of u_n which is calculated is available to us. In practice temperatures will be bounded, and the procedure outlined above may lead to unfeasible design specifications. We shall have to put such considerations off until later chapters, but it suffices to note here that some of the simplicity of the above approach is lost.

1.13 ONE-DIMENSIONAL PROCESSES

Simplified models of several industrial processes may be described by a single equation at each stage

$$x_n - f_n(x_{n-1},u_n) = 0 \qquad n = 1, 2, \ldots, N \quad (1)$$

with the total return for the process the sum of individual stage returns

$$\mathcal{E} = \sum_{n=1}^{N} \mathcal{R}_n(x_{n-1},u_n) \quad (2)$$

The choice of optimal conditions $u_1, u_2, \ldots, u_N$ for such a process is easily determined using the Lagrange multiplier rule, but we shall obtain the design equations as a special case of the development of Sec. 1.11.

We define a new variable y_n by

$$y_n - y_{n-1} - \mathcal{R}_n(x_{n-1},u_n) = 0 \qquad y_0 = 0 \quad (3)$$

It follows then that

$$\mathcal{E} = y_N \quad (4)$$

and substitution of Eqs. (1) and (3) into the generalized Euler equation [Eq. (7), Sec. 1.11] yields

$$\frac{\partial \mathcal{R}_{n-1}/\partial u_{n-1}}{\partial f_{n-1}/\partial u_{n-1}} + \frac{\partial \mathcal{R}_n}{\partial x_{n-1}} - \frac{(\partial f_n/\partial x_{n-1})(\partial \mathcal{R}_n/\partial u_n)}{\partial f_n/\partial u_n} = 0$$

$$n = 1, 2, \ldots, N \quad (5)$$

with the boundary condition

$$\frac{\partial \mathcal{R}_N}{\partial u_N} = 0 \tag{6}$$

The difference equations (1) and (5) are then solved by varying u_1 until Eq. (6) is satisfied.

Fan and Wang† have collected a number of examples of processes which can be modeled by Eqs. (1) and (2), including optimal distribution of solvent in cross-current extraction, temperatures and holding times in continuous-flow stirred-tank reactors with a single reaction, hot-air allocation in a moving-bed grain dryer, heat-exchange and refrigeration systems, and multistage gas compressors.

We leave it as a problem to show that Eq. (5) can also be used when a fixed value x_N is required.

1.14 AN INVERSE PROBLEM

It is sometimes found in a multistage system that the optimal decision policy is identical in each stage. Because of the simplicity of such a policy it would be desirable to establish the class of systems for which such a policy will be optimal. A problem of this type, in which the policy rather than the system is given, is called an *inverse problem*. Such problems are generally difficult, and in this case we shall restrict our attention to the one-dimensional processes described in the previous section and, in fact, to those systems in which the state enters linearly and all units are the same.

We shall consider, then, processes for which Eqs. (1) and (2) of Sec. 1.13 reduce to

$$x_n = f_n(x_{n-1}, u_n) = A(u_n)x_{n-1} + B(u_n) \tag{1}$$
$$\mathcal{R}_n(x_{n-1}, u_n) = C(u_n)x_{n-1} + D(u_n) \tag{2}$$

Since we are assuming that all decisions are identical, we shall simply write u in place of u_n. The generalized Euler equation then becomes

$$\frac{C'x_{n-2} + D'}{A'x_{n-2} + B'} + C - \frac{A(C'x_{n-1} + D')}{A'x_{n-1} + B'} = 0 \tag{3}$$

where the prime denotes differentiation with respect to u. After substi-

† See the bibliographical notes at the end of the chapter.

tution of Eq. (1) and some simplification this becomes

$$
\begin{aligned}
(AA'C' + AA'^2C - A^2A'C')x_{n-2}{}^2 \\
+ (A'BC' + B'C' + A'^2BC + AA'B'C \\
+ A'B'C - AA'BC' - A^2B'C')x_{n-2} \\
+ (A'D'B + B'D' + A'BB'C + B'^2 - ABB'C' - AB'D') = 0 \quad (4)
\end{aligned}
$$

Our problem, then, is to find the functions $A(u)$, $B(u)$, $C(u)$, and $D(u)$ such that Eq. (4) holds as an identity, and so we must require that the coefficient of each power of x_{n-2} vanish, which leads to three coupled differential equations

$$
AA'(C' + A'C - AC') = 0 \tag{5a}
$$

$$
A'BC' + B'C' + A'^2BC + AA'B'C + A'B'C
$$
$$
\qquad\qquad - AA'BC' - A^2B'C' = 0 \tag{5b}
$$

$$
A'D'B + B'D' + A'BB'C + B'^2 - ABB'C' - AB'D' = 0 \tag{5c}
$$

If we assume that $A(u)$ is not a constant, Eq. (5a) has the solution

$$
C(u) = \alpha(A - 1) \tag{6}
$$

where α is a constant of integration. Equation (5b) is then satisfied identically, while Eq. (5c) may be solved by the substitution

$$
D(u) = \alpha B(u) + M(u) \tag{7}
$$

to give

$$
M'(A'B + B' - AB') = 0 \tag{8}
$$

or, if M is not a constant,

$$
B(u) = \beta(A - 1) \tag{9}
$$

where β is a constant of integration. $M(u)$ and $A(u)$ are thus arbitrary, but not constant, functions of u, and the system must satisfy the equations

$$
x_n = A(u_n)x_{n-1} + \beta[A(u_n) - 1] \tag{10}
$$

$$
\Re(x_{n-1}, u_n) = \alpha[A(u_n) - 1]x_{n-1} + \alpha B(u_n) + M(u_n)
$$
$$
\qquad\quad = \alpha(x_n - x_{n-1}) + M(u_n) \tag{11}
$$

Thus, the most general linear one-dimensional system whose optimal policy is identical at each stage is described by Eq. (10), with the objective the minimization of

$$
\mathcal{E} = \sum_{n=1}^{N} \Re_n = \alpha(x_N - x_0) + \sum_{n=1}^{N} M(u_n) \tag{12}
$$

This objective function includes the special cases of minimizing $\displaystyle\sum_{n=1}^{N} M(u_n)$

for fixed conversion $x_N - x_0$, in which case α is a Lagrange multiplier,† and maximizing conversion for fixed total resources, in which case

$$\sum_{n=1}^{N} M(u_n) = \lambda \left(\sum_{n=1}^{N} u_n - U \right) \tag{13}$$

with λ a Lagrange multiplier and U the total available resource. Multistage isentropic compression of a gas, the choice of reactor volumes for isothermal first-order reaction, and multistage cross-current extraction with a linear phase-equilibrium relationship are among the processes which are described by Eqs. (10) and (11).

1.15 MEANING OF THE LAGRANGE MULTIPLIERS

The Lagrange multiplier was introduced in Sec. 1.8 in a rather artificial manner, but it has a meaningful interpretation in many situations, two of which we shall examine in this section. For the first, let us restrict attention to the problem of minimizing a function of two variables, $\mathcal{E}(x_1,x_2)$, subject to a constraint which we shall write

$$g(x_1,x_2) - b = 0 \tag{1}$$

where b is a constant. The lagrangian is then

$$\mathcal{L} = \mathcal{E}(x_1,x_2) + \lambda g(x_1,x_2) - \lambda b \tag{2}$$

and the necessary conditions are

$$\frac{\partial \mathcal{E}}{\partial x_1} + \lambda \frac{\partial g}{\partial x_1} = 0 \tag{3a}$$

$$\frac{\partial \mathcal{E}}{\partial x_2} + \lambda \frac{\partial g}{\partial x_2} = 0 \tag{3b}$$

Let us denote the optimum by $\mathcal{E}^*$ and the optimal values of x_1 and x_2 by x_1^*, x_2^*. If we change the value of the constant b, we shall certainly change the value of the optimum, and so we may write $\mathcal{E}^*$ as a function of b

$$\mathcal{E}^* = \mathcal{E}^*(b) \tag{4a}$$

and

$$x_1^* = x_1^*(b) \qquad x_2^* = x_2^*(b) \tag{4b}$$

Thus

$$\frac{d\mathcal{E}^*}{db} = \left(\frac{\partial \mathcal{E}}{\partial x_1} \right)_{\substack{x_1 = x_1^* \\ x_2 = x_2^*}} \frac{dx_1^*}{db} + \left(\frac{\partial \mathcal{E}}{\partial x_2} \right)_{\substack{x_1 = x_1^* \\ x_2 = x_2^*}} \frac{dx_2^*}{db} \tag{5}$$

† Recall that the lagrangian is minimized for a minimum of a convex objective function if the constraint is linear.

and, differentiating Eq. (1) as an identity in b,

$$\left(\frac{\partial g}{\partial x_1}\right)_{\substack{x_1=x_1^* \\ x_2=x_2^*}} \frac{dx_1^*}{db} + \left(\frac{\partial g}{\partial x_2}\right)_{\substack{x_1=x_1^* \\ x_2=x_2^*}} \frac{dx_2^*}{db} - 1 = 0 \tag{6}$$

or

$$\frac{dx_1^*}{db} = \frac{1 - (\partial g/\partial x_2)_{\substack{x_1=x_1^* \\ x_2=x_2^*}} dx_2^*/db}{(\partial g/\partial x_1)_{\substack{x_1=x_1^* \\ x_2=x_2^*}}} \tag{7}$$

Combining Eqs. (3a), (5), and (7), we obtain

$$\frac{d\mathcal{E}^*}{db} = -\lambda + \frac{dx_2^*}{db}\left[\left(\frac{\partial \mathcal{E}}{\partial x_2}\right)_{\substack{x_1=x_1^* \\ x_2=x_2^*}} + \lambda\left(\frac{\partial g}{\partial x_2}\right)_{\substack{x_1=x_1^* \\ x_2=x_2^*}}\right] \tag{8}$$

and from Eq. (3b),

$$\lambda = -\frac{d\mathcal{E}^*}{db} \tag{9}$$

That is, the Lagrange multiplier represents the rate of change of the optimal value of the objective with respect to the value of the constraint. If $\mathcal{E}$ has units of return and b of production, then λ represents the rate of change of optimal return with production. Because of this economic interpretation the multipliers are often referred to as *shadow prices* or *imputed values*, and sometimes as *sensitivity coefficients*, since they represent the sensitivity of the objective to changes in constraint levels.

A second related interpretation may be developed in the context of the one-dimensional processes considered in Sec. 1.13. We seek the minimum of

$$\mathcal{E} = \sum_{n=1}^{N} \mathcal{R}_n(x_{n-1}, u_n) \tag{10}$$

by choice of $u_1, u_2, \ldots, u_N$, where

$$x_n = f_n(x_{n-1}, u_n) \tag{11}$$

Now we can find the optimum by differentiation, so that

$$\frac{\partial \mathcal{E}}{\partial u_n} = 0 \qquad n = 1, 2, \ldots, N \tag{12}$$

or, since x_n depends on all u_i, $i \leq n$, by the chain rule,

$$\frac{\partial \mathcal{E}}{\partial u_n} = \frac{\partial \mathcal{R}_n}{\partial u_n} + \frac{\partial \mathcal{R}_{n+1}}{\partial x_n}\frac{\partial f_n}{\partial u_n} + \frac{\partial \mathcal{R}_{n+2}}{\partial x_{n+1}}\frac{\partial f_{n+1}}{\partial x_n}\frac{\partial f_n}{\partial u_n} + \cdots$$

$$+ \frac{\partial \mathcal{R}_N}{\partial x_{N-1}} \cdots \frac{\partial f_n}{\partial u_n} = \frac{\partial \mathcal{R}_n}{\partial u_n} + \left(\frac{\partial \mathcal{R}_{n+1}}{\partial x_n} + \frac{\partial \mathcal{R}_{n+2}}{\partial x_{n+1}}\frac{\partial f_{n+1}}{\partial x_n} + \cdots \right.$$

$$\left. + \frac{\partial \mathcal{R}_N}{\partial x_{N-1}} \cdots \frac{\partial f_{n+1}}{\partial x_n}\right)\frac{\partial f_n}{\partial u_n} = 0 \tag{13}$$

and similarly,

$$\frac{\partial \mathcal{E}}{\partial u_{n-1}} = \frac{\partial \mathcal{R}_{n-1}}{\partial u_{n-1}} + \left(\frac{\partial \mathcal{R}_n}{\partial x_{n-1}} + \frac{\partial \mathcal{R}_{n+1}}{\partial x_n} \frac{\partial f_n}{\partial x_{n-1}} + \cdots \right.$$

$$\left. + \frac{\partial \mathcal{R}_N}{\partial x_{N-1}} \cdots \frac{\partial f_n}{\partial x_{n-1}} \right) \frac{\partial f_{n-1}}{\partial u_{n-1}} = 0 \quad (14)$$

$$\frac{\partial \mathcal{E}}{\partial u_N} = \frac{\partial \mathcal{R}_N}{\partial u_N} = 0 \tag{15}$$

If we define a variable λ_n satisfying

$$\lambda_{n-1} = \frac{\partial \mathcal{R}_n}{\partial x_{n-1}} + \frac{\partial f_n}{\partial x_{n-1}} \lambda_n \tag{16a}$$

$$\lambda_N = 0 \tag{16b}$$

we see that Eqs. (13) to (15) may be written

$$\frac{\partial \mathcal{E}}{\partial u_n} = \frac{\partial \mathcal{R}_n}{\partial u_n} + \lambda_n \frac{\partial f_n}{\partial u_n} = 0 \tag{17}$$

which, together with Eq. (16), is the result obtainable from the Lagrange multiplier rule. The multiplier is then seen to be a consequence of the chain rule and, in fact, may be interpreted in Eq. (17) as

$$\lambda_n = \frac{\partial \mathcal{E}}{\partial x_n} \tag{18}$$

the partial derivative of the objective with respect to the state at any stage in the process. This interpretation is consistant with the notion of a sensitivity coefficient.

1.16 PENALTY FUNCTIONS

An alternative approach to the use of Lagrange multipliers for constrained minimization which is approximate but frequently useful is the method of *penalty functions*. Let us again consider the problem of minimizing $\mathcal{E}(x_1, x_2)$ subject to

$$g(x_1, x_2) = 0 \tag{1}$$

We recognize that in practice it might be sufficient to obtain a small but nonzero value of g in return for reduced computational effort, and we are led to consider the minimization of a new function

$$\tilde{\mathcal{E}}(x_1, x_2) = \mathcal{E}(x_1, x_2) + \tfrac{1}{2} K[g(x_1, x_2)]^2 \tag{2}$$

Clearly, if K is a very large positive number, then a minimum is obtainable only if the product Kg^2 is small, and hence minimizing $\tilde{\mathcal{E}}$ will be equivalent to minimizing a function not very different from $\mathcal{E}$ while ensuring that the constraint is nearly satisfied.

Before discussing the method further it will be useful to reconsider the geometric example of Sec. 1.9

$$\mathcal{E} = ax_1 + bx_2 \tag{3}$$
$$g = \alpha(x_1)^2 + 2\beta x_1 x_2 + \gamma(x_2)^2 - \Delta^2 = 0 \tag{4}$$
$$\alpha > 0 \qquad \alpha\gamma - \beta^2 > 0 \tag{5}$$

Equation (2) then becomes

$$\tilde{\mathcal{E}} = ax_1 + bx_2 + \tfrac{1}{2}K[\alpha(x_1)^2 + 2\beta x_1 x_2 + \gamma(x_2)^2 - \Delta^2]^2 \tag{6}$$

with the minimum satisfying

$$\frac{\partial \tilde{\mathcal{E}}}{\partial x_1} = a + 2K(\alpha x_1 + \beta x_2)[\alpha(x_1)^2 + 2\beta x_1 x_2 + \gamma(x_2)^2 - \Delta^2] = 0 \tag{7a}$$

$$\frac{\partial \tilde{\mathcal{E}}}{\partial x_2} = b + 2K(\beta x_1 + \gamma x_2)[\alpha(x_1)^2 + 2\beta x_1 x_2 + \gamma(x_2)^2 - \Delta^2] = 0 \tag{7b}$$

Equations (7) give

$$x_2 = \frac{a\beta - b\alpha}{b\beta - a\gamma} x_1 \tag{8}$$

and we obtain, upon substitution into Eq. (7a),

$$a + 2Kx_1\left(\alpha + \beta\frac{a\beta - b\alpha}{b\beta - a\gamma}\right)\left\{(x_1)^2\left[\alpha + 2\beta\frac{a\beta - b\alpha}{b\beta - a\gamma}\right.\right.$$
$$\left.\left. + \gamma\left(\frac{a\beta - b\alpha}{b\beta - a\gamma}\right)^2\right] - \Delta^2\right\} = 0 \tag{9}$$

Equation (9) may be solved for x_1 in terms of K, with x_2 then obtained from Eq. (8). Since we are interested only in large K, however, our purposes are satisfied by considering the limit of Eq. (9) as $K \to \infty$. In order for the second term to remain finite it follows that either $x_1 = 0$, in which case the constraint equation (4) is not satisfied, or

$$(x_1)^2\left[\alpha + 2\beta\frac{a\beta - b\alpha}{b\beta - a\gamma} + \gamma\left(\frac{a\beta - b\alpha}{b\beta - a\gamma}\right)^2 - \Delta^2\right] = 0 \tag{10}$$

This last equation has the solution

$$x_1 = \pm\Delta\frac{a\gamma - b\beta}{\sqrt{(\alpha\gamma - \beta^2)(\gamma a^2 + \alpha b^2 - 2ab\beta)}} \tag{11a}$$

and, from Eq. (8),

$$x_2 = \pm\Delta\frac{b\alpha - a\beta}{\sqrt{(\alpha\gamma - \beta^2)(\gamma a^2 + \alpha b^2 - 2ab\beta)}} \tag{11b}$$

which are the results obtained from the Lagrange multiplier rule.

In practice we would consider a sequence of problems of the form of

Eq. (2) for $K_{(1)}$, $K_{(2)}$, $K_{(3)}$, . . .

$$\tilde{\mathcal{E}}_{(n)} = \mathcal{E}(x_1,x_2) + \tfrac{1}{2}K_{(n)}[g(x_1,x_2)]^2 \tag{12}$$

where $K_{(n+1)} > K_{(n)}$ and $\lim_{n \to \infty} K_{(n)} \to \infty$. It might be hoped that as $K_{(n)}$ becomes arbitrarily large, the sequence $\{\tilde{\mathcal{E}}_{(n)}\}$ will approach a finite limit, with g vanishing, and that this limit will be the constrained minimum of $\mathcal{E}(x_1,x_2)$. It can in fact be shown that if the sequence converges, it does indeed converge to the solution of the constrained problem. In general the sequence will be terminated when some prespecified tolerance on the constraint is reached.

Finally, we note that the particular form $\tfrac{1}{2}Kg^2$ is only for convenience of demonstration and that any nonnegative function which vanishes only when the constraint is satisfied would suffice, the particular situation governing the choice of function. Consider, for example, an *inequality* constraint of the form

$$|x_1| \leq X_1 \tag{13}$$

The function $[x_1/(X_1 + \epsilon)]^{2N}$ will be vanishingly small for small ϵ and large N when the constraint is satisfied and exceedingly large when it is violated. It is thus an excellent penalty function for this type of constraint. Other functions may be constructed to "smooth" hard constraints as needed.

APPENDIX 1.1 LINEAR DIFFERENCE EQUATIONS

We have assumed that the reader is familiar with the method of solution of linear difference equations, but this brief introduction should suffice by demonstrating the analogy to linear differential equations. The nth-order homogeneous linear difference equation with constant coefficients is written

$$a_n x_{k+n} + a_{n-1} x_{k+n-1} + \cdots + a_1 x_{k+1} + a_0 x_k = 0 \tag{1}$$

with x specified at n values of the (discrete) independent variable k. If, by analogy to the usual procedure for differential equations, we seek a solution of the form

$$x_k = e^{mk} \tag{2}$$

Eq. (1) becomes

$$e^{mk} \sum_{p=0}^{n} a_p e^{mp} = 0 \tag{3}$$

or, letting $y = e^m$ and noting that $e^{mk} \neq 0$, we obtain the *characteristic equation*

$$\sum_{p=0}^{n} a_p y^p = 0 \tag{4}$$

This algebraic equation will have n roots, $y_1, y_2, \ldots, y_n$, which we shall assume to be distinct. The general solution to Eq. (1) is then

$$x_k = C_1 y_1{}^k + C_2 y_2{}^k + \cdots + C_n y_n{}^k \tag{5}$$

where $C_1, C_2, \ldots, C_n$ are evaluated from the n specified values of x.

Consider, for example, the second-order difference equation

$$x_{n+2} + 2\alpha x_{n+1} + \beta x_n = 0 \tag{6}$$

The characteristic equation is

$$y^2 + 2\alpha y + \beta = 0 \tag{7}$$

or

$$y = -\alpha \pm \sqrt{\alpha^2 - \beta} \tag{8}$$

The general solution is then

$$x_k = C_1(-\alpha + \sqrt{\alpha^2 - \beta})^k + C_2(-\alpha - \sqrt{\alpha^2 - \beta})^k \tag{9}$$

If initial conditions x_0 and x_1 are given, the constants C_1 and C_2 are evaluated from the equations

$$x_0 = C_1 + C_2 \tag{10a}$$
$$x_1 = (-\alpha + \sqrt{\alpha^2 - \beta})C_1 - (\alpha + \sqrt{\alpha^2 - \beta})C_2 \tag{10b}$$

The modification for repeated roots is the same as for differential equations. If the right-hand side of Eq. (1) is nonzero, the general solution is the sum of a particular and homogeneous solution, the standard methods of finding particular solutions, such as undetermined coefficients and variation of parameters, carrying over from the theory of ordinary differential equations. For instance, if our example were of the form

$$x_{n+2} + 2\alpha x_{n+1} + \beta x_n = n \tag{11}$$

the solution would be of the form

$$x_k = C_1(-\alpha + \sqrt{\alpha^2 - \beta})^k + C_2(-\alpha - \sqrt{\alpha^2 - \beta})^k + x_k{}^{(p)} \tag{12}$$

The particular solution $x_k{}^{(p)}$ can be found from the method of undetermined coefficients by the choice

$$x_k{}^{(p)} = A + Bk \tag{13}$$

Substituting into Eq. (11) gives

$$A + B(n + 2) + 2\alpha[A + B(n + 1)] + \beta(A + Bn) = n \qquad (14)$$

or, equating coefficients of powers of n on both sides,

$$n^0: \qquad (1 + 2\alpha + \beta)A + 2(1 + \alpha)B = 0 \qquad\qquad (15a)$$
$$n^1: \qquad (1 + 2\alpha + \beta)B = 1 \qquad\qquad\qquad\qquad (15b)$$

Thus,

$$B = \frac{1}{1 + 2\alpha + \beta} \qquad\qquad (16a)$$

$$A = -\frac{2(1 + \alpha)}{(1 + 2\alpha + \beta)^2} \qquad\qquad (16b)$$

and the solution to Eq. (11) is

$$x_k = C_1(-\alpha + \sqrt{\alpha^2 - \beta})^k + C_2(-\alpha - \sqrt{\alpha^2 - \beta})^k$$
$$- \frac{2(1 + \alpha)}{(1 + 2\alpha + \beta)^2} + \frac{k}{1 + 2\alpha + \beta} \qquad (17)$$

The constants C_1 and C_2 are again evaluated from the boundary conditions. If, for example, x_0 and x_1 are given, C_1 and C_2 are found from

$$x_0 = C_1 + C_2 - \frac{2(1 + \alpha)}{(1 + 2\alpha + \beta)^2} \qquad\qquad (18a)$$

$$x_1 = C_1(-\alpha + \sqrt{\alpha^2 - \beta}) - C_2(\alpha + \sqrt{\alpha^2 - \beta})$$
$$- \frac{2(1 + \alpha)}{(1 + 2 + \beta)^2} + \frac{1}{1 + 2\alpha + \beta} \qquad (18b)$$

BIBLIOGRAPHICAL NOTES

Sections 1.2 and 1.3: The elementary theory of maxima and minima is treated in all books on advanced calculus. The fundamental reference on the subject is

H. Hancock: "Theory of Maxima and Minima," Dover Publications, Inc., New York, 1960

Useful discussions in the context of modern optimization problems may be found in

T. N. Edelbaum: in G. Leitmann (ed.), "Optimization Techniques with Applications to Aerospace Systems," Academic Press, Inc., New York, 1962
G. Hadley: "Nonlinear and Dynamic Programming," Addison-Wesley Publishing Company, Inc., Reading, Mass., 1964
D. J. Wilde and C. S. Beightler: "Foundations of Optimization," Prentice-Hall, Inc., Englewood Cliffs, N.J., 1967

Sections 1.4 to 1.7: We shall frequently use problems in control as examples of applications of the optimization theory, and complete references are given in later chapters. A useful introduction to the elements of process dynamics and control is

D. R. Coughanowr and L. B. Koppel: "Process Systems Analysis and Control," McGraw-Hill Book Company, New York, 1965

The demonstration that optimal feedback gains are computable from the solution of a Riccati equation is an elementary special case of results obtained by Kalman. References to this work will be given after the development of the required mathematical groundwork.

Section 1.8: The references on the theory of maxima and minima in Secs 1.2 and 1.3 are also pertinent for Lagrange multipliers and constrained minimization. The generalization of the Lagrange multiplier rule to include inequality constraints is based on a theorem of Kuhn and Tucker, which is discussed in the books by Hadley and by Wilde and Beightler. There is a particularly enlightening development of the Kuhn-Tucker theorem in an appendix of

R. E. Bellman and S. E. Dreyfus: "Applied Dynamic Programming," Princeton University Press, Princeton, N.J., 1962

See also

H. P. Künzi and W. Krelle: "Nonlinear Programming," Blaisdell Publishing Company, Waltham, Mass., 1966

An interesting application of Lagrange multiplier–Kuhn-Tucker theory with process applications, known as geometric programming, is discussed in the text by Wilde and Beightler (cited above) and in

R. J. Duffin, E. L. Peterson, and C. Zener: "Geometric Programming," John Wiley & Sons, Inc., New York, 1967
C. D. Eben and J. R. Ferron: *AIChE J.*, **14**:32 (1968)

An alternative approach, taken by some of these authors, is by means of the theory of inequalities.

Sections 1.11 to 1.14: The generalized Euler equations were derived in

M. M. Denn and R. Aris: *Z. Angew. Math. Phys.*, **16**:290 (1965)

Applications to several elementary one-dimensional design problems are contained in

L. T. Fan and C. S. Wang: "The Discrete Maximum Principle," John Wiley & Sons, Inc., New York, 1964

Section 1.15: The interpretation of Lagrange multipliers as sensitivity coefficients follows the books by Bellman and Dreyfus and Hadley. The chain-rule development for one-dimensional staged processes is due to

F. Horn and R. Jackson: *Ind. Eng. Chem. Fundamentals*, **4**:487 (1965)

Section 1.16: The use of penalty functions appears to be due to Courant:

R. Courant: *Bull. Am. Math. Soc.*, **49**:1 (1943)

The theoretical basis is contained in supplements by H. Rubin and M. Kruskal (1950) and J. Moser (1957) to

R. Courant: "The Calculus of Variations," New York University Press, New York, 1945–1946

See also

A. V. Fiacco and G. P. McCormick: *Management Sci.*, **10**:601 (1964)
H. J. Kelley: in G. Leitmann (ed.), "Optimization Techniques with Applications to Aerospace Systems," Academic Press, Inc., New York, 1962

Appendix 1.1: Good introductions to the calculus of finite differences and difference equations may be found in

T. Fort: "Finite Differences and Difference Equations in the Real Domain," Oxford University Press, Fair Lawn, N.J., 1948
V. G. Jenson and G. V. Jeffreys: "Mathematical Methods in Chemical Engineering," Academic Press, Inc., New York, 1963
W. R. Marshall, Jr., and R. L. Pigford: "The Application of Differential Equations to Chemical Engineering Problems," University of Delaware Press, Newark, Del., 1947
H. S. Mickley, T. K. Sherwood, and C. E. Reed: "Applied Mathematics in Chemical Engineering," 2d ed., McGraw-Hill Book Company, New York, 1957

PROBLEMS

1.1. The chemical reaction $X \rightarrow Y \rightarrow Z$, carried out in an isothermal batch reactor, is described by the equations

$$\dot{x} = -k_1 x$$
$$\dot{y} = k_1 x - k_2 y$$
$$x + y + z = \text{const}$$

If the initial concentrations of X, Y, and Z are x_0, 0, and 0, respectively, and the values per unit mole of the species are c_X, c_Y, and c_Z, find the operating time θ which maximizes the value of the mixture in the reactor

$$\mathcal{P} = c_X[x(\theta) - x_0] + c_Y y(\theta) + c_Z z(\theta)$$

1.2. For the system in Prob. 1.1 suppose that k_1 and k_2 depend upon the temperature u in Arrhenius form

$$k_i = k_{i0} \exp\left(\frac{-E_i'}{u}\right) \qquad i = 1, 2$$

For fixed total operating time θ find the optimal constant temperature. Note the difference in results for the two cases $E_1 < E_2$ (exothermic) and $E_2 < E_1$ (endothermic).

1.3. The feed to a single-stage extractor contains a mass fraction x_0 of dissolved solute, and the extracting solvent contains mass fraction y_0. The mass fraction of solute in the effluent is x and in the exit solvent stream is y. Performance is approximately described by the equations

$$x + \sigma y = x_0 + \sigma y_0$$
$$y = Kx$$

where K is a constant (the distribution coefficient) and σ is the solvent-to-feed ratio. The cost of solvent purification may be taken approximately as

$$C = \frac{c(y - y_0)}{y_0}$$

and the net return for the process is the value of material extracted, $P(x_0 - x)$, less the cost of solvent purification. Find the degree of solvent purification y_0 which maximizes the net return.

1.4. The reversible exothermic reaction $X \rightleftharpoons Y$ in a continuous-flow stirred-tank reactor is described by the equations

$$0 = c_f - c - \theta r(c, c_f, T)$$
$$0 = T_f - T + \theta J r(c, c_f, T) - \theta Q$$

where c denotes the concentration of X, T the temperature, and the subscript f refers to the feed stream. r is the reaction rate, a function of c, c_f, and T, θ the residence time, J a constant, and Q the normalized rate of heat removal through a cooling coil. For fixed feed conditions find the design equations defining the heat-removal rate Q which maximizes the conversion, $c_f - c$. Do not use Lagrange multipliers. (*Hint:* First consider c a function of T and find the optimal temperature by implicit differentiation of the first equation. Then find Q from the second equation.) Obtain an explicit equation for Q for the first-order reaction

$$r = k_{10} \exp\left(\frac{-E_1'}{T}\right)(c_f - c) - k_{20} \exp\left(\frac{-E_2'}{T}\right)c$$

1.5. A set of experimental measurements $y_1, y_2, \ldots, y_n$ is made at points $x_1, x_2, \ldots, x_n$, respectively. The data are to be approximated by the equation

$$y = \alpha f(x) + \beta g(x)$$

where $f(x)$ and $g(x)$ are specified functions and the coefficients α and β are to be chosen to minimize the sum of squares of deviations between predicted and measured values of y

$$\min \varepsilon = \sum_{i=1}^{n} [\alpha f(x_i) + \beta g(x_i) - y_i]^2$$

Obtain explicit equations for α and β in terms of the experimental data. Generalize to relations of the form

$$y = \sum_{k=1}^{N} \alpha_k f_k(x)$$

Find the best values of α and β in the equation

$$y = \alpha + \beta x$$

for the following data:

x	0	1	2	3	4	5	6
y	0	4.5	11.0	15.5	17.0	26.5	30.5

1.6. A sequence of functions $\phi_1(x), \phi_2(x), \ldots, \phi_n(x)$ is called *orthogonal* with weighting $\rho(x)$ over an interval $[a,b]$ if it satisfies the relation

$$\int_a^b \rho(x)\phi_i(x)\phi_j(x)\, dx = 0 \qquad i \neq j$$

A given function $y(x)$ is to be approximated by a sum of orthogonal functions

$$y(x) = \sum_{n=1}^{N} c_n \phi_n(x)$$

Find the coefficients $c_1, c_2, \ldots, c_N$ which are best in the sense of minimizing the weighted integral of the square of the deviations

$$\min \mathcal{E} = \int_a^b \rho(x)[y(x) - \sum_{n=1}^{N} c_n \phi_n(x)]^2 \, dx$$

Show that the sequence $\sin x$, $\sin 2x$, $\sin 3x$, . . . is orthogonal over the interval $0 \le x \le \pi$ with weighting unity. Find the coefficients of the first four terms for approximating the functions

(a) $y = 1$ $0 \le x \le \pi$
(b) $y = x$ $0 \le x \le \pi$

(c) $y = \begin{cases} x & 0 \le x \le \dfrac{\pi}{2} \\[2mm] \pi - x & \dfrac{\pi}{2} \le x \le \pi \end{cases}$

Compare the approximate and exact functions graphically.

1.7 The cost in dollars per year of a horizontal vapor condenser may be approximated by

$$C = \beta_1 N^{-7/6} D^{-1} L^{-4/3} + \beta_2 N^{-0.2} D^{0.8} L^{-1} + \beta_3 N D L + \beta_4 N^{-1.8} D^{-4.8} L$$

where N is the number of tubes, D the average tube diameter in inches, and L the tube length in feet. β_1, β_2, β_3, and β_4 are coefficients that vary with fluids and construction costs. The first two terms represent cost of thermal energy; the third, fixed charges on the heat exchanger; and the fourth, pumping costs. Show that for all values of the coefficients the optimal cost distribution is 43.3 percent thermal energy, 53.3 percent fixed charges, and 3.33 percent pumping cost.
Show that the optimal value of the cost can be written

$$C^* = \left(\frac{\beta_1}{f_1}\right)^{f_1} \left(\frac{\beta_2}{f_2}\right)^{f_2} \left(\frac{\beta_3}{f_3}\right)^{f_3} \left(\frac{\beta_4}{f_4}\right)^{f_4}$$

where f_1, f_2, f_3, f_4 are respectively the fractions of the total cost associated with the first, second, third, and fourth terms in the cost. [*Hint:* If $A = \alpha C$, $B = \beta C$, and $\alpha + \beta = 1$, then $C = (A/\alpha)^\alpha (B/\beta)^\beta$.] Thus obtain explicit results for N, D, and L in terms of the β_i. Solve for $\beta_1 = 1.724 \times 10^5$, $\beta_2 = 9.779 \times 10^4$, $\beta_3 = 1.57$, $\beta_4 = 3.82 \times 10^{-2}$, corresponding to a desalination plant using low-pressure steam. (These results are equivalent to the formalism of geometric programming, but in this case they require only the application of the vanishing of partial derivatives at a minimum. The problem is due to Avriel and Wilde.)

1.8. For the minimization of a function of one variable, $\mathcal{E}(x)$, extend the analysis of Sec. 1.3 to obtain necessary and sufficient conditions for a minimum when both the first and second derivatives vanish. Prove that a point is a minimum if and only if the lowest-order nonvanishing derivative is positive and of even order.

1.9. Prove the converse of Eqs. (13) and (14) of Sec. 1.3, namely, that a quadratic form

$$\alpha x^2 + 2\beta xy + \gamma y^2$$

is positive definite if $\alpha > 0$, $\alpha\beta > \gamma^2$.

1.10. For the system described in Prob. 1.4 suppose that the cost of cooling is equal to ρQ. Find the design equation for the rate of heat removal which maximizes conversion less cost of cooling. Lagrange multipliers may be used.

1.11. Prove that when ε is convex (the hessian is positive definite) the minimum of ε subject to the linear constraints

$$\sum_{j=1}^{n} a_{ij}x_j = b_i \qquad i = 1, 2, \ldots, m < n$$

occurs at the minimum of the lagrangian with respect to $x_1, x_2, \ldots, x_n$.

1.12. Obtain the results of Sec. 1.13 by direct application of the Lagrange multiplier rule rather than by specialization of the results of Sec. 1.13. Extend the analysis to include the following two cases:

(a) x_N specified.

(b) Effluent is recycled, so that x_0 and x_N are related by an equation $x_0 = g(x_N)$.

1.13. The reversible reaction $A \rightleftharpoons B$ is to be carried out in a sequence of adiabatic beds with cooling between beds. Conversion in the nth bed follows the relation

$$\theta_n = \int_{x_{k-1}}^{x_k} \frac{d\xi}{r(T, \xi)}$$

where θ_n is the holding time, x_n the conversion in the stream leaving the nth bed, and $r(T,x)$ the reaction rate. In an adiabatic bed the temperature is a linear function of inlet temperature and of conversion. Thus the conversion can be expressed as

$$\theta_n = \int_{x_{k-1}}^{x_k} \frac{d\xi}{R(T_n, \xi)} \equiv F(x_{n-1}\, x_n, T_n)$$

where T_n is the temperature of the stream entering the nth bed. Obtain design equations and a computational procedure for choosing θ_n and T_n in order to maximize conversion in N beds while maintaining a fixed total residence time

$$\Theta = \sum_{n=1}^{N} \theta_n$$

(This problem has been considered by Horn and Kuchler and Aris.)

2
Optimization with Differential Calculus: Computation

2.1 INTRODUCTION

The previous chapter was concerned with developing algebraic conditions which must be satisfied by the optimal variables in minimizing an objective. The examples considered for detailed study were somewhat extraordinary in that the solutions presented could be obtained without recourse to extensive numerical calculation, but clearly this will rarely be the case in practice. In this chapter we shall consider several methods for obtaining numerical solutions to optimization problems of the type introduced in Chap. 1. An entire book could easily be devoted to this subject, and we shall simply examine representative techniques, both for the purpose of introducing the several possible viewpoints and of laying the necessary foundation for our later study of more complex situations. Two of the techniques which we wish to include for completeness are not conveniently derived from a variational point of view, so that in order to maintain continuity of the development the details are included as appendixes.

2.2 SOLUTION OF ALGEBRAIC EQUATIONS

The condition that the first partial derivatives of a function vanish at the minimum leads to algebraic equations for the optimal variables, and these equations will generally be highly nonlinear, requiring some iterative method of solution. One useful method is the Newton-Raphson method, which we shall derive for the solution of an equation in one unknown.

We seek the solution of the equation

$$f(x) = 0 \tag{1}$$

If our kth approximation to the solution is $x^{(k)}$ and we suppose that the $(k + 1)$st trial will be the exact solution, we can write the Taylor series expansion of $f(x^{(k+1)})$ about $f(x^{(k)})$ as

$$f(x^{(k+1)}) = 0 = f(x^{(k)}) + f'(x^{(k)})(x^{(k+1)} - x^{(k)}) + \cdots \tag{2}$$

Neglecting the higher-order terms and solving for $x^{(k+1)}$, we then obtain the recursive equation

$$x^{(k+1)} = x^{(k)} - \frac{f(x^{(k)})}{f'(x^{(k)})} \tag{3}$$

As an example of the use of Eq. (3) let us find the square root of 2 by solving

$$f(x) = x^2 - 2 = 0 \tag{4}$$

Equation (3) then becomes

$$x^{(k+1)} = x^{(k)} - \frac{x^{(k)2} - 2}{2x^{(k)}} = \frac{x^{(k)2} + 2}{2x^{(k)}} \tag{5}$$

If we take our initial approximation as $x^{(1)} = 1$, we obtain $x^{(2)} = 1.5$, $x^{(3)} = 1.4167$, etc., which converges rapidly to the value 1.4142. On the other hand, a negative initial approximation will converge to -1.4142, while an initial value of zero will diverge immediately.

When it converges, the Newton-Raphson method does so rapidly. In fact, convergence is quadratic, which means that the error $|x^{(k+1)} - x|$ is roughly proportional to the square of the previous error, $|x^{(k)} - x|$. Convergence will generally not occur, however, without a good first approximation. The difficulties which to be anticipated can be visualized from Fig. 2.1. The Newton-Raphson procedure is one of estimating the function by its tangent at the point $x^{(k)}$. Thus, as shown, the next estimate, $x^{(k+1)}$, is closer to the root, $x^{(k+2)}$ closer still, etc. Note, however, that convergence can be obtained only when the slope at $x^{(k)}$ has the same algebraic sign as the slope at the root. The starting point

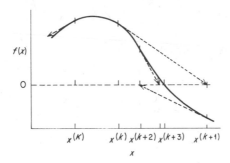

Fig. 2.1 Successive iterations of the Newton-Raphson method.

$x^{(K)}$, where $f'(x)$ is different in sign from $f'(\bar{x})$, will result in divergence from the solution.

For an optimization problem the function $f(x)$ in Eq. (1) is the derivative $\mathcal{E}'(x)$ of the function $\mathcal{E}(x)$ which is being minimized. Equation (3) then has the form

$$x^{(k+1)} = x^{(k)} - \frac{\mathcal{E}'(x^{(k)})}{\mathcal{E}''(x^{(k)})} \tag{6}$$

At the minimum, $\mathcal{E}'' > 0$, so that convergence is possible (but not guaranteed!) only if $\mathcal{E}''$ is positive for each approximation. For the minimization of a function of several variables, $\mathcal{E}(x_1, x_2, \ldots, x_n)$, the iteration formula analogous to Eq. (6) can be demonstrated by an equivalent development to be

$$x_i^{(k+1)} = x_i^{(k)} - \sum_{j=1}^{n} w_{ij} \frac{\partial \mathcal{E}(x_1^{(k)}, x_2^{(k)}, \ldots, x_n^{(k)})}{\partial x_j} \tag{7}$$

where w_{ij} is the *inverse* of the hessian matrix of $\mathcal{E}$, defined as the solution of the n linear algebraic equations

$$\sum_{j=1}^{n} w_{ij} \frac{\partial^2 \mathcal{E}(x_1^{(k)}, x_2^{(k)}, \ldots, x_n^{(k)})}{\partial x_j \, \partial x_p} = \delta_{ip} = \begin{cases} 1 & i = p \\ 0 & i \neq p \end{cases} \tag{8}$$

2.3 AN APPLICATION OF THE NEWTON-RAPHSON METHOD

As a somewhat practical example of the application of the Newton-Raphson method to an optimization problem we shall consider the consecutive-reaction sequence described in Sec. 1.12 and seek the optimal temperature in a single reactor. The reaction sequence is

$$X \rightarrow Y \rightarrow \text{products}$$

and taking the functions F and G as linear, the outlet concentrations of species X and Y are defined by the equations

$$x_0 - x - \theta k_{10}e^{-E_1'/u}x = 0 \tag{1a}$$
$$y_0 - y + \theta k_{10}e^{-E_1'/u}x - \theta k_{20}e^{-E_2'/u}y = 0 \tag{2a}$$

or, equivalently,

$$x = \frac{x_0}{1 + \theta k_{10}e^{-E_1'/u}} \tag{1b}$$

$$y = \frac{y_0}{1 + \theta k_{20}e^{-E_2'/u}} + \frac{\theta k_{10}e^{-E_1'/u}x_0}{(1 + \theta k_{10}e^{-E_1'/u})(1 + \theta k_{20}e^{-E_2'/u})} \tag{2b}$$

The object is to choose u to maximize $y + \rho x$ or to minimize

$$\varepsilon = -y - \rho x = -\frac{y_0}{1 + \theta k_{20}e^{-E_2'/u}}$$
$$- \frac{\theta k_{10}e^{-E_1'/u}x_0}{(1 + \theta k_{10}e^{-E_1'/u})(1 + \theta k_{20}e^{-E_2'/u})} - \frac{\rho x_0}{1 + \theta k_{10}e^{-E_1'/u}} \tag{3}$$

For manipulations it is convenient to define a new variable

$$v = e^{-E_1'/u} \tag{4a}$$
$$u = \frac{E_1'}{\ln v} \tag{4b}$$

so that the function to be minimized may be written

$$\varepsilon(v) = -\frac{y_0}{1 + \theta k_{20}v^\beta} - \frac{\theta k_{10}vx_0}{(1 + \theta k_{10}v)(1 + \theta k_{20}v^\beta)} - \frac{\rho x_0}{1 + \theta k_{10}v} \tag{5}$$

where β is the ratio of activation energies

$$\beta = \frac{E_2'}{E_1'} \tag{6}$$

The iteration procedure, Eq. (6) of the previous section, is then

$$v^{(k+1)} = v^{(k)} - \frac{\varepsilon'(v^{(k)})}{\varepsilon''(v^{(k)})} \tag{7}$$

We shall not write down here the lengthy explicit relations for ε' and ε''.

For purposes of computation the following numerical values were used:

$$x_0 = 1 \qquad\qquad y_0 = 0$$
$$k_{10} = 5.4 \times 10^{10} \qquad k_{20} = 4.6 \times 10^{17}$$
$$E_1' = 9,000 \qquad\qquad E_2' = 15,000$$
$$\rho = 0.3 \qquad\qquad \theta = 10$$

The first estimate of u was taken as 300, with the first estimate of v then calculated from Eq. (4a). Table 2.1 contains the results of the iteration

Table 2.1 Successive approximations by the Newton-Raphson method to the optimal temperature for consecutive reactions

Iteration	$v \times 10^{12}$	u	$-\varepsilon$	$-\varepsilon'$	$\varepsilon'' \times 10^{-22}$
Initial	9.3576224×10^{-2}	300.00000	0.33362777	3.4131804×10^{11}	37.249074
1	1.0098890	325.83690	0.53126091	1.2333063×10^{11}	14.815753
2	1.8423181	333.08664	0.59275666	3.4447925×10^{10}	7.4848782
3	2.3025517	335.85845	0.60156599	5.5884200×10^{9}	5.1916627
4	2.4101939	336.43207	0.60187525	2.3492310×10^{8}	4.7608371
5	2.4151284	336.45780	0.60187583	4.6704000×10^{5}	4.7418880
6	2.4151382	336.45785	0.60187583	8.6400000×10^{2}	4.7418504
7	2.4151382	336.45785	0.60187583	8.6400000×10^{2}	4.7418504

based on Eq. (7), where an unrealistically large number of significant figures has been retained to demonstrate the convergence. Starting rather far from the optimum, convergence is effectively obtained by the fourth correction and convergence to eight significant figures by the sixth.

It is found for this example that convergence cannot be obtained for initial estimates of u smaller distances to the right of the optimum. The reason may be seen in Figs. 2.2 and 2.3, plots of ε and ε' versus u, respectively. There is an inflection point in ε at $u \approx 347$, indicating a change in sign of ε'', which shows up as a maximum in ε'. Thus, care must be taken even in this elementary case to be sure that the initial estimate is one which will lead to convergence.

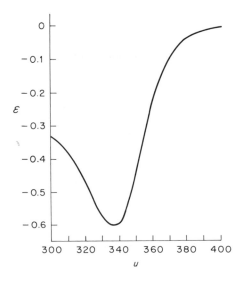

Fig. 2.2 Objective function versus temperature for consecutive reactions.

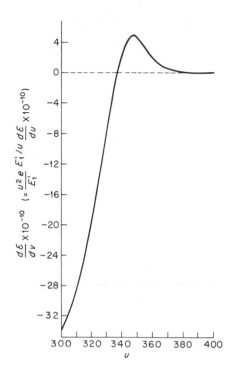

Fig. 2.3 Derivative of objective versus temperature for consecutive reactions.

2.4 FIBONACCI SEARCH

The use of the Newton-Raphson method assumes that the function and its derivatives are continuous and, most importantly, that derivatives are easily obtained. Convergence is not assured, and it may be inconvenient to evaluate derivatives, as would be the case if the function ε were not available in analytical form but only as the outcome of a physical or numerical experiment. Under certain assumptions an appealing alternative is available for functions of a single variable, although we must change our point of view somewhat.

We shall restrict our attention to functions $\varepsilon(x)$ which are unimodal; i.e., they must possess a single minimum and no maximum in the region of interest, but they need not be continuous. An example is shown in Fig. 2.4, where the region of interest is shown as extending from $x = 0$ to $x = L$. An important feature of such functions is the fact that given two observations $\varepsilon(x_1)$ and $\varepsilon(x_2)$ at points x_1 and x_2, respectively, we may say unequivocally that if $\varepsilon(x_1) > \varepsilon(x_2)$, the minimum lies somewhere in the interval $x_1 \leq x \leq L$, while if $\varepsilon(x_2) > \varepsilon(x_1)$, the minimum lies in the interval $0 \leq x \leq x_2$. Note that this is true even if x_1 and x_2 both lie on the same side of the minimum. The Fibonacci search procedure exploits this property of unimodal functions to eliminate in a systematic manner regions of the independent variable in which the minimum *cannot* occur. After N such eliminations there remains an *interval of uncertainty*, which must contain the minimum, and the procedure we shall describe here is the one requiring the minimum number of measurements (evaluations) of the function in order to reach a given uncertainty interval.

The algorithm requires that the function be measured at two symmetric points in the interval of interest, $0.382L$ and $0.618L$. If $\mathcal{E}(0.382L) > \mathcal{E}(0.618L)$, the region to the left of $0.382L$ is excluded, while if $\mathcal{E}(0.618L) > \mathcal{E}(0.382L)$, the region to the right of $0.618L$ is excluded. The process is then repeated for the new interval. Part of the efficiency results from the fact that one of the two previous measurements always lies inside the new interval at either the 38.2 or 61.8 percent location, so that only one new measurement need be made—at the point an equal distance on the other side of the midpoint from the point already in the interval. The proof that this is the *best* such algorithm in the sense defined above is straightforward, but of a very different nature from the variational analysis which we wish to emphasize, and so we bypass it here and refer the interested reader to Appendix 2.1.

To demonstrate the Fibonacci search algorithm we shall again consider the reactor example of the previous section, the minimization of

$$\mathcal{E}(u) = -\frac{y_0}{1 + \theta k_{20}e^{-E_2'/u}} - \frac{\theta k_{10}e^{-E_1'/u}x_0}{(1 + \theta k_{10}e^{-E_1'/u})(1 + \theta k_{20}e^{-E_2'/u})}$$
$$-\frac{\rho x_0}{1 + \theta k_{10}e^{-E_1'/u}} \quad (1)$$

for the values of parameters used previously. The initial interval of interest is $300 \leq u \leq 400$, and we have already seen that Newton-Raphson will converge from starting values in no more than half the region.

In this case $L = 100$, and the points at $0.382L$ and $0.618L$ are $u = 338.2$ and $u = 361.8$, respectively. Here,

$$\mathcal{E}(338.2) = -0.599 \qquad \mathcal{E}(361.8) = -0.193$$

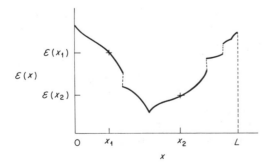

Fig. 2.4 A unimodal function.

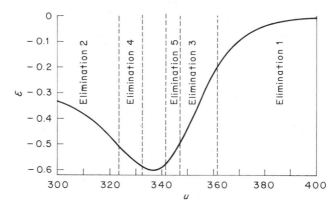

Fig. 2.5 Successive reductions of the interval of uncertainty for the optimal temperature by Fibonacci search.

so that the region $361.8 < u \leq 400$ is excluded. The remaining point is $u = 338.2$, which is at 61.8 percent of the new interval $300 \leq u \leq 361.8$, and the point which is symmetric at 38.2 percent is $u = 323.6076$. The process is then repeated, leading this time to the elimination of the region on the far left. The first several eliminations are shown graphically in Fig. 2.5, and the full sequence of calculations is shown in Table 2.2, where many extra significant figures have again been retained. The

Table 2.2 Successive iterations of the Fibonacci search method to the final interval of uncertainty of the optimal temperature for consecutive reactions

No. of Computations	Interval	$u_{0.382L}$	$\varepsilon(u_{0.382L})$	$u_{0.618L}$	$\varepsilon(u_{0.618L})$
1,2	300.000–400.000	338.20000	0.59912986	361.80000	0.19305488
3	300.000–361.800	323.60760	0.50729013	338.20000	0.59912986
4	323.608–361.800	338.20000	0.59912986	347.21050	0.49250240
5	323.608–347.211	332.62391	0.59023933	338.20000	0.59912986
6	332.624–347.211	338.20000	0.59912986	341.63842	0.57647503
7	332.624–341.638	336.06745	0.60174375	338.20000	0.59912986
8	332.624–338.200	334.75398	0.59943853	336.06745	0.60174375
9	334.754–338.200	336.06745	0.60174375	336.88362	0.60171592
10	334.754–336.884	335.56750	0.60119683	336.06745	0.60174375
11	335.568–336.884	336.06745	0.60174375	336.38086	0.60187065
12	336.067–336.884	336.38086	0.60187065	336.57184	0.60186444
13	336.067–336.572	336.26013	0.60184180	336.38086	0.60187065
14	336.260–336.572	336.38086	0.60187065	336.45277	0.60187581
15	336.381–336.572	336.45277	0.60187581	336.49889	0.60187435
16	336.381–336.499	336.42595	0.60187494	336.45277	0.60187581

rapid convergence can be observed in that 7 evaluations of $\mathcal{E}$ are required to reduce the interval of uncertainty to less than 10 percent of the original, 12 to less than 1 percent, and 16 to nearly 0.1 percent.

2.5 STEEP DESCENT

The Fibonacci search technique does not generalize easily to more than one dimension, and we must look further to find useful alternatives to the solution of the set of nonlinear equations arising from the necessary conditions. The method of steep descent, originated by Cauchy, is one such alternative, and it, too, is a technique for obtaining the solution without recourse to necessary conditions.

Our starting point is again the variational equation

$$\delta\mathcal{E} = \mathcal{E}(\bar{x}_1 + \delta x_1, \bar{x}_2 + \delta x_2) - \mathcal{E}(\bar{x}_1, \bar{x}_2) = \frac{\partial\mathcal{E}}{\partial x_1}\delta x_1 + \frac{\partial\mathcal{E}}{\partial x_2}\delta x_2 + o(\epsilon) \quad (1)$$

where the partial derivatives are evaluated at $\bar{x}_1$, $\bar{x}_2$, but we now suppose that $\bar{x}_1$ and $\bar{x}_2$ are *not* the values which cause $\mathcal{E}(x_1, x_2)$ to take on its minimum. Our problem, then, is to find values δx_1, δx_2 which will bring $\mathcal{E}$ closer to its minimum or, in other words, values δx_1, δx_2 which will ensure that

$$\delta\mathcal{E} < 0 \quad (2)$$

A choice which clearly meets this requirement is

$$\delta x_1 = -w_1\left(\frac{\partial\mathcal{E}}{\partial x_1}\right)_{\bar{x}_1, \bar{x}_2} \quad (3a)$$

$$\delta x_2 = -w_2\left(\frac{\partial\mathcal{E}}{\partial x_2}\right)_{\bar{x}_1, \bar{x}_2} \quad (3b)$$

where w_1 and w_2 are sufficiently small to allow $o(\epsilon)$ to be neglected in Eq. (1). We then have

$$\delta\mathcal{E} = -w_1\left(\frac{\partial\mathcal{E}}{\partial x_1}\right)^2 - w_2\left(\frac{\partial\mathcal{E}}{\partial x_2}\right)^2 < 0 \quad (4)$$

which satisfies Eq. (2), so that if w_1 and w_2 are small enough, the new value $\bar{x}_1 + \delta\bar{x}_1$, $\bar{x}_2 + \delta\bar{x}_2$ will be a better approximation of the minimum than the old.

An obvious generalization is to choose

$$\delta x_1 = -w_{11}\frac{\partial\mathcal{E}}{\partial x_1} - w_{12}\frac{\partial\mathcal{E}}{\partial x_2} \quad (5a)$$

$$\delta x_2 = -w_{21}\frac{\partial\mathcal{E}}{\partial x_1} - w_{22}\frac{\partial\mathcal{E}}{\partial x_2} \quad (5b)$$

where the matrix array $\begin{bmatrix} w_{11} & w_{12} \\ w_{21} & w_{22} \end{bmatrix}$ is positive definite and may depend on the position $\bar{x}_1$, $\bar{x}_2$, guaranteeing that

$$\delta \mathcal{E} = -\sum_{i=1}^{2} \sum_{j=1}^{2} w_{ij} \frac{\partial \mathcal{E}}{\partial x_i} \frac{\partial \mathcal{E}}{\partial x_j} < 0 \tag{6}$$

for small enough w_{ij}. All we have done, of course, is to define a set of directions

$$\delta x_i = -\sum_{j=1}^{2} w_{ij} \frac{\partial \mathcal{E}}{\partial x_j} \tag{7}$$

which will ensure a decrease in the function $\mathcal{E}$ for sufficiently small distances. We have at this point neither a means for choosing the proper *weighting matrix* of components w_{ij} nor one for determining how far to travel in the chosen direction. We simply know that if we guess values x_1, x_2, then by moving a small distance in the direction indicated by Eq. (7) for *any* positive definite matrix, we shall get an improved value.

2.6 A GEOMETRIC INTERPRETATION

It is helpful to approach the method of steep descent from a somewhat different point of view. Let us suppose that we are committed to moving a fixed distance Δ in the $x_1 x_2$ plane and we wish to make that move in such a way that $\mathcal{E}$ will be made as small as possible. That is, neglecting second-order terms, minimize the linear form

$$\delta \mathcal{E} = \left(\frac{\partial \mathcal{E}}{\partial x_1} \right)_{\bar{x}_1, \bar{x}_2} \delta x_1 + \left(\frac{\partial \mathcal{E}}{\partial x_2} \right)_{\bar{x}_1, \bar{x}_2} \delta x_2 \tag{1}$$

by choice of x_1, x_2, subject to the constraint

$$(\delta x_1)^2 + (\delta x_2)^2 - \Delta^2 = 0 \tag{2}$$

This is precisely the problem we solved in Sec. 1.9 using Lagrange multipliers and in Sec. 1.16 using penalty functions, and we may identify terms and write the solution as

$$\delta x_i = -\Delta \frac{\partial \mathcal{E}/\partial x_i}{[(\partial \mathcal{E}/\partial x_1)^2 + (\partial \mathcal{E}/\partial x_2)^2]^{1/2}} \qquad i = 1, 2 \tag{3}$$

That is, in Eqs. (3) of Sec. 2.5 w_i is defined as

$$w_i = \Delta \left[\left(\frac{\partial \mathcal{E}}{\partial x_1} \right)^2 + \left(\frac{\partial \mathcal{E}}{\partial x_2} \right)^2 \right]^{-1/2} \qquad i = 1, 2 \tag{4}$$

where Δ is a step size and w_i is the same for each variable but changes in value at each position.

The ratios

$$\frac{\partial \mathcal{E}/\partial x_i}{[(\partial \mathcal{E}/\partial x_1)^2 + (\partial \mathcal{E}/\partial x_2)^2]^{1/2}}$$

will be recognized as the set of direction cosines for the gradient at the point $(\bar{x}_1, \bar{x}_2)$, and Eq. (3) is simply a statement that the most rapid change in $\mathcal{E}$ will be obtained by moving in the direction of the negative gradient vector. A potential computational scheme would then be as follows:

1. Choose a pair of points $(\bar{x}_1, \bar{x}_2)$ and compute $\mathcal{E}$ and $\partial \mathcal{E}/\partial x_1$, $\partial \mathcal{E}/\partial x_2$.
2. Find the approximate minimum of $\mathcal{E}$ in the negative gradient direction; i.e., solve the single-variable problem

$$\min_{\lambda} \mathcal{E} \left(\bar{x}_1 - \lambda \frac{\partial \mathcal{E}}{\partial x_1}, \bar{x}_2 - \lambda \frac{\partial \mathcal{E}}{\partial x_2} \right)$$

   Call the minimizing point the new $(\bar{x}_1, \bar{x}_2)$. An approximate value of λ might be obtained by evaluating $\mathcal{E}$ at two or three points and interpolating or fitting a cubic in λ. A Fibonacci search could be used, though for an approximate calculation of this nature the number of function evaluations would normally be excessive.
3. Repeat until no further improvement is possible. In place of step 2 it might sometimes be preferable to use some fixed value $w_i = w$ and then recompute the gradient. If the new value of $\mathcal{E}$ is not less than $\mathcal{E}(\bar{x}_1, \bar{x}_2)$, the linearity assumption has been violated and w is too large and must be decreased.

It must be noted that this gradient method will find only a single minimum, usually the one nearest the starting point, despite the possible existence of more than one. Thus, the process must be repeated several times from different starting locations in order to attempt to find all the places where a local minimum exists.

A far more serious reservation exists about the method derived above, which might have been anticipated from the results of the previous section. We have, as is customary, defined distance by the usual euclidean measure

$$\Delta^2 = (\delta x_1)^2 + (\delta x_2)^2 \tag{5}$$

Since we shall frequently be dealing with variables such as temperatures, concentrations, valve settings, flow rates, etc., we must introduce normalizing factors, which are, to a certain extent, arbitrary. The proper distance constraint will then be

$$\Delta^2 = \alpha(\delta x_1)^2 + \gamma(\delta x_2)^2 \tag{6}$$

and the direction of steep descent, *and hence the rate of convergence*, will depend on the scale factors α and γ. Moreover, it is quite presumptuous to assume that the *natural* geometry of, say, a concentration–valve-setting space is even euclidean. A more general definition of distance is

$$\Delta^2 = \alpha(\delta x_1)^2 + 2\beta\,\delta x_1\,\delta x_2 + \gamma(\delta x_2)^2 \tag{7}$$

Here, the matrix array $\begin{bmatrix} \alpha & \beta \\ \beta & \gamma \end{bmatrix}$ is known as the *covariant metric tensor* and must be positive definite, or $\alpha > 0$, $\alpha\gamma - \beta^2 > 0$. It follows, then, from the results of Sec. 1.9 that the coefficients w_{ij} in Eq. (7) of Sec. 2.4 are†

$$w_{11} = \frac{\Delta\gamma}{D} \tag{8a}$$

$$w_{12} = w_{21} = -\frac{\Delta\beta}{D} \tag{8b}$$

$$w_{22} = \frac{\Delta\alpha}{D} \tag{8c}$$

where

$$D = \left\{ (\alpha\gamma - \beta^2) \left[\gamma\left(\frac{\partial\mathcal{E}}{\partial x_1}\right)^2 + \alpha\left(\frac{\partial\mathcal{E}}{\partial x_2}\right)^2 - 2\beta\frac{\partial\mathcal{E}}{\partial x_1}\frac{\partial\mathcal{E}}{\partial x_2} \right] \right\}^{\frac{1}{2}} \tag{8d}$$

There is no general way of determining a suitable geometry for a given problem a priori, and we have thus returned to essentially the same difficulty we faced at the end of the previous section. After an example we shall explore this question further.

2.7 AN APPLICATION OF STEEP DESCENT

As an example of the use of steep descent we again consider the consecutive-reaction sequence with linear kinetics but now with two reactors and, therefore, two temperatures to be chosen optimally. The general relations are

$$x_{n-1} - x_n - \theta k_1(u_n)x_n = 0 \qquad n = 1, 2, \ldots, N \tag{1}$$
$$y_{n-1} - y_n + \theta k_1(u_n)x_n - \theta k_2(u_n)y_n = 0 \qquad n = 1, 2, \ldots, N \tag{2}$$

† The matrix w_{ij} consists of a constant multiplied by the inverse of the covariant metric tensor. The inverse of an array a_{ij} is defined as the array b_{ij} such that

$$\sum_k a_{ik}b_{kj} = \begin{cases} 1 & i = j \\ 0 & i \neq j \end{cases}$$

where

$$k_i(u_n) = k_{i0}e^{-E_1'/u_n} \qquad i = 1, 2 \tag{3}$$

and the objective to be minimized by choice of u_1 and u_2 is

$$\mathcal{E} = -y_N - \rho x_N \tag{4}$$

For $N = 2$ this can be solved explicitly in terms of u_1 and u_2 as

$$\mathcal{E} = -\frac{1}{1 + \theta k_2(u_2)} \left\{ \frac{y_0}{1 + \theta k_2(u_1)} + \frac{\theta k_1(u_1)x_0}{[1 + \theta k_1(u_1)][1 + \theta k_2(u_1)]} \right\}$$
$$-\frac{\theta k_1(u_2)x_0}{[1 + \theta k_1(u_2)][1 + \theta k_2(u_2)][1 + \theta k_1(u_1)]}$$
$$-\frac{\rho x_0}{[1 + \theta k_1(u_2)][1 + \theta k_1(u_1)]} \tag{5}$$

or, for computational simplicity,

$$\mathcal{E} = -\frac{1}{1 + \theta k_{20}v_2^\beta} \left[\frac{y_0}{1 + \theta k_{20}v_1^\beta} + \frac{\theta k_{10}v_1 x_0}{(1 + \theta k_{10}v_1)(1 + \theta k_{20}v_1^\beta)} \right]$$
$$-\frac{\theta k_{10}v_2 x_0}{(1 + \theta k_{10}v_2)(1 + \theta k_{20}v_2^\beta)(1 + \theta k_{10}v_1)}$$
$$-\frac{\rho x_0}{(1 + \theta k_{10}v_2)(1 + \theta k_{10}v_1)} \tag{6}$$

where

$$v_n = e^{-E_1'/u_n} \tag{7}$$

and

$$\beta = \frac{E_2'}{E_1'} \tag{8}$$

Though they are easily computed, we shall not write down the cumbersome expressions for $\partial \mathcal{E}/\partial v_1$ and $\partial \mathcal{E}/\partial v_2$.

The values of the parameters were the same as those used previously in this chapter, except that θ was set equal to 5 in order to maintain comparable total residence times for the one- and two-reactor problems. The simplest form of steep descent was used, in which the correction is based on the relation [Eq. (3) of Sec. 2.5]

$$v_1^{\text{new}} = v_1^{\text{old}} - w_1 \frac{\partial \mathcal{E}}{\partial v_1} \tag{9a}$$

$$v_2^{\text{new}} = v_2^{\text{old}} - w_2 \frac{\partial \mathcal{E}}{\partial v_2} \tag{9b}$$

Since the relative effects of v_1 and v_2 should be the same, no scale factor is needed and w_1 and w_2 were further taken to be the same value w. Based

Table 2.3 Successive approximations to the optimal temperatures in two reactors for consecutive reactions by steep descent

Iteration	$v_1 \times 10^{12}$	u_1	$v_2 \times 10^{12}$	u_2	$-\varepsilon$	$-\partial\varepsilon/\partial v_1$	$-\partial\varepsilon/\partial v_2$
Initial	9.358×10^{-2}	300.0	9.358×10^{-2}	300.0	0.3340	1.749×10^{11}	1.749×10^{11}
1	1.843	333.1	1.842	333.1	0.6360	2.877×10^{10}	2.427×10^{10}
2	2.131	334.6	2.085	334.6	0.6472	1.771×10^{10}	1.354×10^{10}
3	2.308	335.9	2.220	335.4	0.6512	1.192×10^{10}	8.158×10^{9}
4	2.427	336.5	2.302	335.9	0.6530	8.434×10^{9}	5.067×10^{9}
5	2.511	337.0	2.353	336.1	0.6538	6.159×10^{9}	3.168×10^{9}
6	2.573	337.3	2.384	336.3	0.6542	4.602×10^{9}	1.959×10^{9}
7	2.619	337.5	2.404	336.4	0.6544	3.500×10^{9}	1.175×10^{9}
8	2.654	337.6	2.416	336.5	0.6546	2.701×10^{9}	6.623×10^{8}
9	2.681	337.8	2.422	336.5	0.6546	2.111×10^{9}	3.286×10^{8}
10	2.702	337.9	2.426	336.5	0.6547	1.667×10^{9}	1.139×10^{8}
11	2.719	338.0	2.427	336.5	0.6547	1.329×10^{9}	-2.092×10^{7}

upon the values of derivatives computed in the example in Sec. 2.3, this weighting w was initially taken as 10^{-23}, and no adjustment was required during the course of these particular calculations. The initial estimates of u_1 and u_2 were both taken as 300, corresponding to v_1 and v_2 of 9.358×10^{-14}. The full sequence of calculations is shown in Table 2.3.

Examination of the value of the objective on successive iterations shows that the approach to values of ε near the optimum is quite rapid, while ultimate convergence to the optimizing values of v_1 and v_2 is relatively slower. Such behavior is characteristic of steep descent. Some trial and error might have been required to find an acceptable starting value for w had a good a priori estimate not been available, and some improvement in convergence might have been obtained by estimating the optimal value of w at each iteration, but at the expense of more calculation per iteration.

This is an appropriate point at which to interject a comment on the usefulness of the consecutive-chemical-reaction problem as a computational example, for we shall use it frequently in that capacity. The examples done thus far indicate that the objective is relatively insensitive to temperature over a reasonably wide range about the optimum, a fortunate result from the point of view of practical operation. This insensitivity is also helpful in examining computational algorithms, for it means that the optimum lies on a plateau of relatively small values of derivatives, and computational schemes basing corrections upon calculated derivatives will tend to move slowly and have difficulty finding the true optimum. Thus, competitive algorithms may be compared under difficult circumstances.

2.8 THE WEIGHTING MATRIX

We can gain some useful information about a form to choose for the weighting matrix w_{ij} in steep descent by considering the behavior of the function $\mathcal{E}$ near its minimum, where a quadratic approximation may suffice. Since there is no conceptual advantage in restricting $\mathcal{E}$ to depend on only two variables, we shall let $\mathcal{E}$ be a function of n variables x_1, x_2, . . . , x_n and write

$$\mathcal{E}(x_1,x_2, \ldots ,x_n) = \mathcal{E}(\bar{x}_1,\bar{x}_2, \ldots ,\bar{x}_n) + \sum_{i=1}^{n} \frac{\partial \mathcal{E}}{\partial x_i} \delta x_i$$

$$+ \frac{1}{2} \sum_{i=1}^{n} \sum_{j=1}^{n} \frac{\partial^2 \mathcal{E}}{\partial x_i \, \partial x_j} \delta x_i \, \delta x_j + \cdots \quad (1)$$

or, for compactness of notation, we may denote the components of the gradient $\partial \mathcal{E}/\partial x_i$ by G_i and the hessian $\partial^2 \mathcal{E}/(\partial x_i \, \partial x_j)$ by H_{ij}, so that

$$\mathcal{E}(x_1,x_2, \ldots ,x_n) = \mathcal{E}(\bar{x}_1,\bar{x}_2, \ldots ,\bar{x}_n) + \sum_{i=1}^{n} G_i \, \delta x_i$$

$$+ \frac{1}{2} \sum_{i=1}^{n} \sum_{j=1}^{n} H_{ij} \, \delta x_i \, \delta x_j + \cdots \quad (2)$$

If we now minimize $\mathcal{E}$ by setting partial derivatives with respect to each δx_i to zero, we obtain

$$G_i + \sum_{j=1}^{n} H_{ij}\delta x_j = 0 \quad (3)$$

Since the hessian is presumed positive definite at (and hence near) the minimum, its determinant does not vanish and Eq. (3) can be solved by Cramer's rule to give

$$\delta x_i = - \sum_{j=1}^{n} w_{ij}G_j \quad (4)$$

where the weighting matrix satisfies the equations

$$\sum_{k=1}^{n} w_{ik}H_{kj} = \delta_{ij} = \begin{cases} 1 & i = j \\ 0 & i \neq j \end{cases} \quad (5)$$

That is, the proper weighting matrix is the inverse of the hessian. This is equivalent to the Newton-Raphson method described in Sec. 2.2. Note that it will not converge if the hessian fails to be positive definite at the point where the calculation is being made. Thus it will be of use only "near" the solution, but even here its use will require the calculation of *second* derivatives of the function $\mathcal{E}$, which may often be inconvenient or even difficult. It will, however, yield the minimum in a single step for a truly quadratic function and give quadratic convergence (when it converges) for all others.

Several methods have been devised which combine some of the simplicity of the most elementary steep-descent procedure with the rapid ultimate convergence of the Newton-Raphson method. This is done by computing a new weighting matrix w_{ij} at each iteration from formulas requiring only a knowledge of the first derivatives of $\mathcal{E}$, starting at the first iteration with the simple weighting $w_{ij} = w\delta_{ij}$. As the optimum is approached, the weighting matrix w_{ij} approaches that which would be obtained using the Newton-Raphson method but without the necessity of computing second derivatives. In many practical situations such a procedure is needed to obtain satisfactory convergence. A discussion of the basis of the computation of new w_{ij} would be out of place here, however, and we refer the reader to the pertinent literature for details.

2.9 APPROXIMATION TO STEEP DESCENT

In situations where it is inconvenient or impossible to obtain analytical expressions for the derivatives of the function $\mathcal{E}$ required in steep descent some form of numerical approximation must be used. The amount of computation or experimentation required to obtain accurate estimates of the gradient at each iteration is generally excessive in terms of the resulting improvement in the optimum, so that most techniques use crude estimates. A number of such procedures have been developed and tested, and the bibliographical notes will provide a guide for those interested in a comprehensive study. The one technique we shall discuss here is conceptually the simplest yet, surprisingly, one of the most effective.

The procedure can be motivated by reference to Fig. 2.6, where contours of constant $\mathcal{E}$ are drawn for a two-variable problem. The triangle ABC provides data for crudely estimating the gradient, and if A is

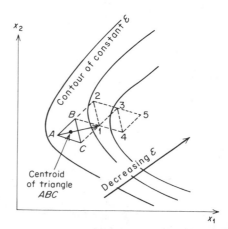

Fig. 2.6 Approximation to steep descent by reflection of triangles.

the worst of the three points, the line with an arrow from A through the centroid of the triangle provides a reasonable first approximation. Thus, we can simply reflect the triangle about the line BC to obtain a new triangle $B1C$ in a region of lower average value of ε. With only a single new calculation the worst point can again be found and the triangle reflected, leading to the triangle $B21$. The process is continually repeated, as shown.

There are several obvious difficulties with this overly simple procedure. First, continuous reflection is not adequate, for it will result in too slow convergence far from the optimum and too much correction and oscillation near the optimum. Thus, the point reflected through the centroid should be moved a fractional distance $\alpha > 1$ on the other side, and if the new point is also the worst, the distance moved should then be reduced by a fractional factor of $1/r < 1$. Hence, the triangle will be distorted in shape on successive iterations. In some cases the distortion will cause the triangle to degenerate to a line, so that more than three starting points will usually be needed. For n variables the number of points would then be greater than $n + 1$. (The coordinate of the centroid of N points is simply the sum of the individual coordinates divided by N.)

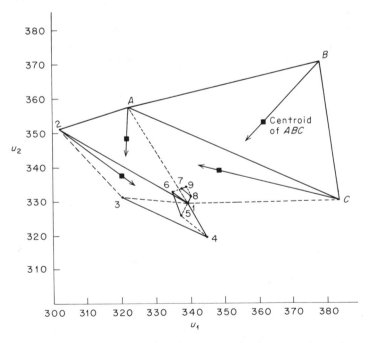

Fig. 2.7 Successive approximations to optimal temperatures in two reactors by the complex method.

Table 2.4 Successive approximations to the optimal temperatures in two reactors for consecutive reactions by the complex method

Iteration	u_1	u_2	$-\varepsilon$	u_1	u_2	$-\varepsilon$	u_1	u_2	$-\varepsilon$
Initial	383.0	330.0	0.0657	322.0	358.0	0.3844	**378.0**	**370.0**	**0.0222**
1	**383.0**	**330.0**	**0.0657**	322.0	358.0	0.3844	338.9	330.1	0.6431
2	302.4	351.6	0.5018	**322.0**	**358.0**	**0.3844**	338.9	330.1	0.6431
3	**302.4**	**351.6**	**0.5018**	319.9	331.7	0.5620	338.9	330.1	0.6431
4	343.8	319.9	0.6055	**319.9**	**331.7**	**0.5620**	338.9	330.1	0.6431
5	**343.8**	**319.9**	**0.6055**	336.4	326.6	0.6229	338.9	330.1	0.6431
6	334.3	332.9	0.6402	**336.4**	**326.6**	**0.6229**	338.9	330.1	0.6431
7	**334.3**	**332.9**	**0.6402**	336.7	334.1	0.6508	338.9	330.1	0.6431
8	339.7	331.7	0.6480	336.7	334.1	0.6508	**338.9**	**330.1**	**0.6431**
9	**339.7**	**331.7**	**0.6480**	336.7	334.1	0.6508	337.9	334.4	0.6529
10	336.0	335.6	0.6518	**336.7**	**334.1**	**0.6508**	337.9	334.4	0.6529
11	**336.0**	**335.6**	**0.6518**	337.0	335.5	0.6534	337.9	334.4	0.6529
12	338.2	334.6	0.6534	337.0	335.5	0.6534	**337.9**	**334.4**	**0.6529**
13	338.2	334.6	0.6534	**337.0**	**335.5**	**0.6534**	337.5	335.4	0.6538
14	**338.2**	**334.6**	**0.6534**	338.3	334.7	0.6537	337.5	335.4	0.6538

The geometrical figure made up by joining $N + 1$ points in an N-dimensional space is called a *simplex*, and the basic procedure is often called the *simplex method*, a name which can cause unfortunate confusion with the unrelated simplex method of linear programming (Appendix 2.2). Box has done a detailed study of the modifications required for systems with constraints and has called the resulting procedure the *complex method*. His paper is highly recommended for computational details.

As an example of the simplex-complex approach the optimal temperature problem for two stages was again solved, the function ε represented by Eq. (5) of Sec. 2.7, and the same parameters used as for steep descent. For simplicity only three points were used, and the starting values were chosen using random numbers in the interval $300 \leq u_1$, $u_2 \leq 400$. The reflection parameter α was taken as 1.3 and r as 2. The first nine iterations are shown graphically in Fig. 2.7, where the sequence of triangles is ABC, $AC1$, $A12$, 213, 314, 145, 561, 671, 781, 798; the first 14 are listed in Table 2.4, with the worst of each set of three points in boldface. The consequence of random starting values is an initial scan over the entire region of interest, followed by rapid approach to one region and systematic descent to the neighborhood of the optimum. It will be observed that convergence is moderately rapid; the final region computed is a small one containing the single value found previously by steep descent and the worst value of the objective close to the minimum found previously.

APPENDIX 2.1 OPTIMALITY OF FIBONACCI SEARCH

In this appendix we wish to prove that the Fibonacci search algorithm described and applied in Sec. 2.4 is optimal for unimodal functions in the sense that for a given final interval of uncertainty it is the sequence of steps requiring the fewest function evaluations. We shall actually prove an equivalent result, that for a given number of function evaluations the Fibonacci algorithm places the minimum within the smallest possible interval of uncertainty or, measuring length in units of the final interval of uncertainty, that it allows the largest possible initial interval $[0,L]$ such that after a given number of observations the minimum can be located within an interval of unit length. We develop the computational scheme by first proving the following result.

Let L_n be any number with the property that the minimum of the unimodal function $\mathcal{E}(x)$ on the interval $0 \leq x \leq L_n$ can be located within an interval of unit length by calculating at most n values and making comparisons. If we define

$$F_n = \sup L_n \tag{1}$$

then

$$F_n = F_{n-1} + F_{n-2} \qquad n \geq 2 \tag{2}$$
$$F_0 = F_1 = 1 \tag{3}$$

The notation sup in Eq. (1) stands for supremum, or least upper bound. We use this instead of "maximum" because while L_n will be able to approach the upper bound F_n arbitrarily closely, it will never in fact be able to take on this value. Our development follows that of Bellman and Dreyfus quite closely.

Clearly if we have made no observations, we can place the minimum within a unit interval only if we have started with a unit interval, so that $F_0 = 1$, and since a single observation is of no use whatsoever in placing the minimum, one observation is no better than none and $F_1 = 1$. For $n = 2$ the minimum may lie either in the interval $[0,x_2]$ or $[x_1,L_2]$, and so neither of these may exceed unity. It is obvious, however, that each of these intervals may be set equal to unity and the value of L_2 maximized by placing x_1 and x_2 equal distances from the center of the interval and as close together as possible; that is,

$$x_1 = 1 - \epsilon \qquad x_2 = 1 \tag{4}$$

in which case

$$L_2 = 2 - \epsilon \tag{5}$$

where ϵ is as small as we wish, and hence

$$F_2 = \sup L_2 = 2 = F_0 + F_1 \tag{6}$$

The remainder of the development proceeds by induction. We assume that

$$F_k = F_{k-1} + F_{k-2} \qquad k = 2, 3, \ldots, n-1 \tag{7}$$

and then prove that this implies Eq. (2) for $k = n$, completing the proof. Referring to Fig. 2.4, setting $L = L_n$, we see that if $\mathcal{E}(x_1) < \mathcal{E}(x_2)$, this implies that the minimum is in the interval $[0, x_2]$. Since we now have only $n - 2$ trials in which to locate the minimum in a unit interval, we must be left with a smaller interval than the largest possible having $n - 1$ trials. Hence

$$x_2 < F_{n-1} \tag{8}$$

Similarly, it is possible that the minimum might in fact be in the interval $[0, x_1]$, but an additional trial would be necessary to establish this, leaving $n - 3$ trials. Thus,

$$x_1 < F_{n-2} \tag{9}$$

On the other hand, if $\mathcal{E}(x_2) < \mathcal{E}(x_1)$, the minimum would lie in the interval $[x_1, L_n]$, and identical reasoning requires

$$L_n - x_1 < F_{n-1} \tag{10}$$

Combining the inequalities (9) and (10), we find

$$L_n < F_{n-1} + x_{n-1} \leq F_{n-1} + F_{n-2} \tag{11}$$

or

$$F_n = \sup L_n \leq F_{n-1} + F_{n-2} \tag{12}$$

Suppose we pick an interval

$$L_n = \left(1 - \frac{\epsilon}{2}\right)(F_{n-1} + F_{n-2}) \tag{13}$$

and place x_1 and x_2 symmetrically,

$$x_1 = \left(1 - \frac{\epsilon}{2}\right)F_{n-2} \qquad x_2 = \left(1 - \frac{\epsilon}{2}\right)F_{n-1} \tag{14}$$

so that the interval remaining after these two trials is as close as possible to the largest possible interval with $n - 2$ evaluations left. Such a placement of points is consistent with our induction hypothesis. It follows, then, that

$$F_n = \sup L_n \geq F_{n-1} + F_{n-2} \tag{15}$$

and combining the inequalities (12) and (15), we obtain the desired result

$$F_n = F_{n-1} + F_{n-2} \tag{2}$$

Let us note, in fact, that the placement of starting points in Eqs. (14) is in fact the optimum, for we have always an optimum interval

$$L_k = \left(1 - \frac{\epsilon}{2}\right) F_k \qquad 2 \le k \le n \tag{16}$$

and an optimum placing of one point, in the position $(1 - \epsilon/2)F_{k-1}$. The procedure is then as follows.

Choose the two starting points symmetrically, a distance $(F_{n-1}/F_n)L$ from each end of the interval $0 \le x \le L$, and make each successive observation at a point which is symmetric in the remaining interval with the observation which already exists in that interval. After n observations the minimum will then be located in the smallest interval possible.

One drawback of this procedure is that it requires advance knowledge of the number n of experiments. This is easily overcome by noting that the sequence defined by Eqs. (2) and (3), known as the *Fibonacci numbers*, may be found explicitly by solving the difference equation (2) by the method outlined in Appendix 1.1. The solution with initial conditions defined by Eq. (3) is

$$F_n = \frac{1 + \sqrt{5}}{2\sqrt{5}} \left(\frac{1 + \sqrt{5}}{2}\right)^n - \frac{1 - \sqrt{5}}{2\sqrt{5}} \left(\frac{1 - \sqrt{5}}{2}\right)^n \tag{17}$$

and for large n this is well approximated by

$$F_n \approx \frac{1 + \sqrt{5}}{2\sqrt{5}} \left(\frac{1 + \sqrt{5}}{2}\right)^n \tag{18}$$

Thus

$$\frac{F_{n-1}}{F_n} \approx \frac{2}{1 + \sqrt{5}} = 0.618 \tag{19}$$

and a near-optimum procedure is to place the first two points symmetrically in the interval $0 \le x \le L$ a distance $0.618L$ from each end and then procede as above. In this way the interval of uncertainty for the location of the minimum can be reduced by a factor of nearly 10,000 with only 20 evaluations of the function $\mathcal{E}(x)$. It is this near-optimum procedure, sometimes referred to as the *golden-section search* and indistinguishable from the true optimum for n greater than about 5, which is described and applied in Sec. 2.4.

APPENDIX 2.2 LINEAR PROGRAMMING

A large number of problems may be cast, exactly or approximately, in the following form:

$$\text{Min } \mathcal{E} = c_1 x_1 + c_2 x_2 + \cdots + c_n x_n = \sum_{j=1}^{n} c_j x_j \qquad (1)$$

with linear constraints

$$\sum_{j=1}^{n} a_{ij} x_j \,(\leq, =, \geq)b_i \qquad i = 1, 2, \ldots, m \qquad (2)$$

$$x_j \geq 0 \qquad j = 1, 2, \ldots, n \qquad (3)$$

where the coefficients c_j and a_{ij} are constants and the notation $(\leq, =, \geq)$ signifies that any of the three possibilities may hold in any constraint. This is the standard linear programming problem. We note that if there should be an inequality in any constraint equation (2), then by defining a new variable $y_i \geq 0$ we may put

$$\sum_{j=1}^{n} a_{ij} x_j \pm y_i = b_i \qquad (4)$$

Thus, provided we are willing to expand the number of variables by introducing "slack" and "surplus" variables, we can always work with equality constraints, and without loss of generality the standard linear programming problem can be written

$$\text{Min } \mathcal{E} = \sum_{j=1}^{n} c_j x_j \qquad (1)$$

$$\sum_{j=1}^{n} a_{ij} x_j = b_i \qquad i = 1, 2, \ldots, m < n \qquad (5)$$

$$x_j \geq 0 \qquad j = 1, 2, \ldots, n \qquad (3)$$

In this formulation we must generally have $m < n$ to prevent the existence of a unique solution or contradictory constraints.

The m linear algebraic equations (5) in the n variables $x_1, x_2, \ldots, x_n$ will generally have an infinite number of solutions if any exist at all. Any solution to Eqs. (5) satisfying the set of nonnegative inequalities [Eq. (3)] will be called a *feasible solution*, and if one exists, an infinite number generally will. A *basic feasible solution* is a feasible solution to Eqs. (5) in which no more than m of the variables $x_1, x_2, \ldots, x_n$ are nonzero. The number of basic feasible solutions is finite and is bounded from above by $n!/[m!(n-m)!]$, the number of combinations of n variables taken m at a time. It can be established without difficulty, although we shall not do so, that the optimal solution which minimizes the linear

form $\mathcal{E}$ subject to the constraints is always a basic feasible solution.
Thus, from among the infinite number of possible combinations of variables satisfying the constraints only a finite number are candidates for the
optimum. The simplex method of linear programming is a systematic
procedure for examining basic feasible solutions in such a way that $\mathcal{E}$ is
decreased on successive iterations until the optimum is found in a finite
number of steps. The number of iterations is generally of order $2m$.
Rather than devoting a great deal of space to the method we shall demonstrate its operation by a single example used previously by Glicksman.
      Let

$$\mathcal{E} = -5x - 4y - 6z \tag{6}$$
$$x + y + z \leq 100 \tag{7a}$$
$$3x + 2y + 4z \leq 210 \tag{7b}$$
$$3x + 2y \leq 150 \tag{7c}$$
$$x, y, z \geq 0 \tag{7d}$$

We first convert to equalities by introducing three nonnegative slack
variables u, v, w, and for convenience we include $\mathcal{E}$ as a variable in the
following set of equations:

$$x + y + z + u \qquad\qquad = 100 \tag{8a}$$
$$3x + 2y + 4z \qquad + v \qquad = 210 \tag{8b}$$
$$3x + 2y \qquad\qquad + w \quad = 150 \tag{8c}$$
$$\overline{5x + 4y + 6z \qquad\qquad + \mathcal{E} = 0} \tag{8d}$$
$$x, y, z, u, v, w \geq 0 \tag{8e}$$

The solid line is meant to separate the "convenience" equation (8d)
from the true constraint equations. A basic feasible solution to Eqs.
(8a), (8b), and (8c) is clearly $u = 100$, $v = 210$, $w = 150$, with x, y, and z
all equal to zero, in which case $\mathcal{E} = 0$.
      Now, computing the gradient of $\mathcal{E}$ from Eq. (6),

$$\frac{\partial \mathcal{E}}{\partial x} = -5 \qquad \frac{\partial \mathcal{E}}{\partial y} = -4 \qquad \frac{\partial \mathcal{E}}{\partial z} = -6 \tag{9}$$

so that improvement in $\mathcal{E}$ can be obtained by increasing x, y, and/or z.
Unlike most steep-descent procedures we choose here to move in only a
single coordinate direction, and since the magnitude of the gradient in
the z direction is greatest, we shall arbitrarily choose that one. From
the point of view of a general computer program this is simply equivalent
to comparing the coefficients in Eq. (8d) and choosing the most positive.
Since we choose to retain x and y as nonbasic (zero), Eqs. (8a) and (8b)
become

$$z + u = 100 \tag{10a}$$
$$4z + v = 210 \tag{10b}$$

[Equation (8c) does not involve z or it, too, would be included.]   As we are interested only in basic feasible solutions, either u or v must go to zero (since z will be nonzero) while the other remains nonnegative. If u goes to zero, $z = 100$ and $v = -190$, while if v goes to zero, $z = 52.5$ and $u = 48.5$.   That is,

$$z = \min \left({}^{100}\!/_1, {}^{210}\!/_4 \right) = 52.5 \tag{11}$$

and v is to be eliminated as z is introduced.   Again, the required calculation for a general program is simply one of dividing the right-hand column by the coefficient of z and comparing.

The Gauss-Jordan procedure is used to manipulate Eqs. (8) so that the new basic variables, u, z, and w each appear in only a single equation.   This is done by dividing the second equation by the coefficient of z, then multiplying it by the coefficient of z in each of the other equations, and subtracting to obtain an equivalent set of equations.   The second equation is chosen because it is the one in which v appears.   The result of this operation is

$$\frac{1}{4}x + \frac{1}{2}y \qquad + u - \frac{1}{4}v \qquad\qquad = 47\frac{1}{2} \tag{12a}$$
$$\frac{3}{4}x + \frac{1}{2}y + z \qquad + \frac{1}{4}v \qquad\qquad = 52\frac{1}{2} \tag{12b}$$
$$3x + 2y \qquad\qquad + w \quad = 150 \tag{12c}$$
$$\overline{\frac{1}{2}x + \quad y \qquad\qquad - \frac{3}{2}v \qquad + \varepsilon = -315} \tag{12d}$$

The basic feasible solution $(x, y, v = 0)$ is $u = 47\frac{1}{2}$, $z = 52\frac{1}{2}$, and $w = 150$.   From Eq. (12d)

$$\varepsilon = -315 - \frac{1}{2}x - y + \frac{3}{2}v \tag{13}$$

so that the value of ε in the basic variables is -315.

Repeating the above procedure, we now find that the largest positive coefficient in Eq. (12d) is that of y, and so y is to be the new basic (nonzero) variable.   The variable to be eliminated is found from examining the coefficients of the basic variables in Eq. (12) in the form

$$y = \min \left(\frac{47\frac{1}{2}}{\frac{1}{2}}, \frac{52\frac{1}{2}}{\frac{1}{2}}, \frac{150}{2} \right) = 75 \tag{14}$$

which corresponds to eliminating w.   Thus, we now use the Gauss-Jordan procedure again to obtain y in the third equation only, the result being

$$-\frac{1}{2}x \qquad + u - \frac{1}{4}v - \frac{1}{4}w \quad = 10 \tag{15a}$$
$$z \qquad + \frac{1}{4}v - \frac{1}{4}w \quad = 15 \tag{15b}$$
$$\frac{3}{2}x + y \qquad\qquad + \frac{1}{2}w \quad = 75 \tag{15c}$$
$$\overline{-x \qquad\qquad - \frac{3}{4}v - \frac{1}{2}w + \varepsilon = -390} \tag{15d}$$

The basic feasible solution is then $x, v, w = 0$, $u = 10$, $z = 15$, $y = 75$, and the corresponding value of $\varepsilon = -390$. There are no positive coefficients in Eq. (15d), and so this is the minimum. In terms of the original variables only, then, $x = 0$, $y = 75$, $z = 15$, and only two of the three original inequality constraints are at equality.

It should be clear from this example how a general computer code using only simple algebraic operations and data comparison could be constructed. The details of obtaining the required starting basic feasible solution for the iterative process under general conditions, as well as other facets of this extensive field, are left to the specialized texts on the subject. The interested reader should establish for himself that the one-at-a-time substitution used in the simplex method is the required result from steep descent when, instead of the quadratic-form definition of distance,

$$\Delta^2 = \sum_{i,j} g_{ij} x_i x_j \qquad (16)$$

a sum-of-absolute-value form is used,

$$\Delta = \sum_j |x_j| \qquad (17)$$

Linear programming can be used to define directions of steep descent in constrained nonlinear minimization problems by linearizing constraints and objective at each iteration and bounding the changes in the variables. The solution to the local linear programming problem will then provide the values at which the linearization for the next iteration occurs. Since gradients must be calculated for the linearization, this is essentially equivalent to finding the weighting matrix in a constrained optimization. The MAP procedure referred to in the bibliographical notes is such a method.

BIBLIOGRAPHICAL NOTES

Section 2.2: Discussions of the convergence properties of the Newton-Raphson and related techniques may be found in such books on numerical analysis as

C. E. Fröberg: "Introduction to Numerical Analysis," Addison-Wesley Publishing Company, Inc., Reading, Mass., 1965
F. B. Hildebrand: "Introduction to Numerical Analysis," McGraw-Hill Book Company, New York, 1956

Section 2.4 and Appendix 2.1: The derivation of the Fibonacci search used here is based on one in

R. E. Bellman and S. E. Dreyfus: "Applied Dynamic Programming," Princeton University Press, Princeton, N.J., 1962 *which contains references to earlier work of Kiefer and Johnson. An alternative approach may be found in*

D. J. Wilde: "Optimum Seeking Methods," Prentice-Hall, Inc., Englewood Cliffs, N.J., 1964
———— and C. S. Beightler: "Foundations of Optimization," Prentice-Hall, Inc., Englewood Cliffs, N.J., 1967

Sections 2.5 and 2.6: The development of steep descent was by Cauchy, although his priority was questioned by Sarrus:

A. Cauchy: *Compt. Rend.*, **25**:536 (1847)
F. Sarrus: *Compt. Rend.*, **25**:726 (1848)

Some discussion of the effect of geometry may be found in the books by Wilde cited above; see also

C. B. Tompkins: in E. F. Beckenbach (ed.), "Modern Mathematics for the Engineer, vol. I," McGraw-Hill Book Company, New York, 1956
T. L. Saaty and J. Bram: "Nonlinear Mathematics," McGraw-Hill Book Company, New York, 1964

Section 2.8: The most powerful of the procedures for computing the weighting matrix is probably a modification of a method of Davidon in

R. Fletcher and M. J. D. Powell: *Computer J.*, **6**:163 (1963)
G. W. Stewart, III: *J. Assoc. Comp. Mach.*, **14**:72 (1967)
W. I. Zangwill: *Computer J.*, **10**:293 (1967)

This has been extended to problems with constrained variables by

D. Goldfarb and L. Lapidus: *Ind. Eng. Chem. Fundamentals*, **7**:142 (1968)

Other procedures leading to quadratic convergence near the optimum are discussed in the books by Wilde and in

H. H. Rosenbrock and C. Storey: "Computational Techniques for Chemical Engineers," Pergamon Press, New York, 1966

Many of the standard computer codes use procedures in which the best direction for descent at each iteration is obtained as the solution to a linear programming problem. The foundations of such procedures are discussed in

G. Hadley: "Nonlinear and Dynamic Programming," Addison-Wesley Publishing Company, Inc., Reading, Mass., 1964

One such technique, known as MAP (method of approximation programming) is developed in

R. E. Griffith and R. A. Stewart: *Management Sci.*, **7**:379 (1961)

MAP has been applied to a reactor design problem in

C. W. DiBella and W. F. Stevens: *Ind. Eng. Chem. Process Design Develop.*, **4**:16 (1965)

Section 2.9:   The references cited for Sec. 2.8 are pertinent for approximate procedures as well, and the texts, in particular, contain extensive references to the periodical literature.   The simplex-complex procedure described here is a simplified version for unconstrained problems of a powerful technique for constrained optimization devised by

M. J. Box: *Computer J.*, **8**:42 (1965)

It is an outgrowth of the more elementary simplex procedure, first described in this section, by

W. Spendley, G. R. Hext, and F. R. Himsworth: *Technometrics*, **4**:441 (1962)

Appendix 2.2:   A delightful introduction to linear programming at a most elementary level is

A. J. Glicksman: "Introduction to Linear Programming and the Theory of Games," John Wiley & Sons, Inc., New York, 1963.

Among the standard texts are

G. B. Dantzig: "Linear Programming and Extensions," Princeton University Press, Princeton, N.J., 1963
S. I. Gass: "Linear Programming: Methods and Applications," 2d ed., McGraw-Hill Book Company, New York, 1964
G. Hadley: "Linear Programming," Addison-Wesley Publishing Company, Inc., Reading, Mass., 1962

Linear programming has been used in the solution of some optimal-control problems; see

G. Dantzig: *SIAM J. Contr.*, **A4**:56 (1966)
G. N. T. Lack and M. Enns: *Preprints 1967 Joint Autom. Contr. Conf.*, 474
H. A. Lesser and L. Lapidus: *AIChE J.*, **12**:143 (1966)
Y. Sakawa: *IEEE Trans. Autom. Contr.*, **AC9**:420 (1964)
H. C. Torng: *J. Franklin Inst.*, **278**:28 (1964)
L. A. Zadeh and B. H. Whalen: *IEEE Trans. Autom. Contr.*, **AC7**:45 (1962)

An extensive review of applications of linear programming in numerical analysis is

P. Rabinowitz: *SIAM Rev.*, **10**:121 (1968)

PROBLEMS

2.1. Solve Prob. 1.3 by both Newton-Raphson and Fibonacci search for the following values of parameters:

$$K = 3 \qquad \sigma = 1 \qquad x_0 = 0.05 \qquad \frac{C}{P} = 0.01$$

2.2. Obtain the optimal heat-removal rate in Prob. 1.4 by the Fibonacci search method, solving the nonlinear equation for T at each value of Q by Newton-Raphson for the following rate and parameters:

$$r = 2.5 \times 10^5 \exp\left(-\frac{20{,}000}{T}\right)(1 - c) - 2.0 \times 10^7 \exp\left(-\frac{40{,}000}{T}\right) c$$
$$J = 10^4 \qquad \theta = 10^{-2}$$

2.3. Derive Eqs. (7) and (8) of Sec. 2.2 for the multidimensional Newton-Raphson method.

2.4. The following function introduced by Rosenbrock is frequently used to test computational methods because of its highly curved contours:

$$\mathcal{E} = 100(x^2 - y^2)^2 + (1 - x)^2$$

Compare the methods of this chapter and that of Fletcher and Powell (cited in the bibliographical notes) for efficiency in obtaining the minimum with initial values $x = -1, y = -1$.

Solve each of the following problems, when appropriate, by steep descent, Newton-Raphson, and the complex method.

2.5. Solve Prob. 1.7 numerically and compare to the exact solution.

2.6. Using the data in Prob. 1.5, find the coefficients α and β which minimize both the sum of squares of deviations and the sum of absolute values of deviations. Compare the former to the exact solution.

2.7. The annual cost of a heavy-water plant per unit yield is given in terms of the flow F, theoretical stages N, and temperature T, as

$$\mathcal{E} = \frac{300F + 4{,}000NA + 80{,}000}{18.3(B - 1)}$$

where

$$A = 2 + 3 \exp\left(16.875 - \frac{T}{14.4}\right)$$

$$B = \frac{\phi(1 - \beta)}{0.6(1 - \beta)(\alpha\beta - 1) + 0.4\phi}$$

$$\phi = \frac{\alpha - 1}{\alpha}(\alpha\beta)^{N+1} + \beta - 1$$

$$\beta = \frac{F}{1{,}400}$$

$$\alpha = \exp\left(\frac{508}{T} - 0.382\right)$$

Find the optimal conditions. For computation the variables may be bounded by

$$250 \leq F \leq 500$$
$$1 \leq N \leq 20$$
$$223 \leq T \leq 295$$

(The problem is due to Rosenbrock and Storey, who give a minimum cost of $\mathcal{E} = 1.97 \times 10^4$.)

2.8. Obtain the optimal heat-removal rate in Prob. 1.4 by including the system equation for temperature in the objective by means of a penalty function. The parameters are given in Prob. 2.2.

2.9. Using the interpretation of Lagrange multipliers developed in Sec. 1.15, formulate a steep-descent algorithm for multistage processes such as those in Sec. 1.13 in which it is not necessary to solve explicitly for the objective in terms of the stage decision variables. Apply this algorithm to the example of Sec. 2.7.

2.10. Solve the following linear programming problem by the simplex method:

$$\min \, \mathcal{E} = 6x_1 + 2x_2 + 3x_3$$
$$30x_1 + 20x_2 + 40x_3 \geq 34$$
$$x_1 + x_2 + x_3 = 1$$
$$10x_1 + 70x_2 \leq 11$$
$$x_1, x_2, x_3 \geq 0$$

Hint: Find a basic feasible solution by solving the linear programming problem

$$\min \, \mathcal{E} = z_1 + z_2$$
$$30x_1 + 20x_2 + 40x_3 - w_1 + z_1 = 34$$
$$x_1 + x_2 + x_3 + z_2 = 1$$
$$10x_1 + 70x_2 + w_3 = 11$$
$$x_1, x_2, x_3, w_1, w_3, z_1, z_2 \geq 0$$

You should be able to deduce from this a general procedure for obtaining basic feasible solutions with which to begin the simplex method.

2.11. Formulate Prob. 1.4 for solution using a linear programming algorithm iteratively.

3
Calculus of Variations

3.1 INTRODUCTION

Until now we have been concerned with finding the optimal values of a finite (or infinite) number of *discrete* variables, x_1, x_2, . . . , x_n, although we have seen that we may use discrete variables to approximate a *function* of a continuous variable, say time, as in the optimal-control problem considered in Secs. 1.4 and 1.6.   In this chapter we shall begin consideration of the problem of finding an optimal function, and much of the remainder of the book will be devoted to this task.   The determination of optimal functions forms a part of the calculus of variations, and we shall be concerned in this first chapter only with the simplest problems of the subject, those which can be solved using the techniques of the differential calculus developed in Chap. 1.

3.2 EULER EQUATION

Let us consider a continuous differentiable function $x(t)$, where t has the range $0 \leq t \leq \theta$, and a function $\mathcal{F}$ which, for each value of t, depends explicitly on the value of $x(t)$, the derivative $\dot{x}(t)$, and t; that is,

$$\mathfrak{F} = \mathfrak{F}(x,\dot{x},t) \tag{1}$$

For each function $x(t)$ we may then define the integral

$$\mathcal{E}[x(t)] = \int_0^\theta \mathfrak{F}(x,\dot{x},t) \, dt \tag{2}$$

The number $\mathcal{E}[x(t)]$ depends not on a discrete set of variables but on an entire function, and is commonly referred to as a *functional*, a function of a function. We shall seek conditions defining the particular function $\bar{x}(t)$ which causes $\mathcal{E}$ to take on its minimum value.

We shall introduce an arbitrary continuous differentiable function $\eta(t)$, with the stipulation that if $x(0)$ is specified as part of the problem, then $\eta(0)$ must vanish, and similarly if $x(\theta)$ is specified, then $\eta(\theta)$ must vanish. If we then define a new function

$$x(t) = \bar{x}(t) + \epsilon\eta(t) \tag{3}$$

we see that $\mathcal{E}[\bar{x}(t) + \epsilon\eta(t)]$ depends only on the constant ϵ, since $\bar{x}(t)$ and $\eta(t)$ are (unknown) specified functions, and we may write

$$\mathcal{E}(\epsilon) = \int_0^\theta \mathfrak{F}(\bar{x} + \epsilon\eta, \, \dot{\bar{x}} + \epsilon\dot{\eta}, \, t) \, dt \tag{4}$$

Since $\bar{x}$ is the function which minimizes $\mathcal{E}$, it follows that the *function* $\mathcal{E}(\epsilon)$ must take on its minimum at $\epsilon = 0$, where its derivative must vanish, so that

$$\left.\frac{d\mathcal{E}}{d\epsilon}\right|_{\epsilon=0} = 0 \tag{5}$$

We can differentiate under the integral sign in Eq. (4), noting that

$$\frac{d\mathfrak{F}}{d\epsilon} = \frac{\partial \mathfrak{F}}{\partial x}\frac{dx}{d\epsilon} + \frac{\partial \mathfrak{F}}{\partial \dot{x}}\frac{d\dot{x}}{d\epsilon}$$

$$= \frac{\partial \mathfrak{F}}{\partial x}\eta + \frac{\partial \mathfrak{F}}{\partial \dot{x}}\dot{\eta} \tag{6}$$

and

$$\left.\frac{d\mathcal{E}}{d\epsilon}\right|_{\epsilon=0} = \int_0^\theta \left[\left(\frac{\partial \mathfrak{F}}{\partial x}\right)_{x=\bar{x}}\eta + \left(\frac{\partial \mathfrak{F}}{\partial \dot{x}}\right)_{x=\bar{x}}\dot{\eta}\right] dt = 0 \tag{7}$$

We integrate the second term by parts to obtain†

$$\int_0^\theta \frac{\partial \mathfrak{F}}{\partial \dot{x}}\dot{\eta} \, dt = \frac{\partial \mathfrak{F}}{\partial \dot{x}}\eta \Big|_0^\theta - \int_0^\theta \eta \frac{d}{dt}\frac{\partial \mathfrak{F}}{\partial \dot{x}} \, dt \tag{8}$$

---

† Note that we are assuming that $\partial \mathfrak{F}/\partial \dot{x}$ is continuously differentiable, which will not be the case if $\dot{x}(t)$ is not continuous. We have thus excluded the possible cases of interest which have cusps in the solution curve.

where we have dropped the notation $(\quad)_{x=\bar{x}}$ for convenience but it is understood that we are always evaluating functions at $\bar{x}$. We thus have, from Eq. (7),

$$\int_0^\theta \left(\frac{\partial \mathfrak{F}}{\partial x} - \frac{d}{dt} \frac{\partial \mathfrak{F}}{\partial \dot{x}} \right) \eta(t) \, dt + \frac{\partial \mathfrak{F}}{\partial \dot{x}} \eta \Big|_0^\theta = 0 \tag{9}$$

for arbitrary functions $\eta(t)$.

If $x(0)$ is specified, then since $\eta(0)$ must be zero, the term $(\partial \mathfrak{F}/\partial \dot{x})\eta$ must vanish at $t = 0$, with an identical condition at $t = \theta$. If $x(0)$ or $x(\theta)$ is not specified, we shall for the moment restrict our attention to functions $\eta(t)$ which vanish, but it will be necessary to return to consider the effect of the term $(\partial \mathfrak{F}/\partial \dot{x})\eta \Big|_0^\theta$. Thus, for arbitrary differentiable functions $\eta(t)$ which vanish at $t = 0$ and θ we must have

$$\int_0^\theta \left(\frac{\partial \mathfrak{F}}{\partial x} - \frac{d}{dt} \frac{\partial \mathfrak{F}}{\partial \dot{x}} \right) \eta(t) \, dt = 0 \tag{10}$$

If we choose the *particular* function

$$\eta(t) = w(t) \left(\frac{\partial \mathfrak{F}}{\partial x} - \frac{d}{dt} \frac{\partial \mathfrak{F}}{\partial \dot{x}} \right) \tag{11}$$

where $w(0) = w(\theta) = 0$ and $w(t) > 0,\ 0 < t < \theta$, we obtain

$$\int_0^\theta w(t) \left(\frac{\partial \mathfrak{F}}{\partial x} - \frac{d}{dx} \frac{\partial \mathfrak{F}}{\partial \dot{x}} \right)^2 dt = 0 \tag{12}$$

The only way in which an integral of a nonnegative function over a positive region can be zero is for the integrand to vanish identically, and so we obtain

$$\frac{\partial \mathfrak{F}}{\partial x} - \frac{d}{dt} \frac{\partial \mathfrak{F}}{\partial \dot{x}} = 0 \qquad 0 < t < \theta \tag{13}$$

which is known as *Euler's differential equation*.

If we now return to Eq. (9), we are left with

$$\left(\frac{\partial \mathfrak{F}}{\partial \dot{x}} \eta \right)_{t=\theta} - \left(\frac{\partial \mathfrak{F}}{\partial \dot{x}} \eta \right)_{t=0} = 0 \tag{14}$$

If $x(\theta)$ is not specified, $\eta(\theta)$ need not be zero and we may choose a function such that

$$\eta(\theta) = \epsilon_1 \left(\frac{\partial \mathfrak{F}}{\partial \dot{x}} \right)_{t=\theta} \tag{15}$$

where $\epsilon_1 > 0$ is nonzero only when $x(\theta)$ is not specified. Similarly, if $x(0)$ is not specified, we may choose $\eta(t)$ such that

$$\eta(0) = -\epsilon_2 \left(\frac{\partial \mathfrak{F}}{\partial \dot{x}} \right)_{t=0} \tag{16}$$

Thus, for these particular choices, we have

$$\epsilon_1 \left(\frac{\partial \mathfrak{F}}{\partial \dot{x}} \right)^2_{t=\theta} + \epsilon_2 \left(\frac{\partial \mathfrak{F}}{\partial \dot{x}} \right)^2_{t=0} = 0 \tag{17}$$

and it follows that $\partial \mathfrak{F} / \partial \dot{x}$ must vanish at an end at which $x(t)$ is not specified.

We have thus obtained a condition for the minimizing function equivalent to the vanishing of the first derivative for the simple calculus problem: *the optimal function $\bar{x}(t)$ must satisfy the Euler second-order differential equation*

$$\frac{d}{dt} \frac{\partial \mathfrak{F}}{\partial \dot{x}} = \frac{\partial^2 \mathfrak{F}}{\partial t \, \partial \dot{x}} + \frac{\partial^2 \mathfrak{F}}{\partial x \, \partial \dot{x}} \dot{x} + \frac{\partial^2 \mathfrak{F}}{\partial \dot{x}^2} \ddot{x} = \frac{\partial \mathfrak{F}}{\partial x} \tag{18}$$

with the boundary conditions

At $t = 0$: $x(0)$ specified   or $\dfrac{\partial \mathfrak{F}}{\partial \dot{x}} = 0$ $\qquad$ (19a)

At $t = \theta$: $x(\theta)$ specified   or $\dfrac{\partial \mathfrak{F}}{\partial \dot{x}} = 0$ $\qquad$ (19b)

It is sometimes convenient to use an alternative formulation for the Euler equation.   We consider

$$\frac{d}{dt} \left(\mathfrak{F} - \dot{x} \frac{\partial \mathfrak{F}}{\partial \dot{x}} \right) = \frac{\partial \mathfrak{F}}{\partial t} + \frac{\partial \mathfrak{F}}{\partial x} \dot{x} + \frac{\partial \mathfrak{F}}{\partial \dot{x}} \ddot{x} - \ddot{x} \frac{\partial \mathfrak{F}}{\partial \dot{x}} - \dot{x} \frac{d}{dt} \frac{\partial \mathfrak{F}}{\partial \dot{x}} \tag{20}$$

and substituting Eq. (18),

$$\frac{d}{dt} \left(\mathfrak{F} - \dot{x} \frac{\partial \mathfrak{F}}{\partial \dot{x}} \right) = \frac{\partial \mathfrak{F}}{\partial t} \tag{21}$$

If $\mathfrak{F}$ does not depend explicitly on t, then $\partial \mathfrak{F} / \partial t$ is zero and we may integrate Eq. (21) to obtain a first integral of the Euler equation

$$\mathfrak{F} - \dot{x} \frac{\partial \mathfrak{F}}{\partial \dot{x}} = \text{const} \tag{22}$$

Finally, if we seek not one but several functions $x_1(t)$, $x_2(t)$, . . . , $x_n(t)$ and $\mathfrak{F}$ is a function $\mathfrak{F}(x_1, x_2, \ . \ . \ . \ , x_n, \dot{x}_1, \dot{x}_2, \ . \ . \ . \ , \dot{x}_n, t)$, then the Euler equations (13) and (18) are written

$$\frac{d}{dt} \frac{\partial \mathfrak{F}}{\partial \dot{x}_i} - \frac{\partial \mathfrak{F}}{\partial x_i} = 0 \qquad i = 1, 2, \ . \ . \ . \ , n \tag{23}$$

$$\mathfrak{F} - \sum_{i=1}^{n} \dot{x}_i \frac{\partial \mathfrak{F}}{\partial \dot{x}_i} = \text{const} \tag{24}$$

It should be noted that the Euler equation is obtained from the necessary condition for a minimum, the setting of a first derivative to zero. Just as in the simple problems of differential calculus, we cannot distinguish from this condition alone between maxima, minima, and other stationary values. We shall refer to all solutions of the Euler equation as *extremals*, recognizing that not all extremals will correspond to minima.

3.3 BRACHISTOCHRONE

Solutions of Euler's equation for various functions yield the answers to many problems of interest in geometry and mechanics. For historical reasons we shall first consider a "control" problem studied (incorrectly) by Galileo and later associated with the Bernoullis, Newton, Leibniz, and L'Hospital. We assume that a particle slides along a wire without frictional resistance, and we seek the shape, or curve in space, which will enable a particle acted upon only by gravitational forces to travel between two points in the minimum time. This curve is called a *brachistochrone*.

The system is shown in Fig. 3.1, where m is the particle mass, s is arc length, t time, g the acceleration due to gravity, x the vertical coordinate, and z the horizontal. The speed is dx/dt, and the acceleration d^2x/dt^2. A force balance at any point along the curve then requires

$$m \frac{d^2s}{dt^2} = mg \sin\left(\frac{\pi}{2} - \alpha\right) = mg \frac{dx}{ds} \tag{1}$$

Dividing out the common factor m and multiplying both sides by ds/dt, we find

$$\frac{1}{2} \frac{d}{dt} \left(\frac{ds}{dt}\right)^2 = g \frac{dx}{dt} \tag{2}$$

or, integrating,

$$\frac{ds}{dt} = \sqrt{2g(x - c)} \tag{3}$$

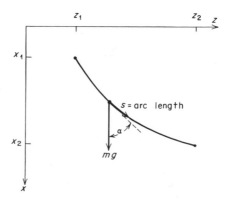

Fig. 3.1 Motion of a particle under the influence of gravity along a frictionless wire between two points.

where the constant c is determined from the initial height and velocity as

$$c = x(0) - \frac{1}{2g}\left(\frac{ds}{dt}\right)^2_{t=0} \tag{4}$$

We note that

$$ds^2 = dx^2 + dz^2 \tag{5}$$

or

$$ds = \sqrt{1 + \left(\frac{dx}{dz}\right)^2}\, dz \tag{6}$$

Thus, substituting Eq. (6) into (3) and integrating, we obtain the total time T to travel the path

$$\sqrt{2g}\, T = \int_{z=z_1}^{z=z_2} \sqrt{\frac{1 + (dx/dz)^2}{x - c}}\, dz \tag{7}$$

This is the integral to be minimized.

Identifying Eq. (7) with our notation in Sec. 3.2, we find

$$\mathfrak{F} = \sqrt{\frac{1 + \dot{x}^2}{x - c}} \tag{8}$$

Because $\mathfrak{F}$ is independent of t (or, as we have used it, z) we may use the first integral of the Euler equation [Eq. (22) of Sec. 3.2], which becomes

$$\frac{1}{\sqrt{(x - c)(1 + \dot{x}^2)}} = \text{const} \equiv \sqrt{\frac{1}{2b}} \tag{9}$$

Equation (9) is a first-order differential equation for x, with the constant b to be determined from the boundary conditions. If we presume that the solution is available and b is known, the control problem is solved, for we have an expression for the steering angle, $\arctan \dot{x}$, as a function of the height x. That is, we have the relation for an optimal nonlinear feedback controller. We do seek the complete solution, however, and to do so we utilize a standard trick and introduce a new variable ζ such that

$$\dot{x} = -\frac{\sin \zeta}{1 + \cos \zeta} \tag{10}$$

It then follows from Eq. (9) that

$$x = c + b(1 + \cos \zeta) \tag{11}$$

and

$$\frac{dz}{d\zeta} = \frac{dz}{dx}\frac{dx}{d\zeta} = b(1 + \cos \zeta) \tag{12}$$

or

$$z = k + b(\zeta + \sin \zeta) \tag{13}$$

Equations (11) and (13) are the parametric representations of the curve of minimum time, which is a cycloid. This curve maps out the locus of a point fixed on the circumference of a circle of radius b as the circle is rolled along the line $x = c$, as shown in Fig. 3.2. When the boundary conditions are applied, the solution is obtained as

$$x = x_1 + (x_2 - x_1) \frac{1 - \cos \zeta}{1 - \cos \zeta_f} \qquad 0 < \zeta < \zeta_f \tag{14a}$$

$$z = z_1 + (z_2 - z_1) \frac{\zeta + \sin \zeta}{\zeta_f + \sin \zeta_f} \qquad 0 < \zeta < \zeta_f \tag{14b}$$

where ζ_f is the solution of the algebraic equation

$$\frac{1 - \cos \zeta_f}{\zeta + \sin \zeta_f} = - \frac{x_2 - x_1}{z_2 - z_1} \tag{15}$$

The existence of such solutions is demonstrated in texts on the calculus of variations.

3.4 OPTIMAL LINEAR CONTROL

As a second and perhaps more useful example of the application of the Euler equation we shall return to the optimal-control problem formulated in Sec. 1.4. We have a system which varies in time according to the differential equation

$$\dot{x} = Ax + w \tag{1}$$

where A is a constant and w a function of time which we are free to choose. The optimal-control problem is to choose $w(t)$ so as to minimize the integral

$$\mathcal{E} = \frac{1}{2} \int_0^\theta (x^2 + c^2 w^2) \, dt \tag{2}$$

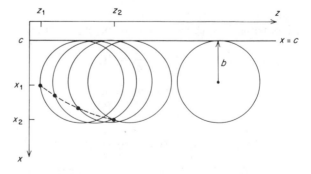

Fig. 3.2 Construction of a cycloid.

If we solve for w in Eq. (1) and substitute into Eq. (2), we have

$$\mathcal{E} = \frac{1}{2} \int_0^\theta [x^2 + c^2(\dot{x} - Ax)^2] \, dt \tag{3}$$

or

$$\mathcal{F} = \frac{1}{2}[x^2 + c^2(\dot{x} - Ax)^2] \tag{4}$$

The Euler equation is then

$$\frac{d}{dt} c^2(\dot{x} - Ax) = x - Ac^2(\dot{x} - Ax) \tag{5a}$$

or

$$\ddot{x} = \frac{1}{c^2}(1 + A^2c^2)x \tag{5b}$$

$x(0)$ is given. Since $x(\theta)$ is not specified, the second boundary condition is obtained from $\partial \mathcal{F}/\partial \dot{x} = 0$ at $t = \theta$, or

$$\dot{x} - Ax = 0 \qquad \text{at } t = \theta \tag{6}$$

Equation (5) may be solved quite easily to obtain $x(t)$, and hence $\dot{x}(t)$ and the optimal control policy $w(t)$. For purposes of implementation, however, it is convenient to have a feedback controller, where w is a function of the present state x, for such a result will be independent of the particular initial condition $x(0)$. Motivated by the results obtained in Chap. 1, we seek a solution in the form of a proportional controller

$$w(t) = M(t)x \tag{7}$$

and Eq. (1) becomes

$$\dot{x} = (A + M)x \tag{8}$$

Differentiating once, we have

$$\ddot{x} = A\dot{x} + M\dot{x} + \dot{M}x = \dot{M}x + (A + M)^2x \tag{9}$$

and this must equal the right-hand side of Eq. (5b). For $x \neq 0$, then, M must satisfy the Riccati ordinary differential equation

$$\dot{M} + 2AM + M^2 - \frac{1}{c^2} = 0 \tag{10}$$

with the boundary condition from Eq. (6)

$$M(\theta) = 0 \qquad \text{or} \qquad x(\theta) = 0 \tag{11}$$

The Riccati equation is solved by use of the standard transformation

$$M(t) = \frac{\dot{y}(t)}{y(t)} \tag{12}$$

with the resulting solution

$$M(t) = -\left\{A + \sqrt{\frac{1 + A^2c^2}{c^2}} \frac{1 - k \exp\left[2\sqrt{(1 + A^2c^2)/c^2}\, t\right]}{1 + k \exp\left[2\sqrt{(1 + A^2c^2)/c^2}\, t\right]}\right\} \tag{13}$$

where the constant k is determined from the condition $M(\theta) = 0$ as

$$k = \frac{\sqrt{1 + A^2c^2} - Ac}{\sqrt{1 + A^2c^2} + Ac} \exp\left(-2\sqrt{\frac{1 + A^2c^2}{c^2}}\,\theta\right) \tag{14}$$

In the limit as $\theta \to \infty$ we have $k \to 0$, and M becomes a constant

$$M = -\left(A + \sqrt{\frac{1 + A^2c^2}{c^2}}\right) \tag{15}$$

which is the result obtained in Sec. 1.5.   Here we do not satisfy $M(\theta) = 0$ but rather $x(\theta) = 0$.   We have again found, then, that *the optimal control for the linear system with quadratic objective function is a proportional feedback control, with the controller gain the solution of a Riccati equation, and a constant for an infinite control time.*

3.5 A DISJOINT POLICY

It sometimes happens that a problem can be formulated so that the integrand is independent of the derivative, $\dot{x}$; that is,

$$\mathcal{E}[x(t)] = \int_0^\theta \mathcal{F}(x,t)\, dt \tag{1}$$

The Euler equation then reduces to

$$0 = \frac{\partial \mathcal{F}}{\partial x} \tag{2}$$

In this case, however, a much stronger result is easily obtained without any considerations of differentiability or constraints on x.

If $\bar{x}(t)$ is the function which minimizes $\mathcal{E}$, then for all allowable functions $x(t)$

$$\int_0^\theta \mathcal{F}[\bar{x}(t),t]\, dt - \int_0^\theta \mathcal{F}[x(t),t]\, dt \leq 0 \tag{3}$$

In particular we shall take $x(t)$ different from $\bar{x}(t)$ only over the small interval $t_1 \leq t \leq t_1 + \Delta$, for arbitrary t_1.   We thus have

$$\int_{t_1}^{t_1+\Delta} \{\mathcal{F}[\bar{x}(t),t] - \mathcal{F}[x(t),t]\}\, dt \leq 0 \tag{4}$$

and using the mean-value theorem,

$$\Delta\{\mathcal{F}[\bar{x}(t),t] - \mathcal{F}[x(t),t]\} \leq 0 \tag{5}$$

where t is somewhere in the interval $t_1 \leq t \leq t_1 + \Delta$. Dividing by Δ and then letting Δ go to zero, we find

$$\mathfrak{F}[\bar{x}(t),t] \leq \mathfrak{F}[x(t),t] \tag{6}$$

for all t, since t_1 was arbitrary. That is, *to minimize the integral in Eq.* (1) *choose* $x(t)$ *to minimize the integrand at each value of* t. Such an optimal policy is called *disjoint*.

An immediate application arises in the design and control of a batch chemical reactor in which a single reaction is taking place. If, for simplicity, we suppose that the reaction is of the form

$$n_1 A \rightleftharpoons n_2 B \tag{7}$$

that is, A reacting to form B with some reverse reaction, then since there are no inflow and outflow streams in a batch reactor, the conservation of mass requires that the rate of change of the concentration of A, denoted by $a(t)$, be equal to the net rate of formation of A. That is

$$\dot{a} = -r(a,b,u) \tag{8}$$

where b is the concentration of B and u is the temperature. Since the amount of B present is simply the initial amount of B plus the amount formed by reaction, we have

$$b(t) = b(0) + \frac{n_1}{n_2}[a(0) - a(t)] \tag{9}$$

where n_1/n_2 represents the number of moles of B formed from a mole of A. Thus, we can write

$$\dot{a} = -r(a,u) \tag{10}$$

where it is understood that $a(0)$ and $b(0)$ enter as parameters.

Suppose now that we wish to achieve a given conversion in the minimum time by a programmed control of the reactor temperature $u(t)$. We can solve Eq. (10) formally for the total time as

$$\int_{a(\theta)}^{a(0)} \frac{da}{r(a,u)} = \theta \tag{11}$$

We wish to choose $u(t)$ to minimize θ, and we note that since $a(t)$ will be a monotonic function of time, when the solution is available, it will be sufficient to know a if we wish to know the time t. Thus, we may consider u as a function of a and write

$$\mathfrak{F}[u(a),a] = \frac{1}{r(a,u)} \tag{12}$$

Since $r > 0$, we shall minimize $1/r$ by maximizing r. The optimal policy is then to *choose the temperature at each instant to maximize the instanta-neous rate of reaction*. This result is probably intuitively obvious.

The kinetics of the reaction will usually be of the form

$$r(a,b,u) = p_1 a^{n_1} \exp\left(-\frac{E_1'}{u}\right) - p_2 b^{n_2} \exp\left(-\frac{E_2'}{u}\right)$$

$$= p_1 a^{n_1} \exp\left(-\frac{E_1'}{u}\right) - p_2 \left(b_0 + \frac{n_2}{n_1} a_0 - \frac{n_2}{n_1} a\right)^{n_2} \exp\left(-\frac{E_2'}{u}\right)$$

$$(13)$$

An internal maximum cannot occur when $E_1' > E_2'$ (an *endothermic* reaction), since $r > 0$, and the maximum is attained at the highest value of u allowable. This, too, is perhaps intuitively obvious. The only case of interest, then, is the exothermic reaction, $E_2' > E_1'$.

The maximum of Eq. (13) can be found by differentiation to give

$$u = \frac{E_2' - E_1'}{\ln\left\{\dfrac{p_2 E_2'}{p_1 E_1'} \dfrac{[b(0) + (n_2/n_1)a(0) - (n_2/n_1)a]^{n_2}}{a^{n_1}}\right\}} \tag{14}$$

By differentiating Eq. (14) we obtain

$$\frac{du}{da} = \frac{u^2}{n_1(E_2' - E_1')}\left(\frac{n_1^2}{a} + \frac{n_2^2}{b}\right) > 0 \tag{15}$$

and since $a(t)$ decreases in the course of the reaction, the optimal temperature is monotone decreasing.

There will, in general, be practical upper and lower limits on the temperature which can be obtained, say $u_* \le u \le u^*$, and it can be seen from Eq. (14) that a small or zero value of $b(0)$ may lead to a temperature which exceeds u^*. The starting temperature is then

$$u = \min\left\{u^*, \frac{E_2' - E_1'}{\ln\left[\dfrac{p_2 E_2'}{p_1 E_1'} \dfrac{b(0)^{n_2}}{a(0)^{n_1}}\right]}\right\} \tag{16}$$

If the starting temperature is u^*, this is maintained until Eq. (14) is satisfied within the allowable bounds. In a similar way, the temperature is maintained at u_* if the solution to Eq. (14) ever falls below this value. The optimal temperature policy will thus have the form shown in Fig. 3.3. The solution for the concentration $a(t)$ is obtained by substituting the relation for the optimal temperature into Eq. (10) and integrating the resulting nonlinear ordinary differential equation. The solutions have been discussed in some detail in the literature.

The monotonicity of the function $a(t)$ means that the policy which provides the minimum time to a given conversion must simultaneously provide the maximum conversion for a given time. This must be true,

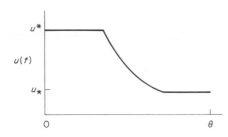

Fig. 3.3 Optimal temperature schedule in a batch reactor.

for were it possible to reach a smaller value of $a(\theta)$ [larger conversion $a(0) - a(\theta)$] in the same time θ, the required value of $a(\theta)$ could have been reached in a shorter time, which contradicts the assumption of a minimum-time solution. Since Eq. (10) can be applied to chemical reaction in a pipeline reactor in which diffusion is negligible when t is interpreted as residence time in the reactor, we have thus obtained at the same time the solution to the following important problem: *find the temperature profile in a pipeline reactor of fixed length which will maximize the conversion.*

3.6 INTEGRAL CONSTRAINTS

In applications of the calculus of variations it frequently happens that the integral

$$\mathcal{E}[x(t)] = \int_0^\theta \mathcal{F}(x,\dot{x},t)\, dt \tag{1}$$

must be minimized subject to a so-called *isoperimetric* constraint,

$$\int_0^\theta G(x,\dot{x},t)\, dt - b = 0 \tag{2}$$

We can obtain a formal solution to this problem in the following way. Let $\bar{x}(t)$ be the solution, and take

$$x(t) = \bar{x}(t) + \epsilon_1 \eta_1(t) + \epsilon_2 \eta_2(t) \tag{3}$$

where ϵ_1 and ϵ_2 are constants and $\eta_1(t)$ is an arbitrary differentiable function. For a given η_1, ϵ_1, and ϵ_2, the function η_2 will be determined by Eq. (2). The minimum of $\mathcal{E}$ subject to the constraint must then occur at $\epsilon_1 = \epsilon_2 = 0$, and we may write

$$\mathcal{E}(\epsilon_1,\epsilon_2) = \int_0^\theta \mathcal{F}(\bar{x} + \epsilon_1 \eta_1 + \epsilon_2 \eta_2,\, \dot{\bar{x}} + \epsilon_1 \dot{\eta}_1 + \epsilon_2 \dot{\eta}_2,\, t)\, dt \tag{4}$$

$$g(\epsilon_1,\epsilon_2) = \int_0^\theta G(\bar{x} + \epsilon_1 \eta_1 + \epsilon_2 \eta_2,\, \dot{\bar{x}} + \epsilon_1 \dot{\eta}_1 + \epsilon_2 \dot{\eta}_2,\, t)\, dt - b = 0 \tag{5}$$

The minimum of $\mathcal{E}$ subject to the constraint equation (5) is found by forming the lagrangian

$$\mathcal{L}(\epsilon_1, \epsilon_2) = \mathcal{E}(\epsilon_1, \epsilon_2) + \lambda g(\epsilon_1, \epsilon_2) \tag{6}$$

and setting partial derivatives with respect to ϵ_1 and ϵ_2 equal to zero at $\epsilon_1 = \epsilon_2 = 0$, we obtain

$$\int_0^\theta \left[\frac{\partial}{\partial x} (\mathcal{F} + \lambda G) - \frac{d}{dt} \frac{\partial}{\partial \dot{x}} (\mathcal{F} + \lambda G) \right] \eta_1(t) \, dt + \frac{\partial}{\partial \dot{x}} (\mathcal{F} + \lambda G) \eta_1 \Big|_0^\theta = 0 \tag{7a}$$

$$\int_0^\theta \left[\frac{\partial}{\partial x} (\mathcal{F} + \lambda G) - \frac{d}{dt} \frac{\partial}{\partial \dot{x}} (\mathcal{F} + \lambda G) \right] \eta_2(t) \, dt + \frac{\partial}{\partial \dot{x}} (\mathcal{F} + \lambda G) \eta_2 \Big|_0^\theta = 0 \tag{7b}$$

Since $\eta_1(t)$ is arbitrary, we obtain, as in Sec. 3.2, the Euler equation

$$\frac{d}{dt} \frac{\partial}{\partial \dot{x}} (\mathcal{F} + \lambda G) - \frac{\partial}{\partial x} (\mathcal{F} + \lambda G) = 0 \tag{8}$$

with the boundary conditions

At $t = 0$: $x(0)$ given   or $\dfrac{\partial}{\partial \dot{x}} (\mathcal{F} + \lambda G) = 0$ (9a)

At $t = \theta$: $x(\theta)$ given   or $\dfrac{\partial}{\partial \dot{x}} (\mathcal{F} + \lambda G) = 0$ (9b)

The constant Lagrange multiplier λ is found from Eq. (2). For n functions $x_1(t)$, $x_2(t)$, . . . , $x_n(t)$ and m constraints

$$\int_0^\theta G_i(x_1, x_2, \ldots, x_n, \dot{x}_1, \dot{x}_2, \ldots, \dot{x}_n, t) \, dt - b_i = 0$$

$$i = 1, 2, \ldots, m < n \tag{10}$$

The Euler equation is

$$\frac{d}{dt} \frac{\partial}{\partial \dot{x}_k} \left(\mathcal{F} + \sum_{j=1}^m \lambda_j G_j \right) - \frac{\partial}{\partial x_k} \left(\mathcal{F} + \sum_{j=1}^m \lambda_j G_j \right) = 0$$

$$k = 1, 2, \ldots, n \tag{11}$$

It is interesting to note that the Euler equation (8) is the same as that which would be obtained for the problem of extremalizing $\int_0^\theta G(x, \dot{x}, t) \, dt$ subject to the constraint $\int_0^\theta \mathcal{F}(x, \dot{x}, t) \, dt = $ const (after dividing by the constant $1/\lambda$). This duality is of the same type as that found in the previous section for the minimum-time and maximum-conversion problems.

3.7 MAXIMUM AREA

The source of the name isoperimetric is the historically interesting problem of finding the curve of fixed length which maximizes a given area.

If, in particular, we choose the area between the function $x(t)$ and the axis $x = 0$ between $t = 0$ and $t = \theta$, as shown in Fig. 3.4, for a curve $x(t)$ of fixed arc length L, we then seek to minimize

$$\mathcal{E} = - \text{area} = - \int_0^\theta x(t) \, dt \tag{1}$$

with

$$\text{Arc length} = \int_0^\theta (1 + \dot{x}^2)^{\frac{1}{2}} \, dt = L \tag{2}$$

The Euler equation (8) of Sec. 3.6 is then

$$\frac{d}{dt} \frac{\partial}{\partial \dot{x}} [-x + \lambda(1 + \dot{x}^2)^{\frac{1}{2}}] = \frac{\partial}{\partial x} [-x + \lambda(1 + \dot{x}^2)^{\frac{1}{2}}] \tag{3}$$

or

$$\lambda \frac{d}{dt} \frac{\dot{x}}{\sqrt{1 + \dot{x}^2}} = -1 \tag{4}$$

This has the solution

$$(x - \xi)^2 + (t - \tau)^2 = \lambda^2 \tag{5}$$

which is the arc of a circle. The constants ξ, τ, and λ can be evaluated from the boundary conditions and Eq. (2).

3.8 AN INVERSE PROBLEM

The second-order linear ordinary differential equation

$$\frac{d}{dt} p(t)\dot{x} + h(t)x + f(t) = 0 \tag{1}$$

appears frequently in physical problems, and, in fact, an arbitrary linear second-order equation

$$a(t)\ddot{y} + b(t)\dot{y} + c(t)y + d(t) = 0 \tag{2}$$

may be put in the *self-adjoint* form of Eq. (1) by introducing the change

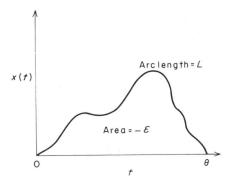

Fig. 3.4 Area enclosed by a curve of length L and the t axis.

of variable

$$x(t) = y(t) \exp \left(\int_0^t \frac{b - \dot{a}}{a} \, dt \right) \tag{3}$$

Because of the wide application of Eq. (1) it is of interest to determine under what conditions it corresponds to the Euler equation for some variational problem.

We shall write $h(t)$ and $f(t)$ as

$$h(t) = \dot{r}(t) - q(t) \qquad f(t) = \dot{m}(t) - n(t) \tag{4}$$

so that Eq. (1) can be rewritten

$$\frac{d}{dt} [p(t)\dot{x} + r(t)x + m(t)] = r(t)\dot{x} + q(y)x + n(t) \tag{5}$$

This will be of the form of an Euler equation

$$\frac{d}{dt} \frac{\partial \mathfrak{F}}{\partial \dot{x}} = \frac{\partial \mathfrak{F}}{\partial x} \tag{6}$$

provided that

$$\frac{\partial \mathfrak{F}}{\partial \dot{x}} = p(t)\dot{x} + r(t)x + m(t) \tag{7a}$$

$$\frac{\partial \mathfrak{F}}{\partial x} = r(t)\dot{x} + q(t)x + n(t) \tag{7b}$$

Integrating, from Eq. (7a)

$$\mathfrak{F} = \tfrac{1}{2}[p(t)\dot{x}^2 + 2r(t)x\dot{x} + 2m(t)\dot{x}] + \text{arbitrary function of } x \tag{8a}$$

while from (7b)

$$\mathfrak{F} = \tfrac{1}{2}[q(t)x^2 + 2r(t)x\dot{x} + 2n(t)x] + \text{arbitrary function of } \dot{x} \tag{8b}$$

so that, within an additive constant,

$$\mathfrak{F}(x,\dot{x},t) = \tfrac{1}{2}[p(t)\dot{x}^2 + 2r(t)x\dot{x} + q(t)x^2] + m(t)\dot{x} + n(t)x \tag{9}$$

with the boundary conditions then becoming

$$\text{At } t = 0 \text{ and } \theta: \qquad x \text{ fixed} \qquad \text{or} \qquad p\dot{x} + rx + m = 0 \tag{10}$$

As shown in the problems, these boundary conditions may be generalized somewhat.

In this next section we shall discuss an approximate method for determining stationary values for the integral $\int_0^\theta \mathfrak{F}(x,\dot{x},t) \, dt$, and as a consequence of the result of this section, such a procedure can always be applied to the solution of a linear second-order differential equation.

3.9 THE RITZ–GALERKIN METHOD

The Ritz-Galerkin method is an approximate procedure for the solution of variational problems. Here a particular functional form is assumed for the solution with only parameters undetermined, and the integral

$$\mathcal{E} = \int_0^\theta \mathfrak{F}(x,\dot{x},t)\, dt \tag{1}$$

which now depends only on the parameters, is minimized (or more generally, made stationary) by choice of the parameters. Suppose, for example, we seek an approximate solution of the form

$$x(t) \approx \psi_0(t) + \sum_{n=1}^{N} C_n\psi_n(t) \tag{2}$$

where ψ_0 satisfies any nonzero boundary specifications and the ψ_n are members of a complete set which vanish at any boundary where $x(t)$ is specified. The minimum of $\mathcal{E}$ is then found approximately by choosing the coefficients $C_1, C_2, \ldots, C_N$.

Substituting Eq. (2) into Eq. (1), we obtain

$$\mathcal{E} \approx \int_0^\theta \mathfrak{F}(\psi_0 + \Sigma C_n\psi_n, \dot{\psi}_0 + \Sigma C_n\dot{\psi}_n, t)\, dt \tag{3}$$

$\mathcal{E}$ is now a function of $C_1, C_2, \ldots, C_N$, and upon differentiating $\mathcal{E}$ with respect to each of the C_n and setting the result to zero

$$\frac{\partial \mathcal{E}}{\partial C_n} \approx \int_0^\theta \left(\frac{\partial \mathfrak{F}_{ap}}{\partial x} \psi_n + \frac{\partial \mathfrak{F}_{ap}}{\partial \dot{x}} \dot{\psi}_n \right) dt = 0 \qquad n = 1, 2, \ldots, N \tag{4}$$

Here the subscript *ap* refers to the fact that the partial derivatives are evaluated for the approximate solution. Equation (4) leads to N algebraic equations for the N coefficients after the integration is carried out, and is generally known as the *Ritz method*.

An alternative form which requires less computation when dealing directly with the solution of Euler differential equations is obtained by integrating Eq. (4) by parts. Thus

$$\int_0^\theta \left(\frac{\partial \mathfrak{F}_{ap}}{\partial x} - \frac{d}{dt}\frac{\partial \mathfrak{F}_{ap}}{\partial \dot{x}} \right) \psi_n(t)\, dt + \frac{\partial \mathfrak{F}_{ap}}{\partial \dot{x}} \psi_n \bigg|_0^\theta = 0 \tag{5}$$

and since either ψ_n is zero (x specified) or $\partial \mathfrak{F}/\partial \dot{x}$ vanishes (*natural* boundary condition), we obtain, finally, the *Galerkin form*

$$\int_0^\theta \left(\frac{\partial \mathfrak{F}_{ap}}{\partial x} - \frac{d}{dt}\frac{\partial \mathfrak{F}_{ap}}{\partial \dot{x}} \right) \psi_n(t)\, dt = 0 \qquad n = 1, 2, \ldots, N \tag{6}$$

Since an exact solution would satisfy the Euler equation,

$$\frac{\partial \mathfrak{F}}{\partial x} - \frac{d}{dt}\frac{\partial \mathfrak{F}}{\partial \dot{x}} = 0 \tag{7}$$

we may consider the quantity in parentheses in Eq. (6) as the *residual* at each value of t which remains when an approximation is used in the left-hand side of the differential equation. Thus, writing

$$\frac{\partial \mathfrak{F}_{ap}}{\partial x} - \frac{d}{dt}\frac{\partial \mathfrak{F}_{ap}}{\partial \dot{x}} \equiv R(C_1, C_2, \ldots, C_N, t) \tag{8}$$

Galerkin's method may be written

$$\int_0^\theta R(C_1, C_2, \ldots, C_N, t)\psi_n(t)\, dt = 0 \qquad n = 1, 2, \ldots, N \tag{9}$$

This procedure is often used for the approximate solution of differential equations which are not Euler equations but without the rigorous justification provided here.

For demonstration, let us consider the approximate solution of the nonhomogeneous Bessel equation

$$\frac{d}{dt}\left(t\frac{dx}{dt}\right) + tx - t = 0 \qquad x(0) \text{ finite} \qquad x(1) = 0 \tag{10}$$

Comparing with Eq. (1) of Sec. 3.8 this corresponds to

$$p(t) = t \qquad q(t) = -t \qquad r(t) = 0 \qquad m(t) = 0 \qquad n(t) = 1 \tag{11}$$

and provided that $\dot{x}(0)$ remains finite, the boundary conditions of Eq. (10) of Sec. 3.9 are satisfied, the condition at $t = 0$ being the natural boundary condition.

We shall use for approximating functions

$$\psi_n = t^{n-1}(1 - t) \qquad n = 1, 2, \ldots, N \tag{12a}$$

which vanish at $t = 1$ and remain finite at $t = 0$. A function ψ_0 is not needed. Thus, substituting the approximation for $x(t)$ into the left-hand side of Eq. (10), the residual is written

$$R(C_1, C_2, \ldots, C_N, t) = \sum_{n=1}^{N} C_n \left\{\frac{d}{dt}\left[t\frac{d}{dt}t^{n-1}(1 - t)\right] + t^n(1 - t)\right\} - t \tag{12b}$$

and the coefficients are found by solving the *linear* algebraic equations

$$\int_0^1 R(C_1, C_2, \ldots, C_N, t)t^{n-1}(1 - t)\, dt = 0$$
$$n = 1, 2, \ldots, N \tag{13}$$

In particular, for a one-term approximation

$$N = 1: \qquad R(C_1, t) = -C_1 + (C_1 - 1)t - C_1 t^2 \tag{14}$$

and the solution of Eq. (13) is $C_1 = -0.4$, or

$$N = 1: \qquad x(t) \approx -0.4(1 - t) \tag{15}$$

Table 3.1 Comparison of the first- and second-order approximations by the Ritz-Galerkin method to the exact solution

t	$-x(t)$		
	$N = 1$	$N = 2$	$Exact$
0.0	0.40	0.310	0.307
0.1	0.36	0.304	0.303
0.2	0.32	0.293	0.294
0.3	0.28	0.276	0.277
0.4	0.24	0.253	0.255
0.5	0.20	0.225	0.226
0.6	0.16	0.191	0.192
0.7	0.12	0.152	0.151
0.8	0.08	0.107	0.106
0.9	0.04	0.056	0.055
1.0	0.0	0.0	0.0

For a two-term approximation,

$$N = 2: \quad R(C_1, C_2, t) = (C_2 - C_1) + (C_1 - 4C_2 - 1)t \\ + (C_2 - C_1)t^2 - C_2 t^3 \quad (16)$$

and Eq. (13) becomes simultaneous algebraic equations for C_1 and C_2, with solutions $C_1 = -0.31$, $C_2 = -0.28$, or

$$N = 2: \quad x(t) \approx -0.31(1 - t) - 0.28t(1 - t) \quad (17)$$

The exact solution is

$$x(t) = 1 - \frac{J_0(t)}{J_0(1)} \quad (18)$$

where J_0 is the Bessel function of zero order and first kind.   Table 3.1 compares the two approximations to the exact solution.

3.10 AN EIGENVALUE PROBLEM

The homogeneous equation

$$\frac{d}{dt}\left(t\frac{dx}{dt}\right) + \lambda tx = 0 \quad (1)$$

with boundary conditions

$$x(0) \text{ finite} \qquad x(1) = 0 \quad (2)$$

clearly has the trivial solution $x(t) = 0$, but for certain *eigenvalues*, or *characteristic numbers*, λ, a nontrivial solution can be obtained.   These numbers are solutions of the equation

$$J_0 (\sqrt{\lambda}) = 0 \tag{3}$$

and the first two values are $\lambda_1 = 5.30$, $\lambda_2 = 30.47$.

The Ritz-Galerkin method may be used to estimate the first several eigenvalues. If we again use the approximation

$$x(t) \approx \sum_{n=1}^{N} C_n t^{n-1}(1 - t) \tag{4}$$

then

$$N = 1: \qquad R(C_1,t) = -C_1 + \lambda C_1 t - C_1 t^2 \tag{5}$$

and Eq. (13) of Sec. 3.9 becomes

$$N = 1: \qquad C_1(\lambda - 6) = 0 \tag{6}$$

Since if C_1 vanishes we obtain only the undesired trivial solution, we can satisfy Eq. (6) only by

$$\lambda = 6 \tag{7}$$

Thus, by setting $N = 1$ we obtain an estimate of the first (smallest) eigenvalue.

Similarly, for a two-term expansion

$$N = 2: \qquad R(C_1,C_2,t) = (C_2 - C_1) + (\lambda C_1 - 4C_2)t \\ + \lambda(C_2 - C_1)t^2 - \lambda C_2 t^3 \tag{8}$$

and we obtain the two equations

$$C_1(5\lambda - 30) + C_2(2\lambda - 10) = 0 \tag{9a}$$
$$C_1(2\lambda - 10) + C_2(\lambda - 10) = 0 \tag{9b}$$

These homogeneous equations have a nontrivial solution only if the determinant of coefficients vanishes, or λ satisfies the quadratic equation

$$(5\lambda - 30)(\lambda - 10) - (2\lambda - 10)^2 = 0 \tag{10}$$

The two roots are

$$\lambda_1 = 5.86 \qquad \lambda_2 = 34.14$$

giving a (better) estimate of the first eigenvalue and an initial estimate of the second. In general, an N-term expansion will lead to an Nth-order polynomial equation whose roots will approximate the first N eigenvalues.

3.11 A DISTRIBUTED SYSTEM

In all the cases which we have considered thus far the Euler equations have been ordinary differential equations. In fact, the extension of the methods of this chapter, which are based only on differential calculus, to the study of distributed systems is straightforward, and this section and the next will be devoted to typical problems. We shall return to such problems again in Chap. 11, and there is no loss in continuity in skipping this section and the next until that time.

Let us suppose that x is a function of *two* independent variables, which we shall call t and z, and that $x(t,z)$ is completely specified when $t = 0$ or 1 for all z and when $z = 0$ or 1 for all t. We seek the function $x(t,z)$ which minimizes the double integral

$$\mathcal{E}[x(t,z)] = \int_0^1 \int_0^1 \left[\frac{1}{2}\left(\frac{\partial x}{\partial t}\right)^2 + \frac{1}{2}\left(\frac{\partial x}{\partial z}\right)^2 + \phi(x) \right] dt\, dz \tag{1}$$

where $\phi(x)$ is any once-differentiable function of x.

If we call $\bar{x}(t,z)$ the optimum, we may write

$$x(t,z) = \bar{x}(t,z) + \epsilon\eta(t,z) \tag{2}$$

where ϵ is a small number and $\eta(t,z)$ is a function which must vanish at $t = 0$ or 1 and $z = 0$ or 1. For a particular function η the integral $\mathcal{E}$ depends only on ϵ and may be written

$$\mathcal{E}(\epsilon) = \int_0^1 \int_0^1 \left[\frac{1}{2}\left(\frac{\partial \bar{x}}{\partial t} + \epsilon\frac{\partial \eta}{\partial t}\right)^2 + \frac{1}{2}\left(\frac{\partial \bar{x}}{\partial z} + \epsilon\frac{\partial \eta}{\partial z}\right)^2 \right.$$
$$\left. + \phi(\bar{x} + \epsilon\bar{x}) \right] dt\, dz \tag{3}$$

The minimum of $\mathcal{E}$ occurs when $\epsilon = 0$ by the definition of $\bar{x}$, and at $\epsilon = 0$ the derivative of $\mathcal{E}$ with respect to ϵ must vanish. Thus,

$$\frac{d\mathcal{E}}{d\epsilon}\bigg|_{\epsilon=0} = \int_0^1 \int_0^1 \left[\frac{\partial \bar{x}}{\partial t}\frac{\partial \eta}{\partial t} + \frac{\partial \bar{x}}{\partial z}\frac{\partial \eta}{\partial z} + \phi'(\bar{x})\eta \right] dt\, dz = 0 \tag{4}$$

But

$$\int_0^1 \frac{\partial \bar{x}}{\partial t}\frac{\partial \eta}{\partial t}\, dt = \frac{\partial \bar{x}}{\partial t}\eta\bigg|_{t=0}^{t=1} - \int_0^1 \frac{\partial^2 \bar{x}}{\partial t^2}\eta\, dt \tag{5}$$

and the first term vanishes by virtue of the restrictions on η. A similar integration may be carried out on the second term with respect to z, and we obtain

$$\int_0^1 \int_0^1 \left[-\frac{\partial^2 \bar{x}}{\partial t^2} - \frac{\partial^2 \bar{x}}{\partial z^2} + \phi'(\bar{x}) \right] \eta(t,z)\, dt\, dz = 0 \tag{6}$$

Since η is arbitrary except for the boundary conditions and obvious differentiability requirements, we may set

$$\eta(t,z) = w(t,x)\left[-\frac{\partial^2 \bar{x}}{\partial t^2} - \frac{\partial^2 \bar{x}}{\partial z^2} + \phi'(x) \right] \tag{7}$$

where

$$w(0,z) = w(1,z) = w(t,0) = w(t,1) = 0$$
$$w(t,z) > 0 \qquad \text{for } t, z \neq 0, 1 \tag{8}$$

Thus,

$$\int_0^1 \int_0^1 w(t,z) \left[\frac{\partial^2 \bar{x}}{\partial t^2} + \frac{\partial^2 \bar{x}}{\partial z^2} - \phi'(\bar{x}) \right]^2 dt\, dz = 0 \tag{9}$$

and it follows that $\bar{x}$ must satisfy the Euler *partial* differential equation

$$\frac{\partial^2 x}{\partial t^2} + \frac{\partial^2 x}{\partial z^2} - \phi'(x) = 0 \qquad x \text{ specified at } \begin{matrix} t = 0, 1 \\ z = 0, 1 \end{matrix} \tag{10}$$

The partial differential equation

$$\frac{\partial^2 x}{\partial t^2} + \frac{\partial^2 x}{\partial z^2} - kF(x) = 0 \tag{11}$$

arises often in applications, as, for example, in two-dimensional mass or heat transfer with nonlinear generation. The method discussed in Sec. 3.9 may then be extended to this case to determine approximate solutions.

3.12 CONTROL OF A DISTRIBUTED PLANT

Many control problems of interest require the control of a system which is distributed in space by the adjustment of a variable operating only at a physical boundary. The complete study of such systems must await a later chapter, but we can investigate a simple situation with the elementary methods of this chapter. One of our reasons for doing so is to demonstrate still another form of an Euler equation.

A prototype of many important problems is that of adjusting the temperature distribution in a homogeneous slab to a desired distribution by control of the fuel flow to the furnace. The temperature in the slab x satisfies the linear heat-conduction equation

$$\frac{\partial x_1}{\partial t} = \frac{\partial^2 x_1}{\partial z^2} \qquad \begin{matrix} 0 < t \leq \theta \\ 0 < z < 1 \end{matrix} \tag{1}$$

with a zero initial distribution and boundary conditions

$$\frac{\partial x_1}{\partial z} = 0 \qquad\qquad \text{at } z = 1$$
$$\frac{\partial x_1}{\partial z} = \rho(x_1 - x_2) \qquad \text{at } z = 0 \tag{2}$$

Here x_2 is the temperature of the furnace. The first boundary condition is a symmetry condition, while the second reflects Newton's law of cooling, that the rate of heat transfer at the surface is proportional to the temperature difference. The furnace temperature satisfies the ordinary differential equation

$$\tau \frac{dx_2}{dt} + x_2 = u(t) \qquad x_2(0) = 0 \tag{3a}$$

where $u(t)$ is the control variable, a normalized fuel feed rate.

The object of control is to obtain a temperature distribution $x_1^*(z)$ in time θ or, more precisely, to minimize a measure of the deviation between the actual and desired profiles. Here we shall use a least-square criterion, so that we seek to minimize

$$\mathcal{E}[u] = \int_0^1 [x_1^*(z) - x_1(\theta,z)]^2 \, dz \tag{3b}$$

In order to use the procedures of this chapter it will be necessary to obtain an explicit representation of $\mathcal{E}[u]$ in terms of u. Later we shall find methods for avoiding this cumbersome (and often impossible) step. Here, however, we may use either conventional Laplace transform or Fourier methods to obtain the solution

$$x_1(\theta,z) = \int_0^\theta K(\theta - t, z)u(t) \, dt \tag{4}$$

where

$$K(\tau,z) = \frac{\alpha^2 \cos \alpha(1 - z)}{\cos \alpha - (\alpha/\rho) \sin \alpha} e^{-\alpha^2 \tau}$$

$$+ 2\alpha^2 \sum_{i=1}^\infty \frac{\cos (1 - z)\beta_i}{(\alpha^2 - \beta_i^2)\left(\dfrac{1}{\rho} + \dfrac{1 + \rho}{\beta_i^2}\right) \cos \beta_i} e^{-\beta_i^2 \tau} \tag{5}$$

with $\alpha = 1/\sqrt{\tau}$ and β_i the real roots of

$$\beta \tan \beta = \rho \tag{6}$$

Thus,

$$\mathcal{E}[u] = \int_0^1 \left[x_1^*(z) - \int_0^\theta K(\theta - t, z)u(t) \, dt \right]^2 dz \tag{7}$$

Now, in the usual manner, we assume that $\bar{u}(t)$ is the optimum, $\eta(t)$ is an arbitrary differentiable function, and ϵ is a small number and let

$$u(t) = \bar{u} + \epsilon\eta \tag{8}$$

The function of ϵ which results by evaluating Eq. (7) is

$$\mathcal{E}[\bar{u} + \epsilon\eta] = \int_0^1 [x_1^*(z)]^2 \, dz - 2 \int_0^1 x_1^*(z) \int_0^\theta K(\theta - t, z)\bar{u}(t) \, dt \, dz$$

$$- 2\epsilon \int_0^1 x_1^*(z) \int_0^\theta K(\theta - t, z)\eta(t) \, dt \, dz$$

$$+ \int_0^1 \left[\int_0^\theta K(\theta - t, z)\bar{u}(t) \, dt \right]^2 dz$$

$$+ 2\epsilon \int_0^1 \left[\int_0^\theta K(\theta - \tau, z)u(\tau) \, d\tau \right] \left[\int_0^\theta K(\theta - t, z)\eta(t) \, dt \right] dz$$

$$+ \epsilon^2 \int_0^1 \left[\int_0^\theta K(\theta - t, z)\eta(t) \, dt \right]^2 dz \quad (9)$$

and, evaluating $d\mathcal{E}/d\epsilon$ at $\epsilon = 0$,

$$\frac{d\mathcal{E}}{d\epsilon}\bigg|_{\epsilon=0} = -2 \int_0^1 x_1^*(\tau) \int_0^\theta K(\theta - t, z)\eta(t) \, dt \, dz$$

$$+ 2 \int_0^1 \left[\int_0^\theta K(\theta - \tau, z)\bar{u}(\tau) \, d\tau \right] \left[\int_0^\theta K(\theta - t, z)\eta(t) \, dt \right] dz = 0 \quad (10)$$

There is no difficulty in changing the order of time and space integration, and the inner two integrals in the second term may be combined to give

$$\int_0^\theta \eta(t) \left[\int_0^1 K(\theta - t, z)x_1^*(z) \, dz \right.$$

$$\left. - \int_0^\theta \int_0^1 K(\theta - t, z)K(\theta - \tau, z)\bar{u}(\tau) \, dz \, d\tau \right] dt = 0 \quad (11a)$$

or, setting the arbitrary function $\eta(t)$ proportional to the quantity in brackets,

$$\int_0^1 K(\theta - t, z)x_1^*(z) \, dz$$

$$- \int_0^\theta \left[\int_0^1 K(\theta - t, z)K(\theta - \tau, z) \, dz \right] \bar{u}(\tau) \, d\tau = 0 \quad (11b)$$

Equation (11) is an integral equation for the function $\bar{u}(t)$. It is simplified somewhat by defining

$$\psi(t) = \int_0^1 K(\theta - t, z)x_1^*(z) \, dz \quad (12a)$$

$$G(t,\tau) = \int_0^1 K(\theta - t, z)K(\theta - \tau, z) \, dz \quad (12b)$$

so that the Euler equation becomes

$$\int_0^\theta G(t,\tau)u(\tau) \, d\tau = \psi(t) \quad (13)$$

This is a Fredholm integral equation of the first kind. Analytical solutions can be obtained in some special cases, or numerical methods may be used. An obvious approach is to divide the interval $0 \le t \le \theta$ into N even increments of size Δt and write

$$G_{ij} = G(i \, \Delta t, j \, \Delta t) \, \Delta t \quad (14a)$$

$$u_j = u(j \, \Delta t) \quad (14b)$$

$$\psi_i = \psi(i \, \Delta t) \quad (14c)$$

We obtain an approximation to $u(t)$ by solving the linear algebraic equations approximating Eq. (13)

$$\sum_{j=1}^{N} G_{ij}u_j = \psi_i \qquad i = 1, 2, \ldots, N \tag{15}$$

BIBLIOGRAPHICAL NOTES

Sections 3.2 and 3.3: An extremely good introduction to the calculus of vairations by means of detailed study of several examples, including the brachistochrone, is

G. A. Bliss: "Calculus of Variations," Carus Mathematical Monograph, Mathematical Association of America, Open Court Publishing Co., La Salle, Ill., 1925

Other good texts on the calculus of variations include

N. I. Akhiezer: "The Calculus of Variations," Blaisdell Publishing Company, Waltham, Mass., 1962

G. A. Bliss: "Lectures on the Calculus of Variations," The University of Chicago Press, Chicago, 1946

O. Bolza: "Lectures on the Calculus of Variations," Dover Publications, Inc., New York, 1961

R. Courant and D. Hilbert: "Methods of Mathematical Physics," vol. 1, Interscience Publishers, Inc., New York, 1953

L. A. Pars: "An Introduction to the Calculus of Variations," John Wiley & Sons, Inc., New York, 1962

Applications specifically directed to a wide variety of engineering problems are found in

R. S. Schechter: "The Variational Method in Engineering," McGraw-Hill Book Company, New York, 1967

Section 3.4: We shall frequently use problems in control as examples of applications of the optimization theory, and complete references are given in later chapters. A useful introduction to the elements of process dynamics and control is

D. R. Coughanowr and L. B. Koppel: "Process Systems Analysis and Control," McGraw-Hill Book Company, New York, 1965

The reduction of the optimal feedback gain to the solution of a Riccati equation is an elementary special case of results due to Kalman, which are discussed in detail in later chapters. An approach to linear feedback control like the one used here, based on the classical calculus-of-variations formulation, is contained in

P. Das: *Automation Remote Contr.*, **27**:1506 (1966)

Section 3.5: The first discussion of the single-exothermic-reaction problem is in

K. G. Denbigh: *Trans. Faraday Soc.*, **40**:352 (1944)

A fairly complete discussion is in

R. Aris: "The Optimal Design of Chemical Reactors," Academic Press, Inc., New York, 1961

The approach taken here, as applied particularly to the exothermic-reaction problem, is credited to Horn in

K. G. Denbigh: "Chemical Reactor Theory," Cambridge University Press, New York, 1965

where there is discussion of the practical significance of the result. Some useful considerations for implementation of the optimal policy when parameters are uncertain can be found in

W. H. Ray and R. Aris: *Ind. Eng. Chem. Fundamentals*, **5**:478 (1966)

Sections 3.6 and 3.7: The texts cited for Secs. 3.2 and 3.3 are pertinent here as well.

Section 3.8: The general inverse problem may be stated: When is a differential equation an Euler equation? This is taken up in the text by Bolza cited above and in

J. Douglas: *Trans. Am. Math. Soc.*, **50**:71 (1941)
P. Funk: "Variationsrechnung und ihr Anwendung in Physik und Technik," Springer-Verlag OHG, Berlin, 1962
F. B. Hildebrand: "Methods of Applied Mathematics," Prentice-Hall, Inc., Englewood Cliffs, N.J., 1952

Sections 3.9 and 3.10: The approximate procedure outlined here is generally known as the Ritz or Rayleigh-Ritz method and as Galerkin's method when expressed in terms of the residual. Galerkin's method is one of a number of related procedures known as methods of weighted residuals for obtaining approximate solutions to systems of equations. A review of such methods with an extensive bibliography is

B. A. Finlayson and L. E. Scriven: *Appl. Mech. Rev.*, **19**:735 (1966)

See also the text by Schechter cited above and others, such as

W. F. Ames: "Nonlinear Partial Differential Equations in Engineering," Academic Press, Inc., New York, 1965
L. Collatz: "The Numerical Treatment of Differential Equations," Springer-Verlag OHG, Berlin, 1960
L. V. Kantorovich and V. I. Krylov: "Approximate Methods of Higher Analysis," Interscience Publishers, Inc., New York, 1958

Sections 3.11 and 3.12: Distributed-parameter systems are considered in some detail in Chap. 11, with particular attention to the control problem of Sec. 3.12. The derivation of the Euler equation used here for that process follows

Y. Sakawa: *IEEE Trans. Autom. Contr.*, **AC9**:420 (1964)

PROBLEMS

3.1. A system follows the equation

$$\dot{x} = -x + u$$

Find the function $u(t)$ which takes x from an initial value x_0 to zero while minimizing

$$\varepsilon = \int_0^\theta (K + u^2)\, dt$$

(time plus cost of control) where θ is unspecified. *Hint:* Solve for fixed θ; then determine the value of θ which minimizes $\mathcal{E}$.

3.2. A body of revolution with axis of symmetry in the x direction may be defined as one which intersects all planes orthogonal to the x axis in a circle. Consider such a body whose surface in any plane containing the x axis is described by the curve $y(x)$,

$$y(0) = 0 \qquad y(L) = R$$

The drag exerted by a gas stream of density ρ and velocity v flowing in the x direction is approximately

$$f = 4\pi \rho v^2 \int_0^L y \left(\frac{dy}{dx}\right)^3 dx$$

Find the function $y(x)$ passing through the required end points which makes the drag a minimum.

3.3. $\mathcal{F}$ is a function of t, x, and the first n derivatives of x with respect to t. Find the Euler equation and boundary conditions for

$$\min \mathcal{E} = \int_0^\theta \mathcal{F}(x, \dot{x}, \ldots, x^{(n)}, t)\, dt$$

3.4. A second-order process described by the equation

$$\ddot{x} + a\dot{x} + bx = u$$

is to be controlled to minimize the error integral

$$\min \mathcal{E} = \int_0^\theta (x^2 + c^2 u^2)\, dt$$

Show that the optimal control can be expressed in the feedback form

$$u = M_1 x + M_2 \dot{x}$$

and find the equations for M_1 and M_2.

3.5. Obtain the Euler equation and boundary conditions for minimization of

$$\mathcal{E} = \frac{1}{2} \int_0^\theta [p(t)\dot{x}^2 + 2r(t)x\dot{x} + q(t)x^2 + 2m(t)x + 2n(t)x]\, dt + ax^2(\theta) + bx^2(0)$$

and relate the result to the discussion of Sec. 3.8.

3.6. Steady-state diffusion with isothermal second-order chemical reaction, as well as other phenomena, can be described by the equations

$$\frac{d^2x}{dz^2} - \frac{k}{D}x^2 = 0$$

$$D\frac{dx}{dz} = h(x - x_0) \qquad \text{at } z = 0$$

$$D\frac{dx}{dz} = 0 \qquad \text{at } z = L$$

where k, D, h, and x_0 are constants. Find the parameters in a cubic approximation to the solution.

3.7. Using a polynomial approximation, estimate the first two eigenvalues of

$$\ddot{x} + \lambda x = 0$$

for the following boundary conditions:

(a) $x = 0$ at $t = 0, \pi$
(b) $x = 0$ at $t = 0$
    $x - \dot{x} = 0$ at $t = \pi$

Compare with the exact values.

3.8. Laminar flow of a newtonian liquid in a square duct is described by the equation

$$\frac{\partial^2 v}{\partial x^2} + \frac{\partial^2 v}{\partial y^2} - \frac{1}{\mu}\frac{\Delta P}{L} = 0 \qquad v = 0 \text{ at } \begin{array}{l} x = \pm a \\ y = \pm a \end{array}$$

Here v is the velocity, μ the viscosity, and $\Delta P/L$ the constant pressure gradient.

(a) Using the Galerkin method, find the coefficients A, B, and C in the approximate form

$$v = (x^2 - a^2)(y^2 - a^2)[A + B(x^2 + y^2) + Cx^2y^2]$$

The solution is most conveniently expressed in terms of the average velocity,

$$\bar{v} = \frac{1}{4a^2}\int_{-a}^{a}\int_{-a}^{a} v(x,y)\,dx\,dy$$

(The form of the approximation is due to Sparrow and Siegal. Numerical values of the exact solution are given in the book by Schechter.)

(b) Formulate the flow problem in terms of minimization of an integral and use the complex method to estimate values for A, B, and C.

4
Continuous Systems: I

4.1 INTRODUCTION

In the previous chapter we investigated the determination of an entire
function which would minimize an objective, and we were led to the
Euler differential equation for the minimizing function. It is rare that
a problem of interest can be formulated in so simple a fashion, and we
shall require a more general theory. Consider, for example, a chemical
reactor which we wish to control in an optimal manner by changing certain flow rates as functions of time. The laws of conservation of mass
and energy in this dynamic situation are represented by ordinary differential equations, and the optimizing function must be chosen consistent
with these constraints.

   We shall assume that the state of our system can be adequately
represented by N variables, which we shall denote by x_1, x_2, . . . , x_N.
In a chemical system these variables might be concentrations of the pertinent species and perhaps temperature or pressure, while for a space
vehicle they would represent coordinates and velocities. In addition,

we suppose that certain control or design variables are at our disposal to adjust as we wish, and we shall denote them by $u_1, u_2, \ldots, u_R$. These might be flow rates, temperatures, accelerations, turning angles, etc. Finally, we suppose that the state variables satisfy ordinary differential equations

$$\dot{x}_i = \frac{dx_i}{dt} = f_i(x_1, x_2, \ldots, x_N, u_1, u_2, \ldots, u_R) \qquad \begin{array}{l} i = 1, 2, \ldots, N \\ 0 < t \leq \theta \end{array} \tag{1}$$

and we wish to choose the functions $u_k(t)$, $k = 1, 2, \ldots, R$ in order to minimize an integral

$$\mathcal{E}[u_1, u_2, \ldots, u_R] = \int_0^\theta \mathcal{F}(x_1, x_2, \ldots, x_N, u_1, u_2, \ldots, u_R)\, dt \tag{2}$$

We shall generally refer to the independent variable t as time, although it might in fact refer to a spatial coordinate, as in a pipeline chemical reactor. The total operating duration θ may or may not be specified in advance, and we may or may not wish to impose conditions on the variables x_i at time θ. A typical measure of performance in a control problem might be a weighted sum of squares of deviations from preset operating conditions, so that $\mathcal{F}$ would be

$$\mathcal{F} = c_1(x_1 - x_{1S})^2 + \cdots + c_N(x_N - x_{NS})^2 \\ + c_{N+1}(u_1 - u_{1S})^2 + \cdots + C_{N+R}(u_R - u_{RS})^2 \tag{3}$$

On the other hand, if the controls were to be set to bring $x_1, x_2, \ldots, x_N$ to fixed values in the minimum time θ, we would wish to minimize $\mathcal{E} = \theta$ or, equivalently,

$$\mathcal{F} = 1 \tag{4}$$

As we shall see, there is no loss of generality in the choice of the performance index, Eq. (2).

4.2 VARIATIONAL EQUATIONS

The development in this section parallels that of Sec. 1.8. For convenience we shall put $N = R = 2$, although it will be clear that the results obtained are valid for any N and R. We thus have the state defined by the two differential equations

$$\dot{x}_1 = f_1(x_1, x_2, u_1, u_2) \tag{1a}$$
$$\dot{x}_2 = f_2(x_1, x_2, u_1, u_2) \tag{1b}$$

and the performance index

$$\mathcal{E}[u_1,u_2] = \int_0^\theta \mathcal{F}(x_1,x_2,u_1,u_2)\,dt \tag{2}$$

Let us suppose that we have specified the decision variables $u_1(t)$, $u_2(t)$ over the entire interval $0 \le t \le \theta$. Call these functions $\bar{u}_1(t)$, $\bar{u}_2(t)$. For given initial conditions x_{10}, x_{20} we may then solve Eqs. (1) for $\bar{x}_1(t)$ and $\bar{x}_2(t)$, $0 \le t \le \theta$, corresponding to the choices $\bar{u}_1(t)$, $\bar{u}_2(t)$. The value of the performance index is completely determined by the choice of decision functions, and we may call the result $\mathcal{E}[\bar{u}_1,\bar{u}_2]$.

We now change u_1 and u_2 at every point by small amounts, $\delta u_1(t)$ and $\delta u_2(t)$, where

$$|\delta u_1(t)|, \ |\delta u_2(t)| \le \epsilon \qquad 0 \le t \le \theta \tag{3}$$

and ϵ is a small positive constant. [If $x_1(0)$ and $x_2(0)$ are not specified, we also make small changes $\delta x_1(0)$, $\delta x_2(0)$.] That is,

$$u_1(t) = \bar{u}_1(t) + \delta u_1(t) \tag{4a}$$
$$u_2(t) = \bar{u}_2(t) + \delta u_2(t) \tag{4b}$$

and as a result we cause small changes $\delta x_1(t)$ and $\delta x_2(t)$ in $\bar{x}_1$ and $\bar{x}_2$, respectively, and a change $\delta\mathcal{E}$ in the performance index. We can obtain expressions for δx_1, δx_2, and $\delta\mathcal{E}$ by evaluating Eqs. (1) and (2) with $(\bar{u}_1,\bar{u}_2)$ and $(\bar{u}_1 + \delta u_1, \ \bar{u}_2 + \delta u_2)$, successively, and then subtracting. Thus,

$$\frac{d}{dt}(\bar{x}_1 + \delta x_1) - \frac{d}{dt}\bar{x}_1 = \delta\dot{x}_1$$
$$= f_1(\bar{x}_1 + \delta x_1,\ \bar{x}_2 + \delta x_2,\ \bar{u}_1 + \delta u_1,\ \bar{u}_2 + \delta u_2) - f_1(\bar{x}_1,\bar{x}_2,\bar{u}_1,\bar{u}_2) \tag{5a}$$
$$\frac{d}{dt}(\bar{x}_2 + \delta x_2) - \frac{d}{dt}\bar{x}_2 = \delta\dot{x}_2$$
$$= f_2(\bar{x}_1 + \delta x_1,\ \bar{x}_2 + \delta x_2,\ \bar{u}_1 + \delta u_1,\ \bar{u}_2 + \delta u_2) - f_2(\bar{x}_1,\bar{x}_2,\bar{u}_1,\bar{u}_2) \tag{5b}$$
$$\delta\mathcal{E} = \mathcal{E}[\bar{u}_1 + \delta u_1,\ \bar{u}_2 + \delta u_2] - \mathcal{E}[\bar{u}_1,\bar{u}_2]$$
$$= \int_0^{\bar{\theta}} [\mathcal{F}(\bar{x}_1 + \delta x_1,\ \bar{x}_2 + \delta x_2,\ \bar{u}_1 + \delta u_1,\ \bar{u}_2 + \delta u_2) - \mathcal{F}(\bar{x}_1,\bar{x}_2,\bar{u}_1,\bar{u}_2)]\,dt$$
$$+ \int_{\bar{\theta}}^{\bar{\theta}+\delta\theta} \mathcal{F}(\bar{x}_1 + \delta x_1,\ \bar{x}_2 + \delta x_2,\ \bar{u}_1 + \delta u_1,\ \bar{u}_2 + \delta u_2)\,dt \tag{6}$$

The last term in Eq. (6) must be added in the event that the total time θ is not specified, and a change in the decisions requires a change in the total process time in order to meet some preset final condition. $\bar{\theta}$ then represents the optimal duration and $\delta\theta$ the change.

If we expand the functions f_1, f_2, and $\mathcal{F}$ at every t about their respective values at that t when the decisions $\bar{u}_1$, $\bar{u}_2$ are used, we obtain

$$\delta\mathcal{E} = \int_0^{\bar{\theta}} \left(\frac{\partial\mathcal{F}}{\partial x_1} \delta x_1 + \frac{\partial\mathcal{F}}{\partial x_2} \delta x_2 + \frac{\partial\mathcal{F}}{\partial u_1} \delta u_1 + \frac{\partial\mathcal{F}}{\partial u_2} \delta u_2 \right) dt$$
$$+ \mathcal{F}[\bar{x}_1(\bar{\theta}),\bar{x}_2(\bar{\theta}),\bar{u}_1(\bar{\theta}),\bar{u}_2(\bar{\theta})]\ \delta\theta + o(\epsilon) \tag{7}$$

$$\delta \dot{x}_1 = \frac{\partial f_1}{\partial x_1} \delta x_1 + \frac{\partial f_1}{\partial x_2} \delta x_2 + \frac{\partial f_1}{\partial u_1} \delta u_1 + \frac{\partial f_2}{\partial u_2} \delta u_2 + o(\epsilon) \qquad (8a)$$

$$\delta \dot{x}_2 = \frac{\partial f_2}{\partial x_1} \delta x_1 + \frac{\partial f_2}{\partial x_2} \delta x_2 + \frac{\partial f_2}{\partial u_1} \delta u_1 + \frac{\partial f_2}{\partial u_2} \delta u_2 + o(\epsilon) \qquad (8b)$$

We now multiply Eqs. (8a) and (8b), respectively, by *arbitrary* continuous functions $\lambda_1(t)$ and $\lambda_2(t)$, integrate from $t = 0$ to $t = \bar{\theta}$, and add the result to Eq. (7). Thus,

$$\delta \mathcal{E} = \int_0^{\bar{\theta}} \left[\left(\frac{\partial \mathcal{F}}{\partial x_1} + \lambda_1 \frac{\partial f_1}{\partial x_1} + \lambda_2 \frac{\partial f_2}{\partial x_1} \right) \delta x_1 - \lambda_1 \, \delta \dot{x}_1 \right.$$

$$+ \left(\frac{\partial \mathcal{F}}{\partial x_2} + \lambda_1 \frac{\partial f_1}{\partial x_2} + \lambda_2 \frac{\partial f_2}{\partial x_2} \right) \delta x_2 - \lambda_2 \, \delta \dot{x}_2$$

$$+ \left(\frac{\partial \mathcal{F}}{\partial u_1} + \lambda_1 \frac{\partial f_1}{\partial u_1} + \lambda_2 \frac{\partial f_2}{\partial u_1} \right) \delta u_1$$

$$\left. + \left(\frac{\partial \mathcal{F}}{\partial u_2} + \lambda_1 \frac{\partial f_1}{\partial u_2} + \lambda_2 \frac{\partial f_2}{\partial u_2} \right) \delta u_2 \right] dt + \mathcal{F} \Big|_{t=\bar{\theta}} \delta \theta + o(\epsilon) \quad (9)$$

Integrating the terms $\lambda_1 \, \delta \dot{x}_1$ and $\lambda_2 \, \delta \dot{x}_2$ by parts gives, finally,

$$\delta \mathcal{E} = \int_0^{\bar{\theta}} \left[\left(\frac{\partial \mathcal{F}}{\partial x_1} + \lambda_1 \frac{\partial f_1}{\partial x_1} + \lambda_2 \frac{\partial f_2}{\partial x_1} + \dot{\lambda}_1 \right) \delta x_1 \right.$$

$$+ \left(\frac{\partial \mathcal{F}}{\partial x_2} + \lambda_1 \frac{\partial f_1}{\partial x_2} + \lambda_2 \frac{\partial f_2}{\partial x_2} + \dot{\lambda}_2 \right) \delta x_2$$

$$+ \left(\frac{\partial \mathcal{F}}{\partial u_1} + \lambda_1 \frac{\partial f_1}{\partial u_1} + \lambda_2 \frac{\partial f_2}{\partial u_1} \right) \delta u_1$$

$$\left. + \left(\frac{\partial \mathcal{F}}{\partial u_2} + \lambda_1 \frac{\partial f_1}{\partial u_2} + \lambda_2 \frac{\partial f_2}{\partial u_2} \right) \delta u_2 \right] dt$$

$$+ \mathcal{F} \Big|_{t=\bar{\theta}} \delta \theta - \lambda_1(\bar{\theta}) \, \delta x_1(\bar{\theta})$$

$$- \lambda_2(\bar{\theta}) \, \delta x_2(\bar{\theta}) + \lambda_1(0) \, \delta x_1(0) + \lambda_2(0) \, \delta x_2(0) + o(\epsilon) \quad (10)$$

Now, just as in Sec. 1.8, we find ourselves with terms which are at our disposal, namely, the decision changes $\delta u_1(t)$, $\delta u_2(t)$, and terms which are not, the state variations $\delta x_1(t)$ and $\delta x_2(t)$. We therefore eliminate these latter terms from the expression for $\delta \mathcal{E}$ by removing some of the arbitrariness from the functions $\lambda_1(t)$, $\lambda_2(t)$ and requiring that they satisfy the differential equations

$$\dot{\lambda}_1 = -\frac{\partial \mathcal{F}}{\partial x_1} - \lambda_1 \frac{\partial f_1}{\partial x_1} - \lambda_2 \frac{\partial f_2}{\partial x_1} \qquad (11a)$$

$$\dot{\lambda}_2 = -\frac{\partial \mathcal{F}}{\partial x_2} - \lambda_1 \frac{\partial f_1}{\partial x_2} - \lambda_2 \frac{\partial f_2}{\partial x_2} \qquad (11b)$$

Note that we have not yet specified boundary conditions. The variation

$\delta\mathcal{E}$ is now

$$\delta\mathcal{E} = \int_0^{\bar{\theta}} \left[\left(\frac{\partial \mathcal{F}}{\partial u_1} + \lambda_1 \frac{\partial f_1}{\partial u_1} + \lambda_2 \frac{\partial f_2}{\partial u_1} \right) \delta u_1 \right.$$

$$\left. + \left(\frac{\partial \mathcal{F}}{\partial u_2} + \lambda_1 \frac{\partial f_1}{\partial u_2} + \lambda_2 \frac{\partial f_2}{\partial u_2} \right) \delta u_2 \right] dt$$

$$+ \left. \mathcal{F} \right|_{t=\bar{\theta}} \delta\theta - \lambda_1(\bar{\theta})\,\delta x_1(\bar{\theta}) - \lambda_2(\bar{\theta})\,\delta x_2(\bar{\theta})$$

$$+ \lambda_1(0)\,\delta x_1(0) + \lambda_2(0)\,\delta x_2(0) + o(\epsilon) \quad (12)$$

It is necessary at this point to distinguish between cases when the total duration θ is specified and when it is not. If θ is specified, then $\delta\theta$ must be zero. If $x_1(\theta)$ is fixed, then $\delta x_1(\theta)$ is zero and the term $\lambda_1(\theta)\,\delta x_1(\theta)$ vanishes. If, on the other hand, $x_1(\theta)$ is free to take on any value, we have no control over $\delta x_1(\theta)$, so that we remove it from the expression for $\delta\mathcal{E}$ by specifying that $\lambda_1(\theta) = 0$. Similar considerations apply to the term $\lambda_2(\theta)\,\delta x_2(\theta)$, and we obtain the boundary conditions:

θ specified:

$x_1(\theta)$ free	$\lambda_1(\theta) = 0$
$x_1(\theta)$ fixed	$\lambda_1(\theta)$ unspecified
$x_2(\theta)$ free	$\lambda_2(\theta) = 0$
$x_2(\theta)$ fixed	$\lambda_2(\theta)$ unspecified

$$(13)$$

If θ is not specified, the variations $\delta\theta$, $\delta x_1(\theta)$, and $\delta x_2(\theta)$ are related. In fact,

$$x_1(\bar{\theta} + \delta\theta) = \bar{x}_1(\bar{\theta}) + \delta x_1(\bar{\theta}) + f_1 \big|_{t=\bar{\theta}} \delta\theta + o(\epsilon) \quad (14)$$

and similarly for x_2. Thus, if x_1 is fixed at the end of the process, we require $x_1(\bar{\theta} + \delta\theta) = \bar{x}_1(\bar{\theta})$, and if x_2 is free, the terms

$$\mathcal{F}\,\delta\theta - \lambda_1\,\delta x_1 - \lambda_2\,\delta x_2 = (\mathcal{F} + \lambda_1 f_1)\,\delta\theta - \lambda_2\,\delta x_2 \quad (15)$$

and similarly for x_2 fixed, or both. If neither x_1 nor x_2 is fixed, of course we may choose $\delta\theta = 0$ and use the previous results. Thus, applying the logic of the previous paragraph, we obtain the following conditions:

θ unspecified:

$x_1(\theta)$ free	$x_2(\theta)$ free	$\lambda_1(\theta) = 0$	$\lambda_2(\theta) = 0$
$x_1(\theta)$ fixed	$x_2(\theta)$ free	$\mathcal{F} + \lambda_1 f_1 = 0$	$\lambda_2 = 0$
$x_1(\theta)$ free	$x_2(\theta)$ fixed	$\mathcal{F} + \lambda_2 f_2 = 0$	$\lambda_1 = 0$
$x_1(\theta)$ fixed	$x_2(\theta)$ fixed	$\mathcal{F} + \lambda_1 f_1 + \lambda_2 f_2 = 0$	

$$(16)$$

We now apply the same approach to the terms $\lambda_1(0)\,\delta x_1(0)$ and $\lambda_2(0)\,\delta x_2(0)$ and obtain the further boundary conditions:

$x_1(0)$ free	$\lambda_1(0) = 0$
$x_1(0)$ fixed	$\lambda_1(0)$ unspecified
$x_2(0)$ free	$\lambda_2(0) = 0$
$x_2(0)$ fixed	$\lambda_2(0)$ unspecified

$$(17)$$

For the problem with fixed θ, Eqs. (13) and (17) provide a total of four boundary conditions for the four differential equations (1a), (1b) and (11a), (11b). When θ is variable, conditions (16) and (17) provide five conditions, four boundary conditions and a *stopping condition*.† With these conditions we finally obtain an expression for $\delta\mathcal{E}$ which depends only on the variations in the decision variables,

$$\delta\mathcal{E} = \int_0^\theta \left[\left(\frac{\partial \mathcal{F}}{\partial u_1} + \lambda_1 \frac{\partial f_1}{\partial u_1} + \lambda_2 \frac{\partial f_2}{\partial u_1} \right) \delta u_1 \right.$$
$$\left. + \left(\frac{\partial \mathcal{F}}{\partial u_2} + \lambda_1 \frac{\partial f_1}{\partial u_2} + \lambda_2 \frac{\partial f_2}{\partial u_2} \right) \delta u_2 \right] dt + o(\epsilon) \quad (18)$$

4.3 FIRST NECESSARY CONDITIONS

We now introduce into the discussion, for the first time, the fact that we are considering an optimization problem. Thus, if the choices $\bar{u}_1(t)$, $\bar{u}_2(t)$ are those which minimize $\mathcal{E}$, then for all variations $\delta u_1(t)$, $\delta u_2(t)$ it is necessary that $\mathcal{E}$ increase, or $\delta\mathcal{E} \geq 0$. We consider in this section only situations in which the optimal decisions are unbounded and we are free to make any (small) variations we wish. In doing so we exclude a large class of problems of interest (indeed, most!), and we shall return shortly to considerations of constraints on the allowable decisions.

As in Chap. 1, we choose a particular set of variations which makes our task easy. We set

$$\delta u_1 = -\epsilon' \left(\frac{\partial \mathcal{F}}{\partial u_1} + \lambda_1 \frac{\partial f_1}{\partial u_1} + \lambda_2 \frac{\partial f_2}{\partial u_1} \right) \tag{1a}$$

$$\delta u_2 = -\epsilon' \left(\frac{\partial \mathcal{F}}{\partial u_2} + \lambda_1 \frac{\partial f_1}{\partial u_2} + \lambda_2 \frac{\partial f_2}{\partial u_2} \right) \tag{1b}$$

where ϵ' is a small positive constant. Thus Eq. (18) of the preceding section becomes

$$\delta\mathcal{E} = -\epsilon' \int_0^\theta \left[\left(\frac{\partial \mathcal{F}}{\partial u_1} + \lambda_1 \frac{\partial f_1}{\partial u_1} + \lambda_2 \frac{\partial f_2}{\partial u_1} \right)^2 \right.$$
$$\left. + \left(\frac{\partial \mathcal{F}}{\partial u_2} + \lambda_1 \frac{\partial f_1}{\partial u_2} + \lambda_2 \frac{\partial f_2}{\partial u_2} \right)^2 \right] dt + o(\epsilon) \geq 0 \quad (2)$$

Since ϵ and ϵ' are of the same order, it follows that

$$\lim_{\epsilon' \to 0} \frac{o(\epsilon)}{\epsilon'} = 0 \tag{3}$$

---

† For the case θ unspecified and both $x_1(\theta)$ and $x_2(\theta)$ unspecified we require an additional condition. We shall find later that this condition is $\mathcal{F} \Big|_{t=\theta} = 0$.

and dividing by ϵ' and taking the limit in Eq. (2), we obtain

$$\int_0^\theta \left[\left(\frac{\partial \mathfrak{F}}{\partial u_1} + \lambda_1 \frac{\partial f_1}{\partial u_1} + \lambda_2 \frac{\partial f_2}{\partial u_1} \right)^2 + \left(\frac{\partial \mathfrak{F}}{\partial u_2} + \lambda_1 \frac{\partial f_1}{\partial u_2} + \lambda_2 \frac{\partial f_2}{\partial u_2} \right)^2 \right] dt \leq 0$$

(4)

Since this is the integral over a positive region of a sum of squares, we can satisfy the inequality only if the integrand vanishes identically (except, perhaps, on a set of discrete points). We therefore conclude that *if u_1 and u_2 are the unconstrained functions which cause $\mathcal{E}$ to take on its minimum value, it is necessary that*

$$\frac{\partial \mathfrak{F}}{\partial u_1} + \lambda_1 \frac{\partial f_1}{\partial u_1} + \lambda_2 \frac{\partial f_2}{\partial u_1} = 0 \tag{5a}$$

$$\frac{\partial \mathfrak{F}}{\partial u_2} + \lambda_1 \frac{\partial f_1}{\partial u_2} + \lambda_2 \frac{\partial f_2}{\partial u_2} = 0 \tag{5b}$$

except, perhaps, on a set of discrete points. These equations represent the extension of the multiplier rule to the problem of minimizing an integral subject to differential-equation side conditions. It is important to note that the optimal *functions* $u_1(t)$, $u_2(t)$ are found from a set of algebraic relations to be satisfied *at each point*.

Let us pause for a moment and contemplate what we have done. In order to obtain conditions for the minimizing functions $u_1(t)$ and $u_2(t)$ we have had to introduce two *additional* functions $\lambda_1(t)$ and $\lambda_2(t)$ which also satisfy a set of differential equations. The four boundary conditions for the total of four differential equations are split between the two ends, $t = 0$ and $t = \theta$, and there is an additional set of algebraic conditions to be satisfied at each point. This is a rather formidable problem but one which we must accept if we are to attack design and control problems of any significance.

It is often convenient, and always elegant, to introduce the *hamiltonian function H*, defined as

$$H = \mathfrak{F} + \lambda_1 f_1 + \lambda_2 f_2 \tag{6}$$

The differential equations for x_1, x_2, λ_1, and λ_2 can then be written in the *canonical* form

$$\dot{x}_1 = \frac{\partial H}{\partial \lambda_1} \qquad \dot{x}_2 = \frac{\partial H}{\partial \lambda_2} \tag{7a}$$

$$\dot{\lambda}_1 = -\frac{\partial H}{\partial x_1} \qquad \dot{\lambda}_2 = -\frac{\partial H}{\partial x_2} \tag{7b}$$

while the necessary conditions (5a) and (5b) can be written

$$\frac{\partial H}{\partial u_1} = 0 \qquad \frac{\partial H}{\partial u_2} = 0 \tag{8}$$

Furthermore, the hamiltonian is a constant along the optimal path, for

$$\frac{dH}{dt} = \frac{\partial H}{\partial x_1} \dot{x}_1 + \frac{\partial H}{\partial x_2} \dot{x}_2 + \frac{\partial H}{\partial \lambda_1} \dot{\lambda}_1 + \frac{\partial H}{\partial \lambda_2} \dot{\lambda}_2 + \frac{\partial H}{\partial u_1} \dot{u}_1 + \frac{\partial H}{\partial u_2} \dot{u}_2 \qquad (9)$$

which equals zero after substitution of Eqs. (7) and (8).

When θ is unspecified, the constant value of the hamiltonian is found from the value at $t = \theta$ to be zero. If $x_1(\theta)$ and $x_2(\theta)$ are free, we can simply apply the calculus to find the optimal stopping time

$$\frac{\partial \mathcal{E}}{\partial \theta} = \frac{\partial}{\partial \theta} \int_0^\theta \mathcal{F} \, dt = \mathcal{F} \Big|_{t=\theta} = 0 \qquad (10)$$

Together with the first boundary condition [Sec. 4.2, Eq. (16)] this gives†

$$H = H \Big|_{t=\theta} = 0 \qquad (11)$$

On the other hand, if x_1, x_2, or both are specified, Eq. (11) follows directly from the remaining boundary conditions [Sec. 4.2, Eq. (16)].

We may summarize the results of this section in the following statements:

> *The unconstrained functions $u_1(t)$ and $u_2(t)$ which minimize $\mathcal{E}$ make the hamiltonian stationary‡ for all t, $0 \le t \le \theta$. The hamiltonian is constant along the optimal path, and the constant has the value zero when the stopping time θ is not specified.*

This is a very weak form of what has come to be called *Pontryagin's minimum principle.*

For the more general problem of N state and R decision variables we require N multipliers, and the hamiltonian has the form

$$H = \mathcal{F} + \sum_{n=1}^{N} \lambda_n f_n(x_1, x_2, \ldots, x_N, u_1, u_2, \ldots, u_R) \qquad (12)$$

with canonical equations

$$\dot{x}_i = \frac{\partial H}{\partial \lambda_i} = f_i \qquad (13a)$$

$$\dot{\lambda}_i = -\frac{\partial H}{\partial x_i} = -\frac{\partial \mathcal{F}}{\partial x_i} - \sum_{n=1}^{N} \lambda_n \frac{\partial f_n}{\partial x_i} \qquad (13b)$$

and the boundary conditions

$$\lambda_i = 0 \qquad\qquad x_i \text{ free} \qquad (14a)$$
$$\lambda_i \text{ unspecified} \qquad x_i \text{ fixed} \qquad (14b)$$

† This is the missing stopping condition which we noted previously.

‡ We shall find later that the hamiltonian is in fact a minimum at these stationary points.

The hamiltonian is made stationary by all decision variables u_k, $k = 1, 2,$ . . . , R, and is a constant, with zero value when θ is unspecified.

4.4 EULER EQUATION

At this point it may be useful to reconsider the problem of Sec. 3.2 from the more general result of the previous section. We seek the function $x(t)$ which will minimize the integral

$$\mathcal{E} = \int_0^\theta \mathfrak{F}(x,\dot{x},t) \, dt \tag{1}$$

First, we must put the problem in the form of Eqs. (1) of Sec. 4.2. To do so, we observe that for a given (or optimal) initial condition, $x(0)$, the function $x(t)$ is uniquely determined by its derivative $\dot{x}(t)$. Thus, $\dot{x}$ may be taken as the decision variable. Furthermore, the explicit dependence of $\mathfrak{F}$ on t may be removed by the simple guise of defining a new state variable. That is, if we let x_1 denote x, the problem can be reformulated as

$$\mathcal{E} = \int_0^\theta \mathfrak{F}(x_1,u,x_2) \, dt \tag{2}$$

with

$$\dot{x}_1 = u \tag{3a}$$
$$\dot{x}_2 = 1 \qquad x_2(0) = 0 \tag{3b}$$

so that $x_2(t) \equiv t$. This is precisely the form which we have studied (with the special case of a single decision function).

The hamiltonian is

$$H = \mathfrak{F} + \lambda_1 u + \lambda_2 \tag{4}$$

with multiplier equations

$$\dot{\lambda}_1 = -\frac{\partial H}{\partial x_1} = -\frac{\partial \mathfrak{F}}{\partial x_1} \left(= -\frac{\partial \mathfrak{F}}{\partial x} \right) \tag{5a}$$

$$\dot{\lambda}_2 = -\frac{\partial H}{\partial x_2} = -\frac{\partial \mathfrak{F}}{\partial x_2} \left(= -\frac{\partial \mathfrak{F}}{\partial t} \right) \tag{5b}$$

Equation (5a) can be integrated to give

$$\lambda_1(t) = \lambda_1(0) - \int_0^t \frac{\partial \mathfrak{F}}{\partial x} \, dt \tag{6}$$

We generally do not need Eq. (5b), but we should note that it requires that λ_2 be constant if $\mathfrak{F}$ is independent of t.

The condition that the hamiltonian be stationary is

$$\frac{\partial H}{\partial u} = \frac{\partial \mathfrak{F}}{\partial u} + \lambda_1 = 0 \left(= \frac{\partial \mathfrak{F}}{\partial \dot{x}} + \lambda_1 \right) \tag{7}$$

or, combining with Eq. (6),

$$\frac{\partial \mathfrak{F}}{\partial \dot{x}} = \int_0^t \frac{\partial \mathfrak{F}}{\partial x} \, dt + c \tag{8}$$

where c denotes a constant.   This is the equation of du Bois-Reymond. If we differentiate with respect to t, we obtain Euler's equation

$$\frac{d}{dt} \frac{\partial \mathfrak{F}}{\partial \dot{x}} = \frac{\partial \mathfrak{F}}{\partial x} \tag{9}$$

The necessary boundary conditions follow from Eq. (13) of Sec. 4.2 as

$$x(0) \text{ given} \qquad \text{or} \qquad \frac{\partial \mathfrak{F}}{\partial \dot{x}} \Big|_{t=0} = 0 \tag{10a}$$

$$x(\theta) \text{ given} \qquad \text{or} \qquad \frac{\partial \mathfrak{F}}{\partial \dot{x}} \Big|_{t=\theta} = 0 \tag{10b}$$

Equations (9) and (10) are identical to Eqs. (18) and (19) of Sec. 3.2.

It may be that the optimal function $x(t)$ has a "corner" at some value of t where the derivative is not continuous.   The integrand $\mathfrak{F}(x,\dot{x},t)$ will not be continuous at this point, and the Euler equation (9) will not be defined.   The integral in Eq. (8) will be continuous, however (recall that the value of an integral is not changed by what happens at a single point), and therefore $\partial \mathfrak{F}/\partial \dot{x}$ must also be continuous at the corner.   This is the *Erdmann-Weierstrass corner condition.*

4.5 RELATION TO CLASSICAL MECHANICS

The reader familiar with classical mechanics may wish to relate the hamiltonian defined here with the function of the same name employed in mechanics.   It will be recalled that in a system of N particles, with masses $m_1, m_2, \ldots, m_N$ and positions in a three-dimensional space x_{11}, $x_{12}, x_{13}, x_{21}, x_{22}, x_{23}, \ldots, x_{N1}, x_{N2}, x_{N3}$, the physical system is the one which minimizes (or makes stationary) the *action integral*

$$\mathcal{E} = \int_0^\theta \left[\sum_{i=1}^N \sum_{j=1}^3 \frac{m_i \dot{x}_{ij}^2}{2} - V(x_{11}, x_{12}, x_{13}, \ldots, x_{N1}, x_{N2}, x_{N3}) \right] dt \tag{1}$$

where the first term is kinetic and the second potential energy.   Renumbering, we may use a single-subscript notation

$$\mathcal{E} = \int_0^\theta \left[\sum_{i=1}^{3N} \frac{m_i \dot{x}_i^2}{2} - V(x_1, x_2, \ldots, x_{3N}) \right] dt \tag{2}$$

with $m_1 = m_2 = m_3, m_4 = m_5 = m_6$, etc.

If we take the decision variable as the velocity,

$$\dot{x}_i = u_i \qquad i = 1, 2, \ldots, 3N \tag{3}$$

and

$$\mathfrak{F} = \sum_{i=1}^{3N} \frac{m_i u_i^2}{2} - V(x_1, x_2, \ldots, x_{3N}) \tag{4}$$

then the hamiltonian is

$$H = \sum_{i=1}^{3N} \frac{m_i u_i^2}{2} - V(x_1, x_2, \ldots, x_{3N}) + \sum_{i=1}^{3N} \lambda_i u_i \tag{5}$$

The multiplier equations are

$$\dot{\lambda}_i = -\frac{\partial H}{\partial x_i} = \frac{\partial V}{\partial x_i} \tag{6}$$

while the stationary condition is

$$\frac{\partial H}{\partial u_i} = 0 = m_i u_i + \lambda_i \tag{7a}$$

or

$$\dot{x}_i = -\frac{\lambda_i}{m_i} \tag{7b}$$

Defining the momentum p_i as $-\lambda_i$, we find

$$\dot{x}_i = \frac{p_i}{m_i} \qquad \dot{p}_i = -\frac{\partial V}{\partial x_i} \tag{8}$$

and

$$H = -\sum_{i=1}^{3N} \frac{p_i^2}{2m_i} - V(x_1, x_2, \ldots, x_{3N}) \tag{9}$$

which is simply the negative of the usual hamiltonian of mechanics. (The fact that we have obtained the negative stems from an unimportant sign convention.) This is the total energy, which is a constant in a conservative system.

4.6 SOME PHYSICAL EQUATIONS AND USEFUL TRANSFORMATIONS

In this and subsequent chapters we shall often find it useful to relate the mathematical development to the particular example of the control of a continuous-flow stirred-tank chemical reactor. This example is chosen because, while simple, it retains the basic features of many practical systems and because the basic equations have received considerable attention in the published literature. In this section we shall develop the equations and consider several transformations of variables, each of which will be useful in various formulations of the control problem. One of our purposes is to derive, through the consideration of the reactor example, the general form of a second-order system, and the reader who is not interested in the particular physical application may wish to begin with Eqs. (10).

The reactor system is shown schematically in Fig. 4.1. For simplicity we assume a single liquid-phase first-order reaction

$$A \rightarrow \text{products} \tag{1}$$

The equation of conservation of mass is then

$$\dot{A} = \frac{q}{V}(A_f - A) - kA \tag{2}$$

where A denotes the concentration of reactant, the subscript f refers to the feed stream, V the volume, and q the volumetric flow rate of the feed. k is a temperature-dependent reaction-rate coefficient of Arrhenius form

$$k = k_0 \exp\left(\frac{-E'}{T}\right) \tag{3}$$

where k_0 and E' are constants and T is temperature. The equation simply states that the net rate of accumulation of A in the reactor is equal to the net rate at which it enters less the rate of reaction.

Similarly, an energy balance leads to

$$\dot{T} = \frac{q}{V}(T_f - T) - \frac{UKq_c}{VC_p\rho(1 + Kq_c)}(T - T_c) + \frac{(-\Delta H)}{C_p\rho}kA \tag{4}$$

Here the subscript c refers to the coolant stream, ρ is the density, C_p the

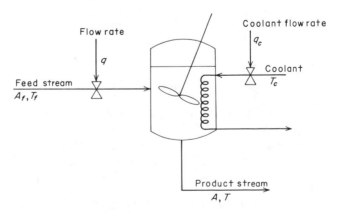

Fig. 4.1 Schematic of a continuous-flow stirred-tank reactor.

specific heat, U the overall heat-transfer coefficient times the cooling area, ΔH the heat of reaction, and K a constant, defined as

$$K = \frac{2C_{pc}\rho_c}{U} \tag{5}$$

Equations (2) and (4) are made dimensionless by defining a dimensionless concentration Z_1 and temperatures Z_2, Z_f, and Z_c, as follows:

$$Z_1 = \frac{A}{A_f} \tag{6a}$$

$$Z_2 = \frac{C_p\rho T}{(-\Delta H)A_f} \tag{6b}$$

$$Z_f = \frac{C_p\rho T_f}{(-\Delta H)A_f} \tag{6c}$$

$$Z_c = \frac{C_p\rho T_c}{(-\Delta H)A_f} \tag{6d}$$

Thus

$$\dot{Z}_1 = \frac{q}{V}(1 - Z_1) - kZ_1 \tag{7a}$$

$$\dot{Z}_2 = \frac{q}{V}(Z_f - Z_2) - \frac{UKq_c}{VC_p\rho(1 + Kq_c)}(Z_2 - Z_c) + kZ_1 \tag{7b}$$

where k may be written in terms of Eqs. (6) as

$$k = k_0 \exp\left[-\frac{E'C_p\rho}{(-\Delta H)A_f}\frac{1}{Z_2}\right] \tag{8}$$

Let us now assume that the reactor has been designed to operate at a stable steady-state condition Z_{1S}, Z_{2S} and that we shall control the reactor in the neighborhood of the steady state by adjusting the flow rates q and q_c. Let x_1 and x_2 denote variations about the steady state in (dimensionless) concentration and temperature, respectively, and u_1 and u_2 in feed and coolant flow rates; that is,

$$x_1 = Z_1 - Z_{1S} \qquad x_2 = Z_2 - Z_{2S} \tag{9a}$$
$$u_1 = q - q_S \qquad u_2 = q_c - q_{cS} \tag{9b}$$

where the subscript S refers to the steady state. Substituting into Eqs. (7) and expanding the right-hand sides in Taylor series about the steady state, exactly as in Sec. 1.4, we obtain the following equations for the dynamic behavior of the reactor in a neighborhood of the steady state $x_1 = x_2 = u_1 = u_2 = 0$:

$$\dot{x}_1 = a_{11}x_1 + a_{12}x_2 + b_{11}u_1 \tag{10a}$$
$$\dot{x}_2 = a_{21}x_1 + a_{22}x_2 + b_{21}u_1 + b_{22}u_2 \tag{10b}$$

where the constants are defined as follows:

$$a_{11} = \frac{qs}{V} + k_S \tag{11a}$$

$$a_{12} = \frac{E'C_p\rho k_S Z_{1S}}{(-\Delta H)A_f Z_{2S}{}^2} \tag{11b}$$

$$a_{21} = -k_S \tag{11c}$$

$$a_{22} = \frac{qs}{V} + \frac{UKq_{cS}}{VC_p\rho(1 + Kq_{cS})} - \frac{E'C_p\rho k_S Z_{1S}}{(-\Delta H)A_f Z_{2S}{}^2} \tag{11d}$$

$$b_{11} = \frac{1}{V}(1 - Z_{1S}) \tag{11e}$$

$$b_{21} = \frac{1}{V}(Z_f - Z_{2S}) \tag{11f}$$

$$b_{22} = -\frac{UK(Z_{2S} - Z_c)}{VC_p\rho(1 + Kq_{cS})^2} \tag{11g}$$

It is clear that Eqs. (10) will apply to a large class of systems besides the reactor problem considered here. Values of the parameters which are used for subsequent numerical calculations are collected in Table 4.1 in consistent (cgs) units.

In some cases we shall choose to hold u_1 at zero and control only with the coolant flow rate. We can simplify the form of the equations by defining new variables y_1 and y_2, as follows:

$$x_1 = -y_1 \tag{12a}$$

$$x_2 = \frac{a_{11}}{a_{12}}y_1 - \frac{1}{a_{12}}y_2 \tag{12b}$$

Substituting into Eqs. (10), we then obtain

$$\dot{y}_1 = y_2 \tag{13a}$$

$$\dot{y}_2 = (a_{12}a_{21} - a_{11}a_{22})y_1 + (a_{11} + a_{22})y_2 + (-a_{12}b_{22}u_2) \tag{13b}$$

or, with obvious notation,

$$\dot{y}_1 = y_2 \tag{14a}$$

$$\dot{y}_2 = -a_2 y_1 - a_1 y_2 + u \tag{14b}$$

Table 4.1 Parameters for continuous-flow stirred-tank reactor

$V = 1{,}000$	$A_f = 0.0065$	$T_f = 350$
$T_c = 340$	$k_0 = 7.86 \times 10^{12}$	$E' = 14{,}000$
$(-\Delta H) = 27{,}000$	$\rho = 1.0$	$C_p = 1.0$
$U = 10$	$K = 0.2$	$qs = 10$
$A_S = 15.31 \times 10^{-5}$	$T_S = 460.91$	$q_{cS} = 5$

Equations (14) are equivalent to the equation of motion of a solid body with viscous drag and a linear restoring force, the general second-order system

$$\ddot{y} + a_1 \dot{y} + a_2 y = u(t) \tag{15}$$

Hence, Eqs. (15) and (10) are completely equivalent when $b_{11} = b_{21} = 0$.

Equations (10) and (15) are also related in another way for this case. We may wish an equation for concentration alone if the temperature is of little importance to us. Differentiating Eq. (10a), we obtain (with $b_{11} = b_{21} = 0$)

$$\ddot{x}_1 = a_{11} \dot{x}_1 + a_{12} \dot{x}_2 \tag{16}$$

and substituting for $\dot{x}_2$ from Eq. (10b) and x_2 from Eq. (10a),

$$\ddot{x}_1 + a_1 \dot{x}_1 + a_2 x_1 = -u(t) \tag{17}$$

where a_1, a_2, and u have the same meanings as above.

Another transformation which we shall find convenient is

$$x_1 = y_1 + y_2 \tag{18a}$$
$$x_2 = -\frac{a_{11} + S_1}{a_{12}} y_1 - \frac{a_{11} + S_2}{a_{12}} y_2 \tag{18b}$$

Here,

$$2S_1 = -(a_{11} + a_{22}) + [(a_{11} + a_{22})^2 - 4(a_{11}a_{22} - a_{12}a_{21})]^{1/2} \tag{19a}$$
$$2S_2 = -(a_{11} + a_{22}) - [(a_{11} + a_{22})^2 - 4(a_{11}a_{22} - a_{12}a_{21})]^{1/2} \tag{19b}$$

For the parameters in Table 4.1, S_1 and S_2 are negative and real. Equations (10) then become

$$\dot{y}_1 = S_1 y_1 - M_{11} u_1 - M_{12} u_2 \tag{20a}$$
$$\dot{y}_2 = S_2 y_2 + M_{21} u_1 + M_{12} u_2 \tag{20b}$$

where

$$M_{11} = \frac{a_{11}b_{11} + a_{12}b_{21} + S_2 b_{11}}{S_1 - S_2} \tag{21a}$$

$$M_{12} = \frac{a_{12}b_{22}}{S_1 - S_2} \tag{21b}$$

$$M_{21} = M_{11} + b_{11} \tag{21c}$$

4.7 LINEAR FEEDBACK CONTROL

We are now in a position to consider the control of the chemical reactor. We shall suppose that the linearized equations are adequate and that the reactor is to be controlled by adjustment of the coolant flow rate. Equations (10) of the preceding section are then

$$\dot{x}_1 = a_{11}x_1 + a_{12}x_2 \tag{1a}$$
$$\dot{x}_2 = a_{21}x_1 + a_{22}x_2 + b_{22}u_2 \tag{1b}$$

where x_1 and x_2 are deviations in concentration and temperature, respectively, and u_2 the variation in coolant flow rate. The system is initially displaced from the equilibrium values, and our control action will be designed to keep fluctuations in all variables small. This objective may be accomplished by minimizing the integral

$$\mathcal{E} = \frac{1}{2} \int_0^\theta (c_1x_1{}^2 + c_2x_2{}^2 + c_3u_2{}^2)\, dt \tag{2}$$

We shall make the transformation

$$x_1 = -y_1 \tag{3a}$$
$$x_2 = \frac{a_{11}}{a_{12}} y_1 - \frac{1}{a_{12}} y_2 \tag{3b}$$

giving

$$\dot{y}_1 = y_2 \tag{4a}$$
$$\dot{y}_2 = -a_2y_1 - a_1y_2 + u \tag{4b}$$

and the objective

$$\mathcal{E} = \frac{1}{2} \int_0^\theta (c_{11}y_1{}^2 + 2c_{12}y_1y_2 + c_{22}y_2{}^2 + c_{33}u^2)\, dt \tag{5}$$

where

$$a_1 = -(a_{11} + a_{22}) \tag{6a}$$
$$a_2 = a_{11}a_{22} - a_{12}a_{21} \tag{6b}$$
$$u = -a_{12}b_{22}u_2 \tag{6c}$$
$$c_{11} = c_1 + c_2 \frac{a_{11}{}^2}{a_{12}{}^2} \tag{6d}$$
$$c_{12} = -c_2 \frac{a_{11}}{a_{12}{}^2} \tag{6e}$$
$$c_{22} = \frac{c_2}{a_{12}{}^2} \tag{6f}$$
$$c_{33} = \frac{c_3}{a_{12}{}^2 b_{22}{}^2} \tag{6g}$$

The hamiltonian is then

$$H = \frac{1}{2}(c_{11}y_1{}^2 + 2c_{12}y_1y_2 + c_{22}y_2{}^2 + c_{33}u^2) \\ + \lambda_1y_2 + \lambda_2(-a_2y_1 - a_1y_2 + u) \tag{7}$$

and the canonical equations for the multipliers

$$\dot{\lambda}_1 = -\frac{\partial H}{\partial y_1} = -c_{11}y_1 - c_{12}y_2 + a_2\lambda_2 \tag{8a}$$

$$\dot{\lambda}_2 = -\frac{\partial H}{\partial y_2} = -c_{12}y_1 - c_{22}y_2 - \lambda_1 + a_1\lambda_2 \tag{8b}$$

with boundary conditions

$$\lambda_1(\theta) = \lambda_2(\theta) = 0 \tag{9}$$

The optimality criterion, that the hamiltonian be stationary, is

$$\frac{\partial H}{\partial u} = 0 = \lambda_2 + c_{33}u \tag{10}$$

or

$$u = -\frac{\lambda_2}{c_{33}} \tag{11}$$

Thus, the problem is one of finding the function $\lambda_2(t)$.

The optimal control for this system can be found in a rather straightforward manner by seeking a solution of the form

$$\lambda_1 = m_{11}y_1 + m_{12}y_2 \tag{12a}$$
$$\lambda_2 = m_{12}y_1 + m_{22}y_2 \tag{12b}$$

From Eqs. (8), then,

$$\dot{\lambda}_1 = -c_{11}y_1 - c_{12}y_2 + a_2(m_{12}y_1 + m_{22}y_2) \tag{13a}$$
$$\dot{\lambda}_2 = -c_{12}y_1 - c_{22}y_2 - (m_{11}y_1 + m_{12}y_2) + a_1(m_{12}y_1 + m_{22}y_2) \tag{13b}$$

By differentiating Eqs. (12) with respect to time we also obtain

$$\dot{\lambda}_1 = \dot{m}_{11}y_1 + m_{11}\dot{y}_1 + \dot{m}_{12}y_2 + m_{12}\dot{y}_2 \tag{14a}$$
$$\dot{\lambda}_2 = \dot{m}_{12}y_1 + m_{12}\dot{y}_1 + \dot{m}_{22}y_2 + m_{22}\dot{y}_2 \tag{14b}$$

and, substituting Eqs. (4),

$$\dot{\lambda}_1 = \dot{m}_{11}y_1 + m_{11}y_2 + \dot{m}_{12}y_2$$
$$+ m_{12}\left(-a_2y_1 - a_1y_2 - \frac{m_{12}}{c_{33}}y_1 - \frac{m_{22}}{c_{33}}y_2\right) \tag{15a}$$

$$\dot{\lambda}_2 = \dot{m}_{12}y_1 + m_{12}y_2 + \dot{m}_{22}y_2$$
$$+ m_{22}\left(-a_2y_1 - a_1y_2 - \frac{m_{12}}{c_{33}}y_1 - \frac{m_{22}}{c_{33}}y_2\right) \tag{15b}$$

If a solution of the form of Eq. (12) is to exist, Eqs. (13) and (15) must be identical for all y_1 and y_2. That is, the coefficients of y_1 in Eqs. (13a) and (15a) must be identical, as they must be in Eqs. (13b) and (15b), and similarly for y_2. Thus, equating coefficients, we obtain the three differential equations

$$\dot{m}_{11} = m_{12}{}^2 + 2a_2 c_{33} m_{12} - c_{33} c_{11} \tag{16a}$$
$$\dot{m}_{22} = m_{22}{}^2 - 2c_{33} m_{12} + 2a_1 c_{33} m_{22} - c_{33} c_{22} \tag{16b}$$
$$\dot{m}_{12} = m_{12} m_{22} - c_{33} m_{11} + a_1 c_{33} m_{12} + a_2 c_{33} m_{22} - c_{33} c_{12} \tag{16c}$$

Equations (16) are the multidimensional generalization of the Riccati equation, which we have considered several times previously. For finite θ a solution can be obtained numerically with boundary conditions $m_{11}(\theta) = m_{12}(\theta) = m_{22}(\theta) = 0$, while as $\theta \to \infty$, a stable constant solution exists. We shall consider this case, and setting the time derivatives in Eqs. (16) to zero, we obtain

$$m_{12} = -c_{33} a_2 + (c_{33}{}^2 a_2{}^2 + c_{33} c_{11})^{1/2} \tag{17a}$$
$$m_{22} = -c_{33} a_1 + [c_{33}{}^2 a_1{}^2 + c_{33} c_{22} - 2c_{33}{}^2 a_2 \\ + c_{33}(c_{33}{}^2 a_2{}^2 + c_{33} c_{11})^{1/2}]^{1/2} \tag{17b}$$

Combining Eqs. (11), (12), and (17), we obtain the optimal control

$$u = \left[a_2 - \left(a_2{}^2 + \frac{c_{11}}{c_{33}} \right)^{1/2} \right] y_1 \\ + \left\{ a_1 - \left[a_1{}^2 - 2a_2 + \frac{c_{22}}{c_{33}} + \left(a_2 + \frac{c_{11}}{c_{33}} \right)^{1/2} \right]^{1/2} \right\} y_2 \tag{18}$$

This is a multivariable linear feedback controller, where the controller gains depend on both the system parameters and relative weights. The resulting system is stable, so that y_1 and y_2 go to zero as $\theta \to \infty$, and hence Eq. (12) for $\lambda_1(\theta)$ and $\lambda_2(\theta)$ satisfies the zero boundary conditions of Eq. (9).

The system equations (4a) and (4b), together with the control (18), are linear, and are easily solved and then transformed back to the Z_1, Z_2 variables defined in the previous section. Figure 4.2 shows the paths in the $Z_1 Z_2$ plane for the linearized reactor equations with the parameters in Table 4.1, using constants

$$c_1 = 84.5 \qquad c_2 = 6.16 \qquad c_3 = 10^{-3} \tag{19}$$

It is interesting to note that from most starting points the equilibrium point is approached along a line in the plane. A weakness of the linearized analyses becomes evident by noting that some of the paths cross into negative concentrations, which are physically meaningless.

4.8 AN APPROXIMATE SOLUTION

The technique used in the previous section to obtain solutions of the multiplier equations is a powerful one when the objective is of the form of Eq. (5), i.e., a quadratic form in the state variables plus a squared cost-

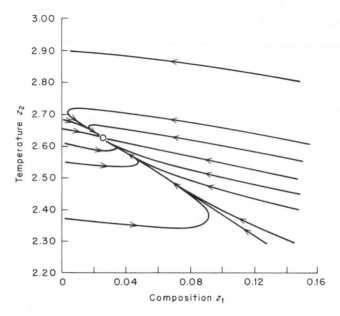

Fig. 4.2 Temperature-concentration paths for the controlled reactor. [*From J. M. Douglas and M. M. Denn, Ind. Eng. Chem.,* **57** (11): 18 (1965). *Copyright 1965 by the American Chemical Society. Reprinted by permission of the copyright owner.*]

of-control term, but it is restricted to linear systems. It is possible, however, to obtain *approximate* solutions to nonlinear systems with the same objective by similar methods. In order to avoid algebraic complexity we shall simplify our reactor example somewhat for the time being, but we shall see later that the result is still of physical as well as mathematical interest.

We shall assume for the reactor model described by Eqs. (6) and (7) of Sec. 4.6 that it is possible to choose the temperature Z_2 at every instant of time, and hence temperature or, equivalently, the reaction rate k [Sec. 4.6, Eq. (9)] is the control variable. The state of the system is described, then, by the single equation

$$\dot{Z}_1 = \frac{q}{V}(1 - Z_1) - kZ_1 \tag{1}$$

and we shall suppose that the objective is again one of keeping fluctuations small, that is,

$$\mathfrak{F} = \tfrac{1}{2}(Z_1 - Z_{1S})^2 + \tfrac{1}{2}c^2(k - k_S)^2 \tag{2}$$

It is convenient to define variables

$$x = Z_1 - Z_{1S} \qquad u = k - k_S \tag{3}$$

and so the system equation becomes, after subtracting out the steady-state terms,

$$\dot{x} = -\left(\frac{q}{V} + k_S\right)x - (Z_{1S} + x)u \tag{4}$$

$$\mathfrak{F} = \tfrac{1}{2}(x^2 + c^2u^2) \tag{5}$$

Note that we have not linearized. The linearized form of Eq. (4) would not contain the term xu.

The hamiltonian for this nonlinear system is

$$H = \tfrac{1}{2}x^2 + \tfrac{1}{2}c^2u^2 - \lambda\left(\frac{q}{V} + k_S\right)x - \lambda(Z_{1S} + x)u \tag{6}$$

and the equation for the multiplier

$$\dot{\lambda} = -\frac{\partial H}{\partial x} = -x + \left(\frac{q}{V} + k_S\right)\lambda + \lambda u \tag{7}$$

The condition that the hamiltonian be stationary is

$$\frac{\partial H}{\partial u} = c^2u - \lambda(Z_{1S} + x) = 0 \tag{8}$$

or

$$u = \frac{\lambda}{c^2}(Z_{1S} + x) \tag{9}$$

so that the problem is again one of finding the solution to the multiplier equation.

For the linear system of the previous section we were able to obtain a solution for λ proportional to x. We might look upon this as the lead term in a series expansion and seek a solution of the form

$$\lambda = mx + px^2 + \cdots \tag{10}$$

Again, for simplicity, we shall take $\theta \to \infty$ so that it will be possible to obtain the solution with $m, p, \ldots$ as constants.

Differentiating Eq. (10), we find, using Eqs. (4) and (9),

$$\dot{\lambda} = -(m + 2px)\left[\left(\frac{q}{V} + k_S\right)x + \frac{(Z_{1S} + x)^2}{c_2}(mx + px^2)\right] + \cdots \tag{11a}$$

while from Eqs. (7), (9), and (10),

$$\dot{\lambda} = -x + \left(\frac{q}{V} + k_S\right)(mx + px^2) + \frac{m^2x^2Z_{1S}}{c^2} + \cdots \tag{11b}$$

The coefficients of each power of x must be equal in these two equations, so that we obtain from the power x^1

$$\frac{m^2 Z_{1S}^2}{c^2} + 2m\left(\frac{q}{V} + k_S\right) - 1 = 0 \tag{12}$$

and from the power x^2

$$p\left(\frac{q}{V} + k_S + \frac{m Z_{1S}^2}{c^2}\right) + \frac{m^2 Z_{1S}}{c^2} = 0 \tag{13}$$

Equation (12) is easily shown to be the steady-state limit of the Riccati equation (10) of Sec. 3.4 for the linearized system when the proper identification of coefficients is made, the stable solution being

$$m = -\frac{c^2}{Z_{1S}^2}\left[\frac{q}{V} + k_S - \sqrt{\left(\frac{q}{V} + k_S\right)^2 + \frac{Z_{1S}^2}{c^2}}\right] \tag{14}$$

while the solution to Eq. (13) is

$$p = -\frac{m^2 Z_{1S}}{c^2(q/V + k_S) + m Z_{1S}^2} \tag{15}$$

Coefficients of x^3, x^4, etc., in the series for λ can be obtained in the same manner. In terms of the original variables, then, we obtain the optimal *nonlinear* feedback control

$$k = k_S + \frac{Z_1}{c^2}[m(Z_1 - Z_{1S}) + p(Z_1 - Z_{1S})^2 + \cdots] \tag{16}$$

Had we not let $\theta \to \infty$, we would have obtained differential equations for $m, p, \ldots$.

Equation (16) defines a curve in the $Z_1 Z_2$ plane. If we assume that we have reached that curve in some way, we can substitute Eq. (16) into Eqs. (7) of Sec. 4.6 to obtain the optimal coolant flow rate q_c to implement this policy, namely, the solution of

$$\frac{q_c}{1 + K q_c} = -\frac{(-\Delta H)VA_f Z_2^2}{kUKE'c^2}\left[\frac{q}{V}(1 - Z_1) - kZ_1\right](2Z_1 - Z_{1S})$$

$$[m + p(Z_1 - Z_{1S}) + \cdots] + \frac{q C_p \rho}{UK}(Z_f - Z_2) + \frac{kV C_p \rho}{UK}Z_1 \tag{17}$$

Hence, had we posed the problem of finding the coolant flow rate to minimize the integral†

$$\mathcal{E} = \int_0^\theta [(Z_1 - Z_{1S})^2 + c^2(k - k_S)^2]\, dt$$

we would have the partial solution that once having reached the line defined by Eq. (16), the optimal control policy is the nonlinear feedback

---

† There is no obvious physical reason why k should be any less meaningful in the integral than Z_2, since it is the reaction rate, rather than the temperature, which affects conversion.

controller defined by Eq. (17).   As we shall see in the next chapter, the overall policy is the one which we intuitively expect, namely, full cooling or no cooling until the line defined by Eq. (16) is reached and then nonlinear control as defined above.

4.9 CONTROL WITH CONTINUOUS DISTURBANCES

The control problems we have considered thus far have dealt with autonomous systems only; i.e., the differential equations are assumed to have no explicit dependence on time.   This will be the case, for example, when the operating conditions have been changed and the system must be brought to a new steady state (a set-point change) or when a large pulse-like disturbance has affected the system but no more disturbances are expected for a time long compared to the system's response time.   In many control situations, however, disturbances of prolonged duration may be expected to enter the system and must be included in a rational control analysis.

For simplicity we shall restrict our attention to linear systems with constant properties in which the state of the system can be described by a single variable.   If the disturbance is a piecewise differentiable function $D(t)$, the system response is described by the differential equation

$$\dot{x} = Ax + u + D(t) \qquad x(0) = x_0 \tag{1}$$

We shall again take the control objective to be the minimization of the integral

$$\mathcal{E} = \frac{1}{2} \int_0^\theta (x^2 + c^2 u^2)\, dt \tag{2}$$

If we follow the procedure of Sec. 4.4 and remove the explicit time dependence by definition of a new variable, we can write Eqs. (1) and (2) in autonomous form as

$$\dot{x}_1 = Ax_1 + D(x_2) + u \qquad x_1(0) = x_0 \tag{3a}$$
$$\dot{x}_2 = 1 \qquad\qquad\qquad\quad x_2(0) = 0 \tag{3b}$$

$$\mathcal{E} = \frac{1}{2} \int_0^\theta (x_1{}^2 + c^2 u^2)\, dt \tag{4}$$

The hamiltonian for this system is

$$H = \tfrac{1}{2}(x_1{}^2 + c^2 u^2) + \lambda_1(Ax_1 + D + u) + \lambda_2 \tag{5}$$

with multiplier equations

$$\dot{\lambda}_1 = -\frac{\partial H}{\partial x_1} = -x_1 - A\lambda_1 \qquad \lambda_1(\theta) = 0 \tag{6a}$$

$$\dot{\lambda}_2 = -\frac{\partial H}{\partial x_2} = -\lambda_1 \frac{dD}{dx_2} = -\lambda_1 \dot{D} \tag{6b}$$

where $\lambda_2(\theta)$ will be zero if θ is unspecified, but unspecified for θ fixed. The condition of optimality is

$$\frac{\partial H}{\partial u} = c^2 u + \lambda_1 = 0 \tag{7a}$$

or

$$u = -\frac{\lambda_1}{c^2} \tag{7b}$$

We see, therefore, that x_2 and λ_2 are in fact extraneous, and we may drop the subscripts on x_1 and λ_1.

Equations (3a) and (6a) form a system

$$\dot{x} = Ax - \frac{\lambda}{c^2} + D(t) \qquad x(0) = x_0 \tag{8a}$$

$$\dot{\lambda} = -x - A\lambda \qquad \lambda(\theta) = 0 \tag{8b}$$

Since we already know that a homogeneous solution $(D = 0)$ for this system can be obtained in the form λ proportional to x, we shall seek a solution

$$\lambda = -c^2[M(t)x + L(t)] \tag{9}$$

From Eq. (8a) we then have

$$\dot{\lambda} = -c^2 M(Ax + D + Mx + L) - c^2\dot{M}x - c^2\dot{L} \tag{10a}$$

while from Eq. (8b)

$$\dot{\lambda} = -x + Ac^2 Mx + Ac^2 L \tag{10b}$$

Equating coefficients of x in Eqs. (10), we obtain an equation for $M(t)$

$$\dot{M} + 2AM + M^2 - \frac{1}{c^2} = 0 \tag{11a}$$

and therefore $L(t)$ must satisfy

$$\dot{L} + [A + M(t)]L + M(t)D(t) = 0 \tag{11b}$$

Equation (11a) is, of course, simply the Riccati equation of Sec. 3.4. In order to establish boundary conditions for Eqs. (11a) and (11b) we shall assume that at some time $t < \theta$ the disturbance vanishes and remains identically zero. During this final period we have the problem we have already solved, namely, a system offset from $x = 0$ with no disturbances, and we know that the solution requires

$$M(\theta) = 0 \qquad \text{or} \qquad x(\theta) = 0 \tag{12a}$$

The solution for $M(t)$ is thus given by Eqs. (13) and (14) of Sec. 3.4. It follows then, from Eqs. (8b) and (9), that the proper boundary condition for Eq. (11b) is

$$L(\theta) = 0 \qquad\qquad (12b)$$

This boundary condition clearly points up the difficulty we face in dealing with nonautonomous systems. While the formal mathematics is straightforward, we must know the future of the disturbance function $D(t)$ in order to obtain a solution to Eq. (11b)' which satisfies the boundary condition [Eq. (12b)] at $t = \theta$.

In practice this difficulty, while serious, may be somewhat less severe on occasion than it first appears. If we let $\theta \to \infty$, the solution of the Riccati equation (11a) is

$$M = -A - \sqrt{\frac{1 + A^2 c^2}{c^2}} \qquad\qquad (13)$$

and Eq. (11b) is

$$\dot{L} - \sqrt{\frac{1 + A^2 c^2}{c^2}}\, L = \left(A + \sqrt{\frac{1 + A^2 c^2}{c^2}} \right) D(t) \qquad L(\infty) = 0 \quad (14)$$

which has the solution

$$L(t) = -\left(A + \sqrt{\frac{1 + A^2 c^2}{c^2}} \right) \int_t^\infty \exp\left[-\sqrt{\frac{1 + A^2 c^2}{c^2}}\, (\tau - t) \right]$$
$$D(\tau)\, d\tau \quad (15)$$

For bounded disturbances the integrand will effectively vanish for future times greater than several time constants, $c(1 + A^2 c^2)^{-\frac12}$, and it is necessary only to know (or estimate) the disturbance for that time into the future. Indeed, disturbances often have the form of step changes,

$$D(t) = D = \text{const} \qquad 0 \le t \le nc(1 + A^2 c^2)^{-\frac12} \qquad (16)$$

in which case Eq. (15) yields

$$L(t) = -\left(1 + \sqrt{\frac{A^2 c^2}{1 + A^2 c^2}} \right)(1 - e^{-n})D \qquad\qquad (17a)$$

and, for n greater than 2 or 3,

$$L \approx -\left(1 + \sqrt{\frac{A^2 c^2}{1 + A^2 c^2}} \right) D \qquad\qquad (17b)$$

That is, the optimal control for a step disturbance is proportional to both the system state and the disturbance

$$u(t) = -\left(A + \sqrt{\frac{1 + A^2 c^2}{c^2}} \right) x(t) - \left(1 + \sqrt{\frac{A^2 c^2}{1 + A^2 c^2}} \right) D \qquad (18)$$

The term $c^2 u^2$ in the objective defined by Eq. (2) is meaningless for many industrial situations, in which a true cost of control is negligible. Such a term might be useful as a penalty function in order to keep the control effort $u(t)$ between bounds, but an immediate disadvantage of such a practice is evident from the substitution of Eq. (18) into Eq. (1). The optimal response to a step-function disturbance is then found to be

$$x(t) = -\frac{Ac^2D}{1 + A^2c^2} + \left(x_0 + \frac{Ac^2D}{1 + A^2c^2}\right)\exp\left(-\frac{c^2}{1 + A^2c^2}t\right) \quad (19)$$

or

$$x(t) \to -\frac{Ac^2D}{1 + A^2c^2} \quad (20)$$

That is, the optimal control with a cost-of-control term does not return the system to $x = 0$ as long as there is a step disturbance present. This is known as *steady-state offset*, and is clearly undesirable in many circumstances.

4.10 PROPORTIONAL PLUS RESET CONTROL

Very often the serious restriction is not the available control effort but the maximum rate at which the control setting may be changed. In such cases the problem of steady-state offset for step disturbances can be resolved, and the result points up an interesting connection between traditional control practice and optimal-control theory.

We shall assume that the system is at equilibrium for $t < 0$, with a step disturbance of magnitude D entering at $t = 0$. We thus have

$$\dot{x} = Ax + D + u \qquad x(0) = 0 \quad (1)$$

with the objective

$$\varepsilon = \frac{1}{2}\int_0^\theta (x^2 + c^2\dot{u}^2)\, dt \quad (2)$$

where the term $c^2\dot{u}^2$ is intended as a penalty function to keep the rate of change of control action within bounds. Because D is a constant, we can differentiate Eq. (1) once to obtain

$$\ddot{x} = A\dot{x} + \dot{u} \qquad x(0) = 0 \qquad \dot{x}(0) = D \quad (3)$$

or defining $x_1 = x$, $x_2 = \dot{x}$, $w = \dot{u}$,

$$\dot{x}_1 = x_2 \qquad\qquad x_1(0) = 0 \quad (4a)$$
$$\dot{x}_2 = Ax_2 + w \qquad x_2(0) = D \quad (4b)$$
$$\varepsilon = \frac{1}{2}\int_0^\theta (x_1{}^2 + c^2w^2)\, dt \quad (5)$$

The problem defined by Eqs. (4) and (5) is precisely the one considered in Sec. 4.7. If we let $\theta \to \infty$, then, by relating coefficients, the optimal solution is obtained from Eq. (18) of Sec. 4.7 as

$$w(t) = \dot{u}(t) = -\frac{1}{c} x_1 - \left(A + \frac{1}{c} \right) x_2 \tag{6}$$

Integrating, the control action is found to be

$$u(t) = -\left(A + \frac{1}{c} \right) x(t) - \frac{1}{c} \int_0^t x(\tau) \, d\tau \tag{7}$$

That is, the optimal control is proportional to both the offset and the integral of the offset. The integral mode is often referred to as *reset*. Most industrial controllers employ proportional and reset modes.

The importance of the reset mode can be clearly seen by substituting the control action into Eq. (3). The system response is determined by the equation

$$\ddot{x} + \frac{1}{c} \dot{x} + \frac{1}{c} x = 0 \qquad x(0) = 0 \qquad \dot{x}(0) = D \tag{8}$$

which has the solution

$$x(t) = \begin{cases} \dfrac{2cD}{\sqrt{1-4c}} e^{-t/2c} \sinh \dfrac{\sqrt{1-4c}}{2c} t & c < \frac{1}{4} \\[3ex] -\dfrac{2cD}{\sqrt{4c-1}} e^{-t/2c} \sin \dfrac{\sqrt{4c-1}}{2c} t & c > \frac{1}{4} \end{cases} \tag{9}$$

That is, the response is overdamped and without oscillation when $c < \frac{1}{4}$, underdamped with oscillations for $c > \frac{1}{4}$, but always decaying exponentially to zero after the initial rise. Thus, there can be no steady-state offset.

4.11 OPTIMAL-YIELD PROBLEMS

In many processes the quantity of interest will not be a cumulative measure of profit or loss in the form

$$\mathcal{E} = \int_0^\theta \mathcal{F}(x_1, x_2, u_1, u_2) \, dt \tag{1}$$

but simply the difference between initial and final values of x_1 and x_2. A particular case would be a chemical-reaction system, where we might seek to maximize some weighted sum of the conversions, with the profit, say $\mathcal{P}$, expressed as

$$\mathcal{P} = c[x_1(\theta) - x_1(0)] + [x_2(\theta) - x_2(0)] \tag{2}$$

This is equivalent, however, to writing

$$\mathcal{P} = \int_0^\theta (c\dot{x}_1 + \dot{x}_2) \, dt = \int_0^\theta (cf_1 + f_2) \, dt \tag{3}$$

which is in the form of Eq. (1) if we obtain a minimization problem by letting $\mathcal{E} = -\mathcal{P}$. Hence, we wish to consider the general problem of two nonlinear state equations, with $\mathcal{F}$ defined as

$$\mathcal{F} = -cf_1 - f_2 \tag{4}$$

For algebraic simplicity we shall assume a single decision function $u(t)$. The canonical equations are

$$\dot{x}_1 = f_1(x_1, x_2, u) \tag{5a}$$
$$\dot{x}_2 = f_2(x_1, x_2, u) \tag{5b}$$
$$\dot{\lambda}_1 = -(\lambda_1 - c)\frac{\partial f_1}{\partial x_1} - (\lambda_2 - 1)\frac{\partial f_2}{\partial x_1} \tag{5c}$$
$$\dot{\lambda}_2 = -(\lambda_1 - c)\frac{\partial f_1}{\partial x_2} - (\lambda_2 - 1)\frac{\partial f_2}{\partial x_2} \tag{5d}$$

with boundary conditions $\lambda_1(\theta) = \lambda_2(\theta) = 0$ and the hamiltonian

$$H = f_1(\lambda_1 - c) + f_2(\lambda_2 - 1) \tag{6}$$

It is convenient to define new variables ψ_1, ψ_2 such that

$$\psi_1 = \lambda_1 - c \tag{7a}$$
$$\psi_2 = \lambda_2 - 1 \tag{7b}$$
$$\psi_1(\theta) = -c \tag{7c}$$
$$\psi_2(\theta) = -1 \tag{7d}$$

We then have

$$\dot{\psi}_1 = -\psi_1\frac{\partial f_1}{\partial x_1} - \psi_2\frac{\partial f_2}{\partial x_1} \tag{8a}$$
$$\dot{\psi}_2 = -\psi_1\frac{\partial f_1}{\partial x_2} - \psi_2\frac{\partial f_2}{\partial x_2} \tag{8b}$$
$$H = \psi_1 f_1 + \psi_2 f_2 \tag{9}$$

and the condition of optimality,

$$\frac{\partial H}{\partial u} = \psi_1\frac{\partial f_1}{\partial u} + \psi_2\frac{\partial f_2}{\partial u} = 0 \tag{10}$$

For this class of problems it is possible to reduce the number of variables which need to be considered. Equation (10) is true for all time, and therefore its derivative with respect to t must vanish. Hence,

$$\frac{d}{dt}\frac{\partial H}{\partial u} = \dot{\psi}_1\frac{\partial f_1}{\partial u} + \psi_1\left(\frac{\partial^2 f_1}{\partial u\, \partial x_1}f_1 + \frac{\partial^2 f_1}{\partial u\, \partial x_2}f_2 + \frac{\partial^2 f_1}{\partial u^2}\dot{u}\right)$$
$$+ \dot{\psi}_2\frac{\partial f_2}{\partial u} + \psi_2\left(\frac{\partial^2 f_2}{\partial u\, \partial x_1}f_1 + \frac{\partial^2 f_2}{\partial u\, \partial x_2}f_2 + \frac{\partial^2 f_2}{\partial u^2}\dot{u}\right) = 0 \tag{11}$$

or, substituting Eqs. (8),

$$\psi_1 \left(-\frac{\partial f_1}{\partial u}\frac{\partial f_1}{\partial x_1} - \frac{\partial f_2}{\partial u}\frac{\partial f_1}{\partial x_2} + \frac{\partial^2 f_1}{\partial u\,\partial x_1}f_1 + \frac{\partial^2 f_1}{\partial u\,\partial x_2}f_2 + \frac{\partial^2 f_1}{\partial u^2}\dot{u} \right)$$

$$+ \psi_2 \left(-\frac{\partial f_1}{\partial u}\frac{\partial f_2}{\partial x_1} - \frac{\partial f_2}{\partial u}\frac{\partial f_2}{\partial x_2} + \frac{\partial^2 f_2}{\partial u\,\partial x_1}f_1 + \frac{\partial^2 f_1}{\partial u\,\partial x_2}f_2 + \frac{\partial^2 f_2}{\partial u^2}\dot{u} \right) = 0 \tag{12}$$

Equations (10) and (12) are linear homogeneous algebraic equations for ψ_1 and ψ_2, and the condition that they have a nontrivial solution (ψ_1 and ψ_2 both not identically zero) is the vanishing of the determinant of coefficients. Thus

$$\frac{\partial f_1}{\partial u} \left(-\frac{\partial f_1}{\partial u}\frac{\partial f_2}{\partial x_1} - \frac{\partial f_2}{\partial u}\frac{\partial f_2}{\partial x_2} + \frac{\partial^2 f_2}{\partial u\,\partial x_1}f_1 + \frac{\partial^2 f_2}{\partial u\,\partial x_2}f_2 + \frac{\partial^2 f_2}{\partial u^2}\dot{u} \right)$$

$$- \frac{\partial f_2}{\partial u} \left(-\frac{\partial f_1}{\partial u}\frac{\partial f_1}{\partial x_1} - \frac{\partial f_2}{\partial u}\frac{\partial f_1}{\partial x_2} + \frac{\partial^2 f_1}{\partial u\,\partial x_1}f_1 + \frac{\partial^2 f_1}{\partial u\,\partial x_2}f_2 + \frac{\partial^2 f_1}{\partial u^2}\dot{u} \right) = 0 \tag{13}$$

or, solving for $\dot{u}$,

$$\dot{u} = \frac{\begin{aligned}\frac{\partial f_2}{\partial u}\left(\frac{\partial^2 f_1}{\partial u\,\partial x_1}f_1 + \frac{\partial^2 f_1}{\partial u\,\partial x_2}f_2 - \frac{\partial f_1}{\partial u}\frac{\partial f_1}{\partial x_1} - \frac{\partial f_2}{\partial u}\frac{\partial f_1}{\partial x_2} \right)\\ - \frac{\partial f_1}{\partial u}\left(\frac{\partial^2 f_2}{\partial u\,\partial x_1}f_1 + \frac{\partial^2 f_2}{\partial u\,\partial x_2}f_2 - \frac{\partial f_1}{\partial u}\frac{\partial f_2}{\partial x_1} - \frac{\partial f_2}{\partial u}\frac{\partial f_1}{\partial x_1} \right)\end{aligned}}{\frac{\partial f_1}{\partial u}\frac{\partial^2 f_2}{\partial u^2} - \frac{\partial f_2}{\partial u}\frac{\partial^2 f_1}{\partial u^2}} \tag{14}$$

Equation (14), which has been called a *generalized Euler equation*, is an ordinary differential equation for the optimal decision function $u(t)$, which must be solved together with Eqs. (5a) and (5b) for x_1 and x_2. A boundary condition is still required, and it is obtained by evaluating Eq. (10) at $t = \theta$ with the values of ψ_1, ψ_2 obtained from Eqs. (7c) and (7d):

$$c\frac{\partial f_1}{\partial u} + \frac{\partial f_2}{\partial u} = 0 \qquad \text{at } t = \theta \tag{15}$$

The problem can then be solved numerically by searching over values of $u(0)$ and solving the three differential equations until the stopping condition, Eq. (15), is satisfied. Equivalently, for each initial value $u(0)$ there is a process running time for which $u(0)$ is optimal, and this correspondence is found when Eq. (15) is satisfied.

The reader will recognize that this section parallels Sec. 1.11 for discrete systems. In each case it has been possible to eliminate the multipliers by use of the optimality condition. It should be clear that the algebraic conditions needed to pass from Eqs. (10) and (12) to (13) require that the number of state variables exceed the number of decision variables by no more than 1, so that there will be at least as many homogeneous equations as multipliers.

4.12 OPTIMAL TEMPERATURES FOR CONSECUTIVE REACTIONS

We now make use of the results of the previous section to consider the continuous analog of the problem of Sec. 1.12. A system of consecutive reactions is assumed to take place,

$$X_1 \rightarrow X_2 \rightarrow \text{decomposition products}$$

and the reaction is to be carried out in a batch reactor. X_2 is the desired product, and we seek to maximize the increase in value of the contents of the reactor after an operating period θ by adjusting the reactor temperature $u(t)$ in time.

The equations describing the course of the reaction are

$$\dot{x}_1 = -k_1(u)F(x_1) \tag{1a}$$

(that is, the rate of reaction of X_1 depends only on temperature and concentration of X_1)

$$\dot{x}_2 = \nu k_1(u)F(x_1) - k_2(u)G(x_2) \tag{1b}$$

(The rate of change of concentration x_2 is the difference between the rate of formation from X_1 and the rate of decomposition. The latter rate depends only on temperature and concentration of X_2.) The coefficient ν is introduced to account for changes in reaction stoichiometry (the number of molecules of X_1 needed to form a molecule of X_2). It should be noted that the equations describing the course of this reaction in a pipeline reactor in which diffusion is negligible are identical if t is interpreted as *residence time,* or length into the reactor divided by fluid velocity. Hence, we may look upon this problem as that of determining the best that could be accomplished in a pipeline reactor if it were possible to specify the temperature at every point, and we shall generally refer to the function $u(t)$ as the *optimal temperature profile.* In the latter case the true reactor design problem would be that of approaching the upper bound represented by the optimum profit with a practical heat-exchange system, since in a real reactor the temperature cannot be specified at every point in space.

The increase in value of the product has precisely the form of Eq. (2) of the preceding section, where c represents the value of the feed X_1 relative to the desired product X_2. Clearly $c < 1$. Equation (14) of Sec. 4.11 for the optimal temperature is

$$\dot{u} = \frac{\nu F(x_1) G'(x_2)[k_1(u) k_2'(u) - k_1'(u) k_2(u)] k_1'(u)}{G(x_2)[k_1''(u) k_2'(u) - k_1'(u) k_2''(u)]} \tag{2}$$

where the prime denotes differentiation with respect to the argument, and the stopping condition, Eq. (15), is

$$k_1'(u) F(x_1)(\nu - c) - k_2'(u) G(x_2) = 0 \qquad t = \theta \tag{3}$$

More specifically, if the functions k_1 and k_2 are of Arrhenius form

$$k_i = k_{i0} \exp\left(-\frac{E_i'}{u}\right) \qquad i = 1, 2 \tag{4}$$

where k_{10}, k_{20}, E_1', and E_2' are constants, then

$$\dot{u} = -\frac{\nu F(x_1) G'(x_2)}{E_2' G(x_2)} k_{10} \exp\left(-\frac{E'}{u}\right) \tag{5}$$

and the final condition on the temperature is

$$u = \frac{E_2' - E_1'}{\ln\left[\dfrac{E_2' k_{20} G(x_2)}{(\nu - c) E_1' k_{10} F(x_1)}\right]} \qquad t = \theta \tag{6}$$

Equation (5) requires that the optimal temperature always decrease in time or with reactor length. When $E_2' > E_1'$, a high temperature favors the second reaction with respect to the first, and a decreasing temperature makes sound physical sense, for it suggests a high temperature initially to encourage the reaction $X_1 \to X_2$ when there is little X_2 to react and then a low temperature in the latter stages in order to prevent the decomposition of the valuable product X_2. On the other hand, if $E_1' > E_2'$, a decreasing temperature profile contradicts the physical intuition that since the reaction $X_1 \to X_2$ is favored with respect to the decomposition of X_2, the highest possible temperature is optimal at all times. In Chap. 6 we shall develop a condition analogous to the second-derivative test of Sec. 1.3 which verifies this physical reasoning and demonstrates that Eq. (5) defines an optimum only when $E_2' > E_1'$.

The procedure for obtaining the optimal temperature profile and optimal profit is as described at the end of Sec. 4.11. The feed compositions $x_1(0)$ and $x_2(0)$ are presumed known, and a value is assumed for $u(0)$. Equations (1) and (5) are then integrated simultaneously until Eq. (6) is satisfied, and $u(0)$ is varied and the procedure repeated until the solution of Eq. (6) occurs at $t = \theta$. Amundson and Bilous have carried out such solutions for several cases.

4.13 OPTIMAL CONVERSION IN A PRESSURE–CONTROLLED REACTION

As a further example of the use of the generalized Euler equation and for purposes of reference in our later discussion of computation we shall consider a second type of optimal-yield problem, the maximization of intermediate conversion in a consecutive-reaction sequence carried out in a pipeline reactor where the reaction rate is dependent not upon temperature but upon pressure. The reaction is

$$X_1 \rightarrow 2X_2 \rightarrow \text{decomposition products}$$

where the first reaction is first-order and the second second-order. Concentrations are denoted by lowercase letters and total pressure by u.

Assuming ideal gases and Dalton's law of additive partial pressures, the state equations may be written

$$\dot{x}_1 = -2k_1 u \frac{x_1}{A + x_2} \qquad x_1(0) = x_{10} \tag{1a}$$

$$\dot{x}_2 = 4k_1 u \frac{x_1}{A + x_2} - 4k_2 u^2 \frac{x_2{}^2}{(A + x_2)^2} \qquad x_2(0) = x_{20} \tag{1b}$$

where k_1 and k_2 are positive constants and $A = 2x_{10} + x_{20}$. To maximize the conversion of intermediate we have

$$\mathcal{P} = x_2(\theta) \tag{2}$$

in which case the parameter c defined in Sec. 4.11 is zero. Performing the required differentiations, Eq. (14) of Sec. 4.11 for the optimal pressure is then found to be

$$\dot{u} = -\frac{4u^2}{x_2(A + x_2)^2} \left[k_1 A x_1 + \frac{k_2 u x_2{}^3}{(A + x_2)} \right] \tag{3}$$

with the boundary condition from Eq. (15) of Sec. 4.11,

$$u = \frac{k_1 x_1 (A + x_2)}{2k_2 x_2{}^2} \qquad \text{at } t = \theta \tag{4}$$

Equation (3) indicates that the optimal pressure decreases with reactor length, and if x_{20} is small, very steep gradients may be required near $t = 0$. It is, of course, impossible to specify the pressure at each point in a pipeline reactor, so that the optimal conversion calculated by the solution of Eqs. (1), (3), and (4) provides an upper bound for evaluating the results of a practical reactor design.

BIBLIOGRAPHICAL NOTES

Sections 4.2 and 4.3: The derivation follows

J. M. Douglas and M. M. Denn: *Ind. Eng. Chem.*, **57**(11): 18 (1965)

The results obtained here are a special case of much more general ones derived in subsequent chapters, and a complete list of references will be included later. A fundamental source for Chaps. 4 to 6 is

L. S. Pontryagin, V. G. Boltyanskii, R. V. Gamkrelidze, and E. F. Mishchenko: "Mathematical Theory of Optimal Processes," John Wiley & Sons, Inc., New York, 1962

Section 4.4: Any of the texts on the calculus of variations noted in the bibliographical notes for Chap. 3 will contain a discussion of the corner condition.

Section 4.5: This section is based on

L. I. Rozenoer: in Automation and Remote Control, I, *Proc. 1st IFAC Congr., Moscow, 1960*, Butterworth & Co. (Publishers), Ltd., London, 1961

The principal of least action is discussed in books on classical mechanics, such as

H. Goldstein: "Classical Mechanics," Addison-Wesley Publishing Company, Inc., Reading, Mass., 1950
L. D. Landau and E. M. Lifshitz: "Mechanics," Addison-Wesley Publishing Company, Inc., Reading, Mass., 1960

Section 4.6: The model of a stirred-tank reactor and an analysis of its transient behavior are contained in

R. Aris: "Introduction to the Analysis of Chemical Reactors," Prentice-Hall Inc., Englewood Cliffs, N.J., 1965

This is also an excellent source of details on other reactor models used as examples throughout this book.

Section 4.7: This section follows the paper by Douglas and Denn cited above. The basic work is

R. E. Kalman: *Bol. Soc. Mat. Mex.*, **5**:102 (1960)

A more general discussion is included in Chap. 8, and an extensive survey of optimal linear control is contained in

M. Athans and P. Falb: "Optimal Control," McGraw-Hill Book Company, New York, 1966

The reader unfamiliar with the conventional approach to process control may wish to consult a text such as

D. R. Coughanowr and L. B. Koppel: "Process Systems Analysis and Control," McGraw-Hill Book Company, New York, 1965
D. D. Perlmutter: "Chemical Process Control," John Wiley & Sons, Inc., New York, 1965
J. Truxal: "Automatic Feedback Control System Synthesis," McGraw-Hill Book Company, New York, 1955

Section 4.8: The expansion technique for obtaining nonlinear feedback controls is due to Merriam; see

C. W. Merriam: "Optimization Theory and the Design of Feedback Control Systems," McGraw-Hill Book Company, New York, 1964

A. R. M. Noton: "Introduction to Variational Methods in Control Engineering," Pergamon Press, New York, 1965

Sections 4.9 and 4.10: The references by Kalman and Athans and Falb cited above are pertinent here also, and the discussion is expanded in Chap. 8. The consequences of the use of $\dot{u}^2$ as the cost term in relating optimal-control theory to conventional feedback control practice is the subject of research being carried on in collaboration with G. E. O'Connor; see

G. E. O'Connor: "Optimal Linear Control of Linear Systems: An Inverse Problem," M. Ch. E. Thesis, Univ. of Delaware, Newark, Del., 1969

Sections 4.11 to 4.13: The generalized Euler equation was obtained in

M. M. Denn and R. Aris: Z. Angew. Math. Phys., **16**:290 (1965)

Prior derivations specific to the optimal-temperature-profile problem are in

N. R. Amundson and O. Bilous: Chem. Eng. Sci., **5**:81, 115 (1956)

R. Aris: "The Optimal Design of Chemical Reactors," Academic Press, Inc., New York (1961)

F. Horn: Chem. Eng. Sci., **14**:77 (1961)

Both the optimal temperature- and pressure-profile problems were studied in

E. S. Lee: AIChE J., **10**:309 (1964)

PROBLEMS

4.1. The pressure-controlled chemical reaction $A \rightleftharpoons 2B$, carried out in a tubular reactor, is described by the equation for the concentration of A

$$\dot{x} = -k_1 u \frac{x}{2x_0 - x} + k_2 u^2 \frac{4(x_0 - x)^2}{(2x_0 - x)^2}$$

where x_0 is the initial value of x, u is the pressure, and k_1 and k_2 are constants. $x(\theta)$ is to be minimized. Obtain an algebraic equation for the theoretical minimum value of $x(\theta)$ in terms of θ, k_1, k_2, and x_0. For comparison in ultimate design obtain the equation for the best yield under constant pressure. (The problem is due to Van de Vusse and Voetter.)

4.2. Batch binary distillation is described by the equations

$$\dot{x}_1 = -u$$

$$\dot{x}_2 = \frac{u}{x_1} [x_2 - F(x_2, u)]$$

with initial conditions x_{10}, x_{20}. Here x_1 denotes the total moles remaining in the still, x_2 the mole fraction of more volatile component in the still, u the product withdrawal rate, and F the overhead mole fraction of more volatile component, a known function of x_2 and u which depends upon the number of stages. The withdrawal rate is to be found so as to maximize the total output

$$\max \mathcal{P} = \int_0^\theta u(t)\, dt$$

while maintaining a specified average purity

$$\bar{F} = \frac{\int_0^\theta Fu\,dt}{\int_0^\theta u\,dt}$$

Formulate the problem so that it can be solved by the methods of this chapter and obtain the complete set of equations describing the optimum. Describe a computational procedure for efficient solution. (The problem is due to Converse and Gross.)

4.3. Consider the linear system

$$\ddot{x} = u$$
$$x(0) = x_0 \qquad \dot{x}(0) = y_0$$
$$x(\theta) = 0 \qquad \dot{x}(\theta) = 0$$

and the objective to be minimized,

$$\min \varepsilon = \frac{1}{2} \int_0^\theta u^2(t)\,dt$$

(a) Find the unconstrained function u which minimizes ε for a fixed θ.

(b) Examine the nature of the minimum ε in part (a) as a function of θ graphically. Comment on the sensitivity of the solution to changes in θ.

(c) Find the unconstrained function u which minimizes ε for unspecified θ. Comment on the significance of the solution in terms of the results of part (b). (The problem is due to Gottlieb.)

4.4. Solve the control problem of Sec. 3.4 by the method of Sec. 4.7.

4.5. Consider the nonlinear system

$$\dot{x} = f(x) + b(x)u$$
$$x(0) = x_0 \qquad x(\theta) = 0$$

where f and b are continuous differentiable functions of x. Show that the optimum unconstrained function u which minimizes

$$\varepsilon = \int_0^\theta [g(x) + c^2u^2]\,dt$$

with c a constant and g a nonnegative continuous differentiable function of x, has the feedback form

$$u = -\frac{f(x)}{b(x)} - \frac{x}{b(x)}\left\{\left[\frac{f(x)}{x}\right]^2 + \frac{[b(x)]^2[g(x) + \beta]}{x^2c^2}\right\}^{\frac{1}{2}}$$

where β is a constant depending on x_0 and θ. Suppose θ is unspecified? Compare the solution of Sec. 3.4. (The problem is due to Johnson.)

4.6. Extend the analysis of Sec. 4.7 to the case

$$\dot{x}_1 = a_{11}x_1 + a_{12}x_2 + b_1u$$
$$\dot{x}_2 = a_{21}x_1 + a_{22}x_2 + b_2u$$
$$\min \varepsilon = \tfrac{1}{2}[C_1e_1{}^2(\theta) + C_2e_2{}^2(\theta) + \int_0^\theta (c_1e_1{}^2 + c_2e_2{}^2 + c_3u^2)\,dt]$$

where

$$e_1 = x_1 - x_1^* \qquad e_2 = x_2 - x_2^*$$

x_1^* and x_2^* are some desired values. Obtain equations for the gains in the optimal controller.

4.7. The kinetics of a one-delayed-neutron group reactor with temperature feedback proportional to flux are

$$\dot{n} = \frac{1}{\Lambda}(un - \alpha n^2 - \beta n) + \gamma c$$

$$\dot{c} = \frac{\beta}{\Lambda} n - \gamma c$$

Here n is the neutron density, c the precursor concentration, u the reactivity, and the constants Λ, γ, α, and β, respectively, the neutron generation time, the decay constant, the power coefficient of reactivity, and the fraction of neutrons given off but not emitted instantaneously. Initial conditions n_0 and c_0 are given, and it is desired to bring the neutron density from n_0 to σn_0 in time θ, with the further condition that $\dot{n}(\theta)$ be zero and the effort be a minimum,

$$\min \varepsilon = \frac{1}{2}\int_0^\theta u^2\,dt$$

Obtain the equations needed for solution. (The problem is due to Rosztoczy, Sage, and Weaver.)

4.8. Reformulate Prob. 4.7 to include the final constraints as penalty functions,

$$\min \varepsilon = \frac{1}{2}\left\{ C_1[n(\theta) - \sigma n_0]^2 + C_2[c(\theta) - c^*]^2 + \int_0^\theta u^2\,dt \right\}$$

(What is c^*?) Obtain the equations needed for solution. Normalize the equations with respect to σn_0 and obtain an approximate solution with the approach of Sec. 4.8, utilizing the result of Prob. 4.6.

4.9. Let I be an inventory, P production rate, and S sales rate. Then

$$\dot{I} = P - S$$

Assuming quadratic marginal costs of manufacturing and holding inventories, the excess cost of production for deviating from desired values is

$$\varepsilon = \int_0^\theta \{C_I[I(t) - \bar{I}]^2 + C_P[P(t) - \bar{P}]^2\}\,dt$$

where θ is fixed, $\bar{I}$ and $\bar{P}$ are the desired levels, and C_I and C_P are constant costs. If the sales forecast $S(t)$ is known, determine the optimal production schedule $P(t)$, $0 \leq t \leq \theta$. Would a feedback solution be helpful here? (The problem is due to Holt, Modigliani, Muth, and Simon.)

4.10. Let x denote the CO_2 concentration in body tissue and u the pulmonary ventilation. An equation relating the two can be written[†]

$$\alpha_0\ddot{x} + (\alpha_1 + \alpha_2 u)\dot{x} + \alpha_3 u x = \alpha_4 + \alpha_5 u$$

where the α_i are constants. Find an approximate feedback solution for the "control" u which regulates the CO_2 level by minimizing

$$\varepsilon = \frac{1}{2}\int_0^\theta (x^2 + c^2 u^2)\,dt$$

It is commonly assumed that $u = a + bx$.

[†] The model equation is due to Grodins et al., *J. Appl. Phys.*, **7**:283 (1954).

5

Continuous Systems: II

5.1 INTRODUCTION

We now generalize our discussion of systems described by ordinary differential equations somewhat by relaxing the requirement that the optimal decision functions be unconstrained. Complete generality must await the next chapter, but most situations of interest will be included within the scope of this one, in which we presume that the optimal decision functions may be bounded from above and below by constant values. Typical bounds would be the open and shut settings on valves, safety limitations on allowable temperatures and pressures, or conditions describing the onset of unfavorable reaction products.

We again assume that the state of the system is described by the N ordinary differential equations

$$\dot{x}_i = f_i(x_1, x_2, \ldots, x_N, u_1, u_2, \ldots, u_R) \qquad \begin{array}{l} i = 1, 2, \ldots, N \\ 0 < t \leq \theta \end{array} \qquad (1)$$

where we wish to choose the R functions $u_1(t)$, $u_2(t)$, $\ldots$, $u_R(t)$ in order to minimize an integral

$$\mathcal{E}[u_1, u_2, \ldots, u_R] = \int_0^\theta \mathcal{F}(x_1, x_2, \ldots, x_N, u_1, u_2, \ldots, u_R) \, dt \qquad (2)$$

We now assume, however, that the functions $u_k(t)$ are bounded

$$u_{k*} \leq u_k(t) \leq u_k^* \qquad k = 1, 2, \ldots, R \qquad (3)$$

where the bounds u_{k*}, u_k^* are constants. The absence of a lower bound simply implies that $u_{k*} \to -\infty$, while the absence of an upper bound implies that $u_k^* \to +\infty$.

5.2 NECESSARY CONDITIONS

For simplicity of presentation we again restrict our attention to the special case $N = R = 2$, although the results will clearly be general. Thus, we consider the state equations

$$\dot{x}_1 = f_1(x_1, x_2, u_1, u_2) \qquad (1a)$$
$$\dot{x}_2 = f_2(x_1, x_2, u_1, u_2) \qquad (1b)$$

with constraints

$$u_{1*} \leq u_1 \leq u_1^* \qquad (2a)$$
$$u_{2*} \leq u_2 \leq u_2^* \qquad (2b)$$

and objective

$$\mathcal{E}[u_1, u_2] = \int_0^\theta \mathcal{F}(x_1, x_2, u_1, u_2) \, dt \qquad (3)$$

In Sec. 4.2 we derived an equation for the change in $\mathcal{E}$ which results from small changes in u_1 and u_2. In doing so we were never required to specify those changes, $\delta u_1(t)$ and $\delta u_2(t)$. Thus, as long as we stipulate that those variations be *admissible*—i.e., that they not be such that u_1 or u_2 violates a constraint—then Eq. (18) of Sec. 4.2 remains valid, and we can write

$$\delta\mathcal{E} = \int_0^{\bar{\theta}} \left(\frac{\partial H}{\partial u_1} \, \delta u_1 + \frac{\partial H}{\partial u_2} \, \delta u_2 \right) dt + o(\epsilon) \geq 0 \qquad (4)$$

Here we have used the hamiltonian notation of Sec. 4.3

$$H = \mathcal{F} + \lambda_1 f_1 + \lambda_2 f_2 \qquad (5)$$

where the multipliers satisfy the canonical equations

$$\dot{\lambda}_1 = -\frac{\partial H}{\partial x_1} \qquad \dot{\lambda}_2 = -\frac{\partial H}{\partial x_2} \qquad (6)$$

with boundary conditions from Eq. (13) or (16) of Sec. 4.2, depending upon whether or not θ is specified.

We shall presume that in some way we know the optimal functions $\bar{u}_1(t)$, $\bar{u}_2(t)$ for this constrained problem. Each may be only piecewise continuous, with segments along upper and lower bounds, as well as within the bounds, as, for example, in Fig. 5.1. It is then necessary that for all allowable variations $\delta\mathcal{E} \geq 0$. We first choose $\delta u_1 = 0$ whenever u_1 is equal to either u_{1*} or u_1^*, and similarly for δu_2. It then follows that since we are not at a constraint, we may vary u_1 and u_2 either positively or negatively (Fig. 5.1). Thus, whenever H is differentiable, we may choose the particular variations

$$\delta u_1 = -\epsilon' \frac{\partial H}{\partial u_1} \qquad \delta u_2 = -\epsilon' \frac{\partial H}{\partial u_2} \tag{7}$$

for sufficiently small ϵ' and obtain the same result as before:

When the optimal decision u_i ($i = 1$ or 2) lies between the constraints u_{i*} and u_i^* and the hamiltonian is differentiable with respect to u_i, it is necessary that the hamiltonian be stationary with respect to u_i ($\partial H/\partial u_i = 0$).

Let us now consider what happens when $\bar{u}_1 = u_1^*$. For convenience we set $\delta u_2 = 0$ for all t and $\delta u_1 = 0$ whenever $u_1 \neq u_1^*$. Because of the constraint all changes in u_1 must be negative (Fig. 5.1), and so we have

$$\delta u_1 \leq 0 \tag{8}$$

Let us make the particular choice

$$\delta u_1 = \epsilon_1 \frac{\partial H}{\partial u_1} \leq 0 \tag{9}$$

where we cannot set the algebraic sign of ϵ_1 since we do not know the sign

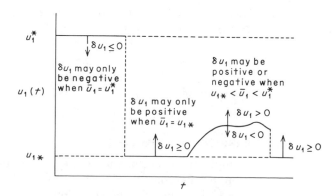

Fig. 5.1 Allowable variations about the optimum decision function.

of $\partial H/\partial u_1$.   From Eq. (4) we then have

$$\delta\mathcal{E} = \epsilon_1 \int \left(\frac{\partial H}{\partial u_1}\right)^2 dt + o(\epsilon) \geq 0 \tag{10}$$

Thus, $\epsilon_1 > 0$, and from Eq. (9),

$$\frac{\partial H}{\partial u_1} \leq 0 \tag{11}$$

Since u_1 decreases as we move into the interior of the allowable region, it follows from the sign of the derivative in Eq. (11) that H is increasing or that *the hamiltonian is a minimum relative to u_1 when $u_1 = u_1^*$.*   (In an exceptional case it might be stationary at u_1^*, and the nature of the stationary point cannot be determined.)

We can repeat this analysis for sections where $\bar{u}_1 = u_{1*}$, and therefore $\delta u_1 \geq 0$ (Figure 5.1) and we obtain the same result. The symmetry of the problem establishes the result for u_2 as well.   If $\bar{u}_1$ or $\bar{u}_2$ lies along an interior interval at a constant value where H is not differentiable but one-sided derivatives exist, then an identical proof establishes that at these points also H is a minimum.   Furthermore, the hamiltonian is still a constant, for whenever the term $\partial H/\partial u_1$ (or $\partial H/\partial u_2$) in Eq. (9) of Sec. 4.3 does not vanish along the optimal path for a finite interval, the optimal decision $\bar{u}_1$ (or $\bar{u}_2$) lies at one of its bounds or at a nondifferentiable point and is a constant.   In that case $du_1/dt = 0$ (or du_2/dt) and the products $\dfrac{\partial H}{\partial u_1}\dfrac{du_1}{dt}$ and $\dfrac{\partial H}{\partial u_2}\dfrac{du_2}{dt}$ always vanish, leading to the result that $dH/dt = 0$.

We summarize the results of this section in the following weak form of the minimum principle:

> *Along the minimizing path the hamiltonian is made stationary by an optimal decision which lies at a differentiable value in the interior of the allowable region, and it is a minimum (or stationary) with respect to an optimal decision which lies along a constraint boundary or at a nondifferentiable interior point. The hamiltonian is constant along the optimal path, and the constant has the value zero when the stopping time θ is not specified.*

5.3 A BANG–BANG CONTROL PROBLEM

As the first example of the use of the necessary conditions derived in the previous section let us consider the control of a particularly simple dynamical system, described by the equation

$$\ddot{x} = u \tag{1}$$

or, equivalently,

$$\dot{x}_1 = x_2 \tag{2a}$$
$$\dot{x}_2 = u \tag{2b}$$

We shall suppose that the system is initially at some state $x_1(0)$, $x_2(0)$ and that we wish to choose the function $u(t)$, subject to the boundedness constraints

$$u_* = -1 \le u \le +1 = u^* \tag{3a}$$

or

$$|u| \le 1 \tag{3b}$$

in order to reach the origin ($x_1 = x_2 = 0$) in the minimum time; that is,

$$\mathcal{E} = \int_0^\theta 1\, dt = \theta \tag{4a}$$
$$\mathcal{F} = 1 \tag{4b}$$

The hamiltonian for this system is

$$H = 1 + \lambda_1 x_2 + \lambda_2 u \tag{5}$$

and since θ is unspecified, the constant value of H along the optimal path is zero. The canonical equations for the multipliers are

$$\dot{\lambda}_1 = -\frac{\partial H}{\partial x_1} = 0 \tag{6a}$$

$$\dot{\lambda}_2 = -\frac{\partial H}{\partial x_2} = -\lambda_1 \tag{6b}$$

and since four boundary conditions are given on the state variables, the boundary conditions for the multipliers are unspecified. Equation (6a) has the solution

$$\lambda_1 = c_1 = \text{const} \tag{7a}$$

and Eq. (6b) the solution

$$\lambda_2 = -c_1 t - c_2 \tag{7b}$$

where c_1 and c_2 are unknown constants of integration resulting from the unspecified boundary conditions.

It is of interest first to investigate whether u may take on values in the interior of the allowable region. In that case the condition for optimality is

$$\frac{\partial H}{\partial u} = 0 = \lambda_2 = -c_1 t - c_2 \tag{8}$$

Equation (8) cannot be satisfied for any finite interval of time unless both c_1 and c_2, the slope and intercept of the straight line, vanish. In that case λ_1 is also zero [Eq. $(7a)$] and the hamiltonian, Eq. (5), has the value unity. Since we have already noted that the optimal value of H must be zero, it follows that the necessary conditions for a minimum can never be satisfied by a control function which is in the interior of the allowable region for any finite time interval.

The only possibilities for the optimum, then, are $u = +1$ and $u = -1$. A control system of this type, which is always at one of its extreme settings, is known as a *bang-bang* or *relay* controller. A typical example is a thermostat-controlled heating system. We note that the question of when to use $u = +1$ or $u = -1$ depends entirely upon the algebraic sign of λ_2, for when λ_2 is positive, the hamiltonian is made a minimum by using $u = -1$ $(-1\lambda_2 < +1\lambda_2,\ \lambda_2 > 0)$, while when λ_2 is negative, the hamiltonian is minimized by $u = +1$ $(+1\lambda_2 < -1\lambda_2,\ \lambda_2 < 0)$. Thus, the optimal policy is

$$u = -\operatorname{sgn} \lambda_2 = \operatorname{sgn} (c_1 t + c_2) \tag{9}$$

Here the sgn (signum) function is defined as

$$\operatorname{sgn} y = \frac{y}{|y|} = \begin{cases} +1 & y > 0 \\ -1 & y < 0 \end{cases} \tag{10}$$

and is undefined when $y = 0$.

We now have sufficient information to solve the system differential equations, starting at $x_1 = x_2 = 0$ and integrating in reverse time, i.e., calling the final time $t = 0$ and the initial time $-\theta$. The condition $H = 0$ establishes that $c_2 = 1$, and for each value of c_1, $-\infty < c_1 < \infty$, we shall define a trajectory in the $x_1 x_2$ plane, thus flooding the entire plane with optimal trajectories and defining a feedback control law. In this case, however, the entire problem can be solved more simply by analytical methods.

We note first that the argument of the signum function in Eq. (9) can change sign at most once. Thus the optimal solution may switch from one extreme value to the other at most once.† During an interval in which the optimal control policy is $u = +1$ the system equations (2) become

$$\dot{x}_1 = x_2 \tag{11a}$$
$$\dot{x}_2 = 1 \tag{11b}$$

or

$$x_2 = t + c_3 \tag{12a}$$
$$x_1 = \tfrac{1}{2} t^2 + c_3 t + c_4 = \tfrac{1}{2}(t + c_3)^2 + \left(c_4 - \frac{c_3{}^2}{2}\right) \tag{12b}$$

---

† It can be demonstrated that for the time-optimal control of an nth-order dynamical system with all real characteristic roots the number of switches cannot exceed 1 less than the order of the system.

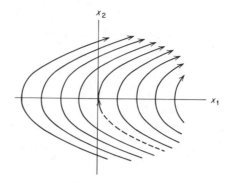

Fig. 5.2   Possible responses for $u = +1$.

Thus, a first integral is

$$x_1 = \tfrac{1}{2}x_2{}^2 + c \tag{13}$$

which defines the family of parabolas shown in Fig. 5.2, the arrows indicating the direction of motion.   Note that the origin can be reached only along the dashed line $x_1 = \tfrac{1}{2}x_2{}^2$, $x_2 \leq 0$, so that if $u = +1$ forms the last part of an optimal trajectory, this must be the path taken.   In a similar way, when $u = -1$, we obtain the family of parabolas

$$x_1 = -\tfrac{1}{2}x_2{}^2 + c \tag{14}$$

shown in Fig. 5.3, with the only possible approach to the origin along the dashed line $x_1 = -\tfrac{1}{2}x_2{}^2$, $x_2 \geq 0$.

When the two sets of trajectories are superimposed, as in Fig. 5.4, the optimal policy becomes obvious at once.   The approach to the origin must be along the dashed line, which has the equation

$$x_1 + \tfrac{1}{2}x_2|x_2| = 0 \tag{15}$$

and at most one switch is possible.   The only way in which initial states below the dashed line can be brought to the origin in this manner is to

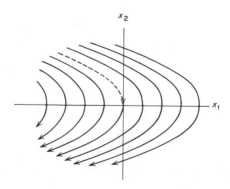

Fig. 5.3   Possible responses for $u = -1$.

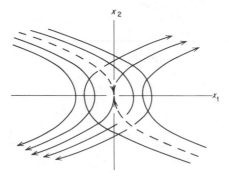

Fig. 5.4 Superposition of all possible responses with bang-bang control.

use the control $u = +1$ until the resulting trajectory intersects the line of Eq. (15) and then to switch to $u = -1$ for the remainder of the control time. Similarly, initial states above the dashed line are brought to the origin by employing the control action $u = -1$ until intersection with the dashed line [Eq. (15)] followed by $u = -1$. This defines the optimal feedback control policy, and only the switching curve [Eq. (15)] is required for implementation. The optimal trajectories are then as shown in Fig. 5.5.

5.4 A PROBLEM OF NONUNIQUENESS

The simple dynamical system considered in the previous section may be used to illustrate another feature of solutions employing the minimum principle. We now suppose that we wish to solve the minimum-time problem to drive x_1 to zero, but we do not choose to specify $x_2(\theta)$. The analysis is essentially unchanged, but because $x_2(\theta)$ is unspecified, we now must invoke the boundary condition

$$\lambda_2(\theta) = 0 \tag{1}$$

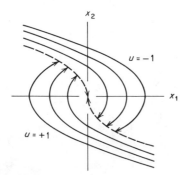

Fig. 5.5 Time-optimal paths to the origin.

or, from Eq. (7b) of the preceding section,

$$\lambda_2(\theta) = -c_1\theta - c_2 = 0 \tag{2}$$

Thus,

$$\lambda_2(t) = c_1(\theta - t) \tag{3}$$

which cannot change algebraic sign, and therefore the optimal control function, defined by Eq. (9) of Sec. 5.3, must always be $+1$ or -1, with no switching possible.

Figure 5.6 shows the trajectories in the right half-plane. For starting values above the dashed line $x_1 + \frac{1}{2}x_2|x_2| = 0$ the line $x_1 = 0$ can be reached without switching only by using the policy $u = -1$. For starting points below the line $x_1 + \frac{1}{2}x_2|x_2| = 0$, however, the x_2 axis can be reached without switching by using either $u = +1$ or $u = -1$. Thus, even in this simplest of problems, the minimum principle does not lead to a unique determination, and the true optimum must be distinguished between the two candidates by other considerations.

In this case the true optimum can be determined analytically. Setting $u = \pm 1$ and dividing Eq. (2a) of Sec. 5.3 by (2b), we obtain the equation for the tangent to each trajectory passing through a point

$$\tan \alpha = \frac{dx_1}{dx_2} = x_2 \qquad u = +1 \tag{4a}$$

$$\tan \beta = \frac{dx_1}{dx_2} = -x_2 \qquad u = -1 \tag{4b}$$

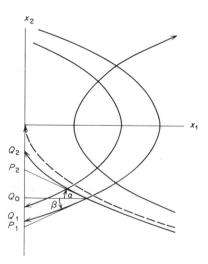

Fig. 5.6 Two possible paths to the x_2 axis satisfying the necessary conditions.

Thus, referring to Fig. 5.6, the line segments $\overline{Q_0P_2}$ and $\overline{Q_0P_1}$ are equal in magnitude. But integrating Eq. (2a) of Sec. 5.3,

$$\theta = x_2(\theta) - x_2(0) = \overline{Q_0Q_2} \qquad u = +1 \tag{5a}$$
$$\theta = x_2(0) - x_2(\theta) = \overline{Q_0Q_1} \qquad u = -1 \tag{5b}$$

and, by inspection,

$$\overline{Q_0Q_2} > \overline{Q_0P_2} = \overline{Q_0P_1} > \overline{Q_0Q_1} \tag{6}$$

Thus, $u = -1$ leads to the shorter time in the entire right-hand plane. By similar reasoning, when $x_1 < 0$, the optimal policy is $u = +1$.

5.5 TIME-OPTIMAL CONTROL OF A STIRRED-TANK REACTOR

We shall now return to the problem of the control of a stirred-tank chemical reactor introduced in Sec. 4.6. The dynamical equations for the reactor after an upset, linearized about the desired steady-state operating conditions, were shown to be of the form of the general second-order system

$$\dot{x}_1 = a_{11}x_1 + a_{12}x_2 + b_{11}u_1 \tag{1a}$$
$$\dot{x}_2 = a_{21}x_1 + a_{22}x_2 + b_{21}u_1 + b_{22}u_2 \tag{1b}$$

where x_1 and x_2 are the deviations from steady state in reduced concentration and temperature, respectively, while u_1 and u_2 are the variations in process and coolant flow rates. It was also shown that after a linear transformation of the dependent variables the system could be represented by the equations

$$\dot{y}_1 = S_1y_1 - M_{11}u_1 - M_{12}u_2 \tag{2a}$$
$$\dot{y}_2 = S_2y_2 + M_{21}u_1 + M_{12}u_2 \tag{2b}$$

where the parameters S_1, S_2, M_{11}, M_{12}, and M_{21} are defined in Sec. 4.6 by Eqs. (19) and (21). In this section we shall consider the problem of returning the system from some initial state $y_1(0)$, $y_2(0)$ to the steady state $y_1 = y_2 = 0$ in the minimum time by choice of the functions $u_1(t)$, $u_2(t)$, subject to the operating constraints on the flow rates

$$u_{1*} \leq u_1 \leq u_1^* \tag{3a}$$
$$u_{2*} \leq u_2 \leq u_2^* \tag{3b}$$

For the minimum-time problem the function $\mathfrak{F}$ is equal to unity, and so the hamiltonian is

$$H = 1 + \lambda_1(S_1y_1 - M_{11}u_1 - M_{12}u_2) + \lambda_2(S_2y_2 + M_{21}u_1 + M_{12}u_2) \tag{4}$$

and the equations for the multipliers are

$$\dot{\lambda}_1 = -\frac{\partial H}{\partial y_1} = -S_1\lambda_1 \tag{5a}$$

$$\dot{\lambda}_2 = -\frac{\partial H}{\partial y_2} = -S_2\lambda_2 \tag{5b}$$

These last equations may be integrated directly to give

$$\lambda_1(t) = \lambda_{10}e^{-S_1t} \tag{6a}$$
$$\lambda_2(t) = \lambda_{20}e^{-S_2t} \tag{6b}$$

although the initial conditions λ_{10}, λ_{20} are unknown.

The possibility that u_1 may lie somewhere between its bounds is considered by setting $\partial H/\partial u_1$ to zero

$$\frac{\partial H}{\partial u_1} = -M_{11}\lambda_1 + M_{21}\lambda_2 = 0 \tag{7}$$

Substitution of Eqs. (6) and (7) demonstrates that this equality can hold for more than a single instant only if $S_1 = S_2$, a degenerate case which we exclude. Thus, the optimal u_1 must always lie at a bound, u_{1*} or u_1^*, and the same may easily be shown true for u_2. The coefficient of u_1 in Eq. (4) is $-\lambda_1 M_{11} + \lambda_2 M_{21}$, so that the hamiltonian is minimized with respect to u_1 by setting u_1 equal to the smallest possible value when the coefficient is positive and the largest possible value when the coefficient is negative, and similarly for u_2:

$$u_1 = \begin{cases} u_1^* & -M_{11}\lambda_1 + M_{21}\lambda_2 < 0 \\ u_{1*} & -M_{11}\lambda_1 + M_{21}\lambda_2 > 0 \end{cases} \tag{8a}$$

$$u_2 = \begin{cases} u_2^* & M_{12}(\lambda_2 - \lambda_1) < 0 \\ u_{2*} & M_{12}(\lambda_2 - \lambda_1) > 0 \end{cases} \tag{8b}$$

For the parameters listed in Table 4.1 we have $S_1 < S_2 < 0$, $M_{21} > M_{11} > 0$, $M_{12} < 0$. Using Eqs. (6), we can then rewrite the optimal control policy in Eqs. (8) as

$$u_1 = \begin{cases} u_1^* & \lambda_{20}\left(r\frac{\lambda_{10}}{\lambda_{20}}e^{(S_2-S_1)t} - 1\right) > 0 \\ \\ u_{1*} & \lambda_{20}\left(r\frac{\lambda_{10}}{\lambda_{20}}e^{(S_2-S_1)t} - 1\right) < 0 \end{cases} \tag{9a}$$

$$u_2 = \begin{cases} u_2^* & -\lambda_{20}\left(\frac{\lambda_{10}}{\lambda_{20}}e^{(S_2-S_1)t} - 1\right) > 0 \\ \\ u_{2*} & -\lambda_{20}\left(\frac{\lambda_{10}}{\lambda_{20}}e^{(S_2-S_1)t} - 1\right) < 0 \end{cases} \tag{9b}$$

where

$$r = \frac{M_{11}}{M_{21}} < 1 \tag{10}$$

The structure of the optimal control policy may now be deduced in a manner similar to that in the two previous sections.

If λ_{10} and λ_{20} have opposite signs or have the same sign with $\lambda_{10}/\lambda_{20} < 1$, the quantities $r(\lambda_{10}/\lambda_{20})e^{(S_2-S_1)t} - 1$ and $(\lambda_{10}/\lambda_{20})e^{(S_2-S_1)t} - 1$ are both always negative, and depending on the sign of λ_{20}, it follows from Eqs. (9) that the optimal control must always be either the pair (u_1^*,u_{2*}) or (u_{1*},u_2^*), with no switching possible. These pairs may also occur when λ_{10} and λ_{20} have the same algebraic sign with $\lambda_{10}/\lambda_{20} > 1$ but only when t is sufficiently large for $(\lambda_{10}/\lambda_{20})e^{(S_1-S_2)t}$ to be less than unity.

If $1 < \lambda_{10}/\lambda_{20} < 1/r$, the initial policy is (u_{1*},u_{2*}) if $\lambda_{20} > 0$, followed by (u_{1*},u_2^*) if the origin has not been reached after a time t_2 such that

$$\frac{\lambda_{10}}{\lambda_{20}} e^{(S_1-S_2)t_2} - 1 = 0 \tag{11a}$$

which is the criterion of switching in Eq. (9b), or

$$t_2 = \frac{1}{S_2 - S_1} \ln \frac{\lambda_{20}}{\lambda_{10}} \tag{11b}$$

with no further switching possible. Similarly, if $\lambda_{20} < 0$, the sequence is (u_1^*,u_2^*), (u_1^*,u_{2*}), with the switching time defined by Eq. (11).

The remaining possibility is that $\lambda_{10}/\lambda_{20} > 1/r$. If $\lambda_{20} > 0$, the initial control policy defined by Eqs. (9) is (u_1^*,u_{2*}), followed, if the origin has not been reached after a time

$$t_1 = \frac{1}{S_2 - S_1} \ln \frac{\lambda_{20}}{\lambda_{10}} \frac{1}{r} \tag{12}$$

by the policy (u_{1*},u_{2*}). A further switch will occur at time t_2 defined by Eq. (11b) to the policy (u_{1*},u_2^*), and no further switching is possible. Thus, the total duration for which the system may be controlled by the policy (u_{1*},u_{2*}) is given as

$$t_2 - t_1 = \frac{1}{S_2 - S_1} \left(\ln \frac{\lambda_{20}}{\lambda_{10}} - \ln \frac{\lambda_{20}}{\lambda_{10}} \frac{1}{r} \right) = \frac{1}{S_2 - S_1} \ln r \tag{13}$$

which depends only on the system parameters. In a similar way, if $\lambda_{20} < 0$, the sequence of optimal policies is (u_{1*},u_2^*), (u_1^*,u_2^*), and (u_1^*,u_{2*}), with the same switching time. This exhausts all possibilities, and it should be noted that no single control variable may switch more than once between limits in this second-order system.

With the sequence of possible control actions available to us we are now in a position to construct the optimal feedback control policy. Since the control action must be piecewise constant, an integral of Eqs. (2) is

$$
\left[\frac{y_1 - (M_{11}/S_1)u_1 - (M_{12}/S_1)u_2}{y_{1r} - (M_{11}/S_1)u_1 - (M_{12}/S_1)u_2}\right]^{S_2}
$$
$$
= \left[\frac{y_2 + (M_{21}/S_2)u_1 + (M_{12}/S_2)u_2}{y_{2r} + (M_{21}/S_2)u_1 + (M_{12}/S_2)u_2}\right]^{S_1} \quad (14)
$$

where y_{1r}, y_{2r} are values of y_1, y_2 somewhere on the path. Thus the paths leading to the origin may be obtained by setting $y_{1r} = y_{2r} = 0$ and putting the appropriate control policy into Eq. (14).

The line marked γ_{+-} in Fig. 5.7 is a plot of Eq. (14) passing through the origin with the control policy (u_1^*, u_{2*}). Since we have found that this policy must always be preceded by the policy (u_1^*, u_2^*), the line γ_{+-} must be a switching curve for trajectories with the control (u_1^*, u_2^*), for

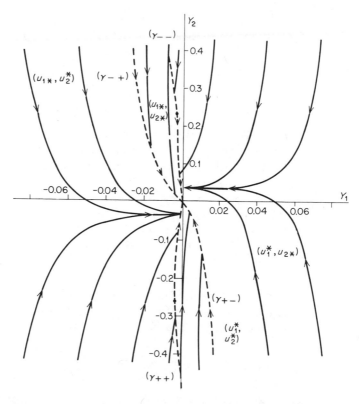

Fig. 5.7   Time-optimal paths to the origin in transformed coordinates for the controlled reactor. [*From J. M. Douglas and M. M. Denn, Ind. Eng. Chem.,* **57**(11):18 (1965). *Copyright 1965 by the American Chemical Society. Reprinted by permission of the copyright owner.*]

otherwise these trajectories could not reach the origin by an optimal sequence. Similarly, the line γ_{-+}, corresponding to (u_{1*}, u_2^*), must be the switching curve for trajectories with control (u_{1*}, u_{2*}).

By choosing points on the γ_{+-} curve and solving Eqs. (2) with the constant control policy (u_1^*, u_2^*) for a time interval $1/(S_2 - S_1) \ln r$ we obtain the curve where the optimal control must have switched from the policy (u_{1*}, u_2^*), the γ_{++} switching curve, and similarly for the γ_{--} curve. We obtain in this way a line for the γ_{++} (or γ_{--}) curve which stops short a finite distance from the origin, for we have seen that we can reach the origin along an optimal trajectory prior to switching from (u_1^*, u_2^*). We obtain the remainder of the γ_{++} curve by setting $x_{1r} = x_{2r} = 0$ and $(u_1, u_2) = (u_1^*, u_2^*)$ in Eq. (14), and similarly for γ_{--}. These switching curves may fail to be smooth at the intersection of the two segments.

We have now divided the $x_1 x_2$ plane into four sections, in each of which the control action is completely specified, with the change in control indicated by reaching a boundary. We have, therefore, by determining the switching curves, constructed the optimal feedback control for the time-optimal problem. The curves in Fig. 5.7 are calculated for the values of the parameters given in Table 4.1, together with the constraints

$$u_{1*} = -8 \le u_1 \le +10 = u_1^* \tag{15a}$$

$$u_{2*} = -5 \le u_2 \le 15 = u_2^* \tag{15b}$$

while Fig. 5.8 shows the trajectories after transformation to the original dimensionless concentration (Z_1) and temperature (Z_2) coordinates. Only one switching was required for most trajectories, and initial conditions for trajectories requiring more than one switching generally fall too far from the origin in the $Z_1 Z_2$ plane for a linearized solution to be useful, in some cases generating trajectories which lead to negative concentrations. It is interesting to observe that many of the optimal trajectories which approach the γ_{++} and γ_{--} curves do so with a common tangent.

At this point it is useful to note again an alternative method of solution which is well suited to automatic digital computation. We make use of the fact that for problems with unspecified total operating times the hamiltonian has the value zero. Thus, when the origin has been reached, from Eq. (4),

$$\lambda_2(\theta) = \frac{\lambda_1(\theta)[M_{11}u_1(\theta) + M_{12}u_2(\theta)] - 1}{M_{21}u_1(\theta) + M_{12}u_2(\theta)} \tag{16}$$

If we specify the final values $u_1(\theta)$ and $u_2(\theta)$, Eq. (16) defines a unique relation between $\lambda_1(\theta)$ and $\lambda_2(\theta)$; for some range of values of $\lambda_1(\theta)$ this relation will be consistant with the requirements of Eq. (9) for the choice of u_1, u_2. For example, the final policy (u_1^*, u_2^*) requires, after some

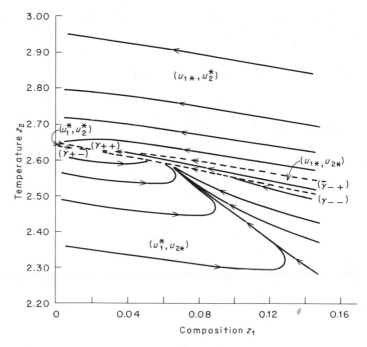

Fig. 5.8 Time-optimal temperature-concentration paths for the controlled reactor. [*From J. M. Douglas and M. M. Denn, Ind. Eng. Chem.,* **57**(11):18 (1965). *Copyright 1965 by the American Chemical Society. Reprinted by permission of the copyright owner.*]

algebra, the satisfaction of the two inequalities

$$M_{21}u_1^* + M_{12}u_2^* > 0 \tag{17a}$$

which is a limitation imposed by the physical properties of the system, and

$$\lambda_1(\theta) > \frac{1}{u_1^*(M_{11} - M_{21})} \tag{17b}$$

Similar relations can be found for other policies.

For a given $\lambda_1(\theta)$ we then have values $\lambda_1(\theta)$, $\lambda_2(\theta)$, $y_1(\theta) = 0$, $y_2(\theta) = 0$, and we can integrate the four differential equations (2) and (5) in the negative time direction, monitoring the combinations in Eq. (9) at all times. When the sense of an inequality changes, we need simply make the appropriate change in the control and continue. In this way we shall map out the switching curves and optimal trajectories as we vary the values of $\lambda_1(\theta)$ over the range $-\infty < \lambda_1(\theta) < \infty$. This backward tracing technique will clearly be of the greatest use in nonlinear systems, where the analytical methods of this section cannot be employed.

5.6 NONLINEAR TIME-OPTIMAL CONTROL

The practical design of a time-optimal control system for the stirred-tank reactor for any but very small upsets will require the use of the full nonlinear equations, and we shall build on the observations of the previous section by following a recent paper of Douglas and considering this more general problem. The nonlinear equations describing the reactor, Eqs. (7) of Sec. 4.6, are

$$\dot{Z}_1 = \frac{q}{V}(1 - Z_1) - kZ_1 \tag{1a}$$

$$\dot{Z}_2 = \frac{q}{V}(Z_f - Z_2) - \frac{UKq_c}{VC_p\rho(1 + Kq_c)}(Z_2 - Z_c) + kZ_1 \tag{1b}$$

where Z_1 is dimensionless concentration, Z_2 dimensionless temperature, and k has the form

$$k = k_0 \exp\left[-\frac{E'C_p\rho}{(-\Delta H)A_f}\frac{1}{Z_2}\right] \tag{2}$$

The constants are defined in Sec. 4.6, with numerical values given in Table 4.1.

In order to avoid complication we shall assume that the flow rate q is fixed at q_S and that control is to be carried out only by varying the coolant flow rate q_c subject to the bounds

$$q_{cS} + u_{2*} \le q_c \le q_{cS} + u_2^* \tag{3}$$

where the numerical values of u_{2*} and u_2^* are the same as those in the previous section. The hamiltonian for time-optimal control is then

$$H = 1 + \lambda_1\left[\frac{q}{V}(1 - Z_1) - kZ_1\right] + \lambda_2\left[\frac{q}{V}(Z_f - Z_2)\right.$$
$$\left. - \frac{UKq_c}{VC_p\rho(1 + Kq_c)}(Z_2 - Z_c) + kZ_1\right] \tag{4}$$

with multiplier equations

$$\dot{\lambda}_1 = -\frac{\partial H}{\partial Z_1} = \left(\frac{q}{V} + k\right)\lambda_1 - k\lambda_2 \tag{5a}$$

$$\dot{\lambda}_2 = -\frac{\partial H}{\partial Z_2} = \frac{E'C_p\rho kZ_1}{(-\Delta H)A_f Z_2{}^2}\lambda_1$$
$$+ \left[\frac{q}{V} + \frac{UKq_c}{VC_p\rho(1 + Kq_c)} - \frac{E'C_p\rho kZ_1}{(-\Delta H)A_f Z_2{}^2}\right]\lambda_2 \tag{5b}$$

We first consider the possibility of intermediate control by setting $\partial H/\partial q_c$ to zero:

$$\frac{\partial H}{\partial q_c} = -\lambda_2 \frac{UK}{VC_{p}\rho} (Z_2 - Z_c) \frac{1}{1 + Kq_c} = 0 \tag{6}$$

This equation can hold for a finite time interval only if $Z_2 = Z_c$ or $\lambda_2 = 0$. In the former case, Z_2 constant implies that Z_1 is also constant, which is clearly impossible except at the steady state. On the other hand, if λ_2 vanishes for a finite time interval then so must its derivative, which, from Eq. (5b), implies that λ_1 also vanishes. But if λ_1 and λ_2 are both zero, then, from Eq. (4), H is equal to unity, which contradicts the necessary condition that $H = 0$ when θ is unspecified. Thus, the control for this nonlinear problem is also bang-bang, and we shall have the solution by construction of the switching surfaces.

The structure of the optimal control function is

$$q_c = \begin{cases} q_c s + u_2^* & (Z_2 - Z_c)\lambda_2 > 0 \\ q_c s + u_{2*} & (Z_2 - Z_c)\lambda_2 < 0 \end{cases} \tag{7}$$

Because of the nonlinear nature of Eqs. (1) and (5) analytical solutions of the type employed in the previous section cannot be used to determine the maximum number of switches or the switching curves. It is to be expected that in a region of the steady state the behavior of the nonlinear system will approximate that of the linearized system, so that a first approximation to the switching curve can be obtained by setting q_c equal, in turn, to its upper and lower limits and obtaining, respectively, the curves γ_+ and γ_-, shown in Fig. 5.9. The optimal trajectories can then be computed, using the policy u_2^* above the switching curve, u_{2*} below, and switching upon intersecting γ_+ or γ_-. Because of the manner of construction of these curves no more than one switch will ever be made. It should be noted that many of the trajectories approach the γ_- switching curve along a common tangent, as in the linearized solution, although this is not true of γ_+. No trajectories can enter the region of negative concentration.

The verification of this solution must be carried out by the backward tracing procedure described in the previous section. When steady state has been reached, Eqs. (4) and (7) become

$$H = -1 + \lambda_1(\theta) \left[\frac{q}{V}(1 - Z_{1S}) - kZ_{1S} \right] + \lambda_2(\theta) \left[\frac{q}{V}(Z_f - Z_{2S}) \right.$$

$$\left. - \frac{UKq_c}{VC_{p}\rho(1 + Kq_c)}(Z_{2S} - Z_c) + kZ_{1S} \right] = 0 \tag{8}$$

$$q_c(\theta) = \begin{cases} q_c s + u_2^* & (Z_{2S} - Z_c)\lambda_2(\theta) > 0 \\ q_c s + u_{2*} & (Z_{2S} - Z_c)\lambda_2(\theta) < 0 \end{cases} \tag{9}$$

A choice of $\lambda_2(\theta)$ uniquely determines $q_c(\theta)$ from Eq. (9), while Eq. (8) determines $\lambda_1(\theta)$. Thus, since $x_1(\theta) = x_2(\theta) = 0$, the four equations (1) and (5) can be integrated simultaneously in the reverse time direction, always monitoring the algebraic sign of $(Z_{2S} - Z_c)\lambda_2$. When this sign changes, q_c is switched to the other extreme of the range and the process continued. This is done for a range of values of $\lambda_2(\theta)$, and the locus of switching points is then the switching curve for the feedback control system. The trajectories and switching curves in Fig. 5.9 were verified by Douglas in this way, except in the region where trajectories approach γ_- along a common tangent, where extremely small changes in $\lambda_2(\theta)$ (of the order of 10^{-6}) are required to generate new trajectories.

It is important to recognize that the nonlinear differential equations (1) may admit more than one steady-state solution; in fact, the parameters listed in Table 4.1 are such that three solutions are possible. Thus, there is a separatrix in the x_1x_2 plane which is approximately the line $Z_2 = 2.4$, below which no trajectory can be forced to the desired steady state with the given control parameters. The design steady state is controllable only subject to certain bounded upsets, a fact which would not be evident from a strictly linearized analysis.

5.7 TIME–OPTIMAL CONTROL OF UNDERDAMPED SYSTEMS

The systems we have studied in the previous several sections had the common property that in the absence of control, the return to the steady state is nonoscillatory, or overdamped. This need not be the case, and,

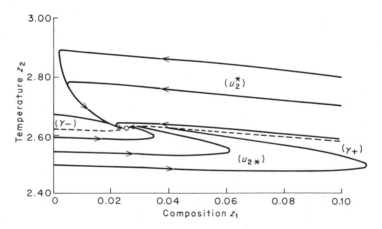

Fig. 5.9 Time-optimal temperature-concentration paths for the nonlinear model of the controlled reactor. [*From J. M. Douglas, Chem. Eng. Sci.*, **21**:519 (1965). *Copyright 1965 by Pergamon Press. Reprinted by permission of the copyright owner.*]

indeed, oscillatory behavior is often observed in physical processes.   Different choices of parameters for the stirred-tank-reactor model would lead to such oscillatory, or underdamped, response, and, since the structure of the time-optimal control is slightly changed, we shall briefly examine this problem.

The prototype of an oscillatory system is the undamped forced harmonic oscillator,

$$\ddot{x} + x = u \tag{1}$$

or

$$\dot{x}_1 = x_2 \tag{2a}$$
$$\dot{x}_2 = -x_1 + u \tag{2b}$$

We shall suppose that u is bounded

$$|u| \leq 1 \tag{3}$$

and seek the minimum time response to the origin.   The hamiltonian is then

$$H = 1 + \lambda_1 x_2 - \lambda_2 x_1 + \lambda_2 u \tag{4}$$

with multiplier equations

$$\dot{\lambda}_1 = -\frac{\partial H}{\partial x_1} = \lambda_2 \tag{5a}$$

$$\dot{\lambda}_2 = -\frac{\partial H}{\partial x_2} = -\lambda_1 \tag{5b}$$

The vanishing of λ_1 or λ_2 over a finite interval implies the vanishing of the derivative, and hence of the other multiplier as well, leading to the contradiction $H = 1$ for the time-optimal case.   Thus, intermediate control, which requires $\lambda_2 = 0$, is impossible, and the solution is again bang-bang, with

$$u = -\operatorname{sgn} \lambda_2 \tag{6}$$

The solution of Eqs. (5) may be written

$$\lambda_2 = -A \sin (t + \phi) \tag{7}$$

where the constants of integration A and ϕ may be adjusted so that $A > 0$, in which case Eq. (6) becomes

$$u = \operatorname{sgn} [\sin (t + \phi)] \tag{8}$$

That is, rather than being limited to a single switch, the controller changes between extremes after each time interval of duration π.

The construction of the optimal feedback control proceeds in the same way as in Secs. 5.3 and 5.4.   When $u = +1$, the first integral of

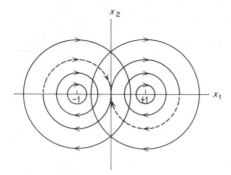

Fig. 5.10 Possible responses of an undamped system with bang-bang control.

Eqs. (2) is

$$u = +1: \qquad (x_1 - 1)^2 + (x_2)^2 = R^2 \tag{9a}$$

a series of concentric circles centered at $x_1 = 1$, $x_2 = 0$, while for $u = -1$ the integral is a series of circles centered at $x_1 = -1$, $x_2 = 0$:

$$u = -1: \qquad (x_1 + 1)^2 + (x_2)^2 = R^2 \tag{9b}$$

All trajectories must then lie along segments of the curves shown in Fig. 5.10, and since the control action changes after every π time units, a trajectory can consist of at most semicircles. Thus, the approach to the origin must be along one of the dashed arcs, which must also form part of the switching curves, γ_+ and γ_-.

We can complete the construction of the switching curves by considering, for example, any point on γ_+. The trajectory leading to it must be a semicircle with center at -1, and so the corresponding point on the γ_- curve can be constructed, as shown in Fig. 5.11. In this fashion the switching curve shown in Fig. 5.12 is built up, with some typical trajectories shown. Above the switching curve and on γ_+ the optimal control is $u = +1$, while below and on γ_- the optimum is $u = -1$.

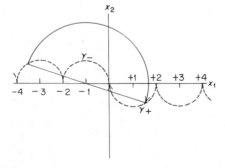

Fig. 5.11 Construction of the switching curve for time-optimal control of an undamped system.

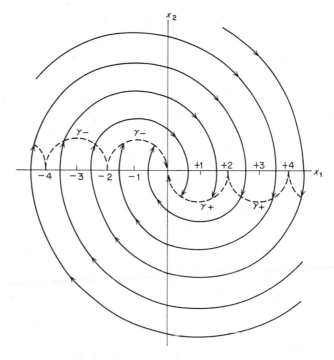

Fig. 5.12 Time-optimal paths to the origin for an undamped system.

The general second-order system,

$$\ddot{x} + a_1\dot{x} + a_2 x = u \tag{10}$$

which is represented by the chemical reactor, may be thought of as a damped forced harmonic oscillator whenever the parameters a_1 and a_2 are such that the characteristic equation

$$m^2 + a_1 m + a_2 = 0 \tag{11}$$

has complex roots. In that case a similar construction to the one above leads to a switching curve of the type shown in Fig. 5.13 for the time-optimal problem. We leave the details of the construction to the interested reader.

5.8 A TIME–AND–FUEL–OPTIMAL PROBLEM

Although rarely of importance in process applications, aerospace problems often involve consideration of limited or minimum fuel expenditures to achieve an objective. If the control variable is assumed to be the

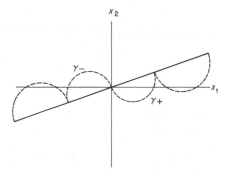

Fig. 5.13 Switching curve for time-optimal control of an underdamped system.

thrust, then to a reasonable approximation the fuel expenditure may be taken as proportional to the magnitude of the thrust, so that the total fuel expenditure is proportional to $\int_0^\theta |u|\, dt$.   Besides the obvious physical importance, optimization problems which involve the magnitude of the decision function introduce a new mathematical structure, and so we shall consider one of the simplest of such problems as a further example of the use of the minimum principle.

The system is again the simple one studied in Sec. 5.3

$$\dot{x}_1 = x_1 \tag{1a}$$
$$\dot{x}_2 = u \tag{1b}$$
$$|u| \le 1 \tag{2}$$

and we shall assume that the total operating time for control to the origin is unspecified but that there is a premium on both time and fuel.   The objective which we wish to minimize is then

$$\mathcal{E} = \rho\theta + \int_0^\theta |u|\, dt = \int_0^\theta (\rho + |u|)\, dt \tag{3}$$

where ρ represents the relative value of time to fuel.   The hamiltonian is

$$H = \rho + |u| + \lambda_1 x_2 + \lambda_2 u \tag{4}$$

with multiplier equations

$$\dot{\lambda}_1 = -\frac{\partial H}{\partial x_1} = 0 \tag{5a}$$

$$\dot{\lambda}_2 = -\frac{\partial H}{\partial x_2} = -\lambda_1 \tag{5b}$$

or

$$\lambda_1 = c_1 = \text{const} \tag{6a}$$
$$\lambda_2 = -c_1 t + c_2 \tag{6b}$$

As usual, we first consider the possibility of a stationary solution by setting $\partial H/\partial u$ to zero, noting that $\partial |u|/\partial u = \text{sgn } u, u \neq 0$:

$$\frac{\partial H}{\partial u} = \text{sgn } u + \lambda_2 = \pm 1 + \lambda_2 = 0 \tag{7}$$

But if λ_2 is a constant for a finite interval of time, $\dot{\lambda}_2 = -\lambda_1$ is zero, so that Eq. (4) becomes

$$H = \rho + |u| - u \text{ sgn } u = \rho + |u| - |u| = \rho > 0 \tag{8}$$

which contradicts the necessary condition that $H = 0$ for unspecified θ. We note, however, that the hamiltonian is not differentiable at $u = 0$ because of the presence of the absolute-value term, and so we may still have an intermediate solution if H is minimized by $u = 0$.

Whenever $\lambda_2 < -1$, the term $|u| + \lambda_2 u$ is less than zero when $u = +1$, while it is zero for $u = 0$ and greater than zero for $u = -1$. Thus the hamiltonian is minimized by $u = +1$. Similarly, when $\lambda_2 > +1$, the hamiltonian is minimized by $u = -1$. For $-1 < \lambda_2 < +1$, however, $|u| + \lambda_2 u$ is zero for $u = 0$ and greater than zero for $u = \pm 1$. Thus, the optimal solution is

$$u = \begin{cases} +1 & \lambda_2 < -1 \\ 0 & -1 \leq \lambda_2 \leq +1 \\ -1 & +1 < \lambda_2 \end{cases} \tag{9}$$

Since λ_2 is a linear function of time, the only possible control sequences are $+1, 0, -1$ and $-1, 0, +1$, or part of either, with a maximum of two switches. This is a *bang-coast-bang* situation, or a relay controller with a dead zone.

We can show further that the final optimal control action must be $u = +1$ or $u = -1$. If $u = 0$ at $t = \theta$, when $x_1 = x_2 = 0$, then, again, $H = \rho \neq 0$, which contradicts a necessary condition for optimality. Thus, the approach to the origin and the switching curve from coasting operation must be the time-optimal switching curve

$$x_1 + \tfrac{1}{2} x_2 |x_2| = 0 \tag{10}$$

shown as the segments γ_{0+} and γ_{0-} in Fig. 5.14. The switching curves γ_{-0} from $u = -1$ to $u = 0$ and γ_{+0} from $u = +1$ to $u = 0$ can be constructed analytically in the following way.

Let t_2 denote the time that a trajectory intersects the switching curve γ_{0+} and t_1 the prior time of intersection with γ_{-0}. From Eqs. (6b) and (9) we write

$$\lambda_2(t_1) = +1 = -c_1 t_1 - c_2 \tag{11a}$$
$$\lambda_2(t_2) = -1 = -c_1 t_2 - c_2 \tag{11b}$$

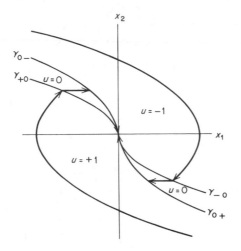

Fig. 5.14 Time-and-fuel-optimal switching curves and paths to the origin.

or, solving for c_1,

$$c_1 = \frac{2}{t_2 - t_1} \tag{12}$$

Evaluating the hamiltonian at $t = t_1$, where $u = 0$,

$$H = \rho + c_1 x_2(t_1) = 0 \tag{13}$$

or, eliminating c_1 in Eqs. (12) and (13),

$$t_2 - t_1 = -\frac{2x_2(t_1)}{\rho} \tag{14}$$

In the interval $t_1 < t < t_2$ the solution of Eqs. (1), with $u = 0$, is

$$x_2(t_2) = x_2(t_1) \tag{15a}$$
$$x_1(t_2) = x_1(t_1) + x_2(t_1)(t_2 - t_1) \tag{15b}$$

But, from Eq. (10), the curve γ_{0+} has the equation

$$x_1(t_2) = \frac{1}{2} x_2{}^2(t_2) \tag{16}$$

and combining Eqs. (14) to (16), we obtain the equation for the switching curve γ_{-0}

$$x_1(t_1) = \frac{\rho + 4}{2\rho} x_2{}^2(t_1) \tag{17}$$

We obtain the γ_{+0} curve in a similar way, with the entire switching curve represented by

$$x_1 + \frac{\rho + 4}{2\rho} x_2|x_2| = 0 \tag{18}$$

The two switching curves, together with typical trajectories, are shown in Fig. 5.14.

Note that as $\rho \to \infty$, in which case only time is important, the two switching curves coincide and the coast period vanishes, giving, as expected, the time-optimal solution. As $\rho \to 0$, the γ_{+0} and γ_{-0} lines tend to the x_1 axis, which we anticipate will be the fuel-optimal solution (bang-coast). In fact the minimum fuel can be obtained by more than one control action, and the one obtained here represents the solution requiring the minimum time. Furthermore, the reader will note that we are asking that the system move at zero velocity between states, which is impossible, so that the limiting solution clearly does not exist, but we can come arbitrarily close to implementing a fuel-optimal solution as ρ is allowed to become arbitrarily large.

5.9 A MINIMUM–INTEGRAL–SQUARE–ERROR CRITERION AND SINGULAR SOLUTIONS

In Sec. 4.7 we studied the optimal control of the linear second-order system, representing the stirred-tank chemical reactor, for an objective in the form of an integral of squares of the state deviations and the square of the control function, leading to a linear feedback control. The u^2 term in the objective may be rationalized as a penalty function to keep the control action within bounds, but this goal can also be accomplished by the methods of this chapter. We thus return to the second-order system with quadratic objective, but we shall now eliminate the cost-of-control term from the objective and include bounds on the control action, the coolant flow rate.

After an appropriate change of variables the reactor equations become Eqs. (4) of Sec. 4.7

$$\dot{y}_1 = y_2 \tag{1a}$$
$$\dot{y}_2 = -a_2 y_1 - a_1 y_2 + u \tag{1b}$$

and the objective

$$\varepsilon = \frac{1}{2} \int_0^\theta (c_{11} y_1{}^2 + 2 c_{12} y_1 y_2 + c_{22} y_2{}^2)\, dt \tag{2}$$

where we have set c_{33} to zero, but we now seek the optimal function u subject to the restriction

$$u_* \leq u \leq u^* \tag{3}$$

The hamiltonian may then be written

$$H = \tfrac{1}{2}(c_{11} y_1{}^2 + 2 c_{12} y_1 y_2 + c_{22} y_2{}^2)$$
$$+ \lambda_1 y_2 - a_2 \lambda_2 y_1 - a_1 \lambda_2 y_2 + \lambda_2 u \tag{4}$$

and the multiplier equations

$$\dot{\lambda}_1 = -\frac{\partial H}{\partial y_1} = a_2\lambda_2 - c_{11}y_1 - c_{12}y_2 \tag{5a}$$

$$\dot{\lambda}_2 = -\frac{\partial H}{\partial y_2} = -\lambda_1 + a_1\lambda_2 - c_{12}y_1 - c_{22}y_2 \tag{5b}$$

Because of the linearity of H in u the minimum principle then implies the optimal solution in part as

$$u = \begin{cases} u^* & \lambda_2 < 0 \\ u_* & \lambda_2 > 0 \end{cases} \tag{6}$$

It will now become clear why we have always been careful to examine each system for the possibility of an intermediate solution. Setting $\partial H/\partial u$ to zero, we obtain

$$\frac{\partial H}{\partial u} = \lambda_2 = 0 \tag{7}$$

which is the situation not covered by Eq. (6). But if λ_2 is zero for a finite time interval, so must be its derivative and Eq. (5b) becomes

$$0 = -\lambda_1 - c_{12}y_1 - c_{22}y_2 \tag{8}$$

This in turn implies that

$$\dot{\lambda}_1 = -c_{12}\dot{y}_1 - c_{22}\dot{y}_2 = a_2c_{22}y_2 + (a_1c_{22} - c_{12})y_2 - c_{22}u \tag{9}$$

which, when compared to Eq. (5a), leads to a solution for u

$$u = \left(a_2 + \frac{c_{11}}{c_{22}}\right)y_1 + a_1y_2 \tag{10}$$

That is, if intermediate control is possible, the form is a linear feedback controller.

We must still satisfy the requirement that the hamiltonian be zero along the optimal path. For the intermediate control defined by Eqs. (7), (8), and (10) the hamiltonian, Eq. (4), becomes

$$\begin{aligned} H &= (-c_{12}y_1 - c_{22}y_2)y_2 + \tfrac{1}{2}(c_{11}y_1^2 + 2c_{12}y_1y_2 + c_{22}y_2^2) \\ &= \tfrac{1}{2}(c_{11}y_1^2 - c_{22}y_2^2) = 0 \end{aligned} \tag{11}$$

so that the intermediate solution represented by Eq. (10) is possible only when the system lies along one of two straight lines in the y_1y_2 state space

$$\sqrt{c_{11}}\, y_1 + \sqrt{c_{22}}\, y_2 = 0 \tag{12a}$$

$$\sqrt{c_{11}}\, y_1 - \sqrt{c_{22}}\, y_2 = 0 \tag{12b}$$

We can further eliminate the second of these possibilities by substituting

Eq. (10) into Eq. (11), giving

$$\dot{y}_1 = y_2 \tag{13q}$$

$$\dot{y}_2 = \frac{c_{11}}{c_{22}} y_1 \tag{13b}$$

and, differentiating Eqs. (12),

$$\dot{y}_2 = \pm \sqrt{\frac{c_{11}}{c_{22}}} \, \dot{y}_1 \tag{14}$$

where the positive sign corresponds to Eq. (12b).   Combining Eqs. (13b) and (14) and solving, we obtain

$$y_1 = y_1(\tau) \exp \pm \sqrt{\frac{c_{11}}{c_{22}}} (t - \tau) \tag{15}$$

with the positive exponential corresponding to Eq. (12b).   Thus the controller is unstable and clearly not a candidate for the optimum along line (12b), since the objective will grow without bound, while the solution is stable along the line (12a) and, hence, perhaps optimal.   At all other points in the $y_1 y_2$ phase plane except this single line, however, the optimal solution must be at one of the extremes, u_* or u^*.   Indeed, only the finite segment of the line (12a) satisfying

$$u_* \leq \left(a_2 + \frac{c_{11}}{c_{22}} \right) y_1 + a_1 y_2 \leq u^* \tag{16}$$

may be optimal.

Next we must establish that when the intermediate, or *singular*, control defined by Eq. (10) is possible, it is in fact optimal.   That is, when λ_2 vanishes at a point y_1', y_2' on the line segment described by Eqs. (12a) and (16), we do not shift to the other extreme of control but rather operate in such a way as to keep λ_2 identically zero.   Because a bang-bang trajectory would again intersect the line segment at some other point y_1'', y_2'', it suffices to show that the singular path between these two points (along the straight line) leads to a smaller value of the objective than any other path.   That is, if we let $\sigma_1(t)$ and $\sigma_2(t)$ denote the values of y_1 and y_2 along the singular path, for all paths between the two points

$$\int_{\substack{y_1 = y_1' \\ y_2 = y_2'}}^{\substack{y_1 = y_1'' \\ y_2 = y_2''}} (c_{11} y_1^2 + 2c_{12} y_1 y_2 + c_{22} y_2^2) \, dt$$

$$- \int_{\substack{y_1 = y_1' \\ y_2 = y_2'}}^{\substack{y_1 = y_1'' \\ y_2 = y_2''}} (c_{11} \sigma_1^2 + 2c_{12} \sigma_1 \sigma_2 + c_{22} \sigma_2^2) \, dt \geq 0 \tag{17}$$

This relation is most easily verified by a method due to Wonham and

Johnson.   We note that from Eq. (1a)

$$y_1 y_2 = y_1 \dot{y}_1 = \frac{1}{2} \frac{d}{dt} (y_1{}^2) \tag{18}$$

The integrand in Eq. (2) may then be written

$$c_{11} y_1{}^2 + 2 c_{12} y_1 y_2 + c_{22} y_2{}^2 = (\sqrt{c_{11}} \, y_1 + \sqrt{c_{22}} \, y_2)^2$$
$$+ (c_{12} - \sqrt{c_{11} c_{22}}) \frac{d}{dt} (y_1{}^2) \quad (19a)$$

and, along the singular path, from Eq. (12a),

$$c_{11} \sigma_1{}^2 + 2 c_{12} \sigma_1 \sigma_2 + c_{22} \sigma_2{}^2 = (c_{12} - \sqrt{c_{11} c_{22}}) \frac{d}{dt} (\sigma_1{}^2) \tag{19b}$$

Integrating Eqs. (19a) and (19b) and subtracting, we make use of the fact that σ_1 and y_1 are identical at both end points (we are comparing the value of the objective along different paths between the same end points) to establish Eq. (17):

$$\int_{\substack{y_1 = y_1' \\ y_2 = y_2'}}^{\substack{y_1 = y_1'' \\ y_2 = y_2''}} (c_{11} y_1{}^2 + 2 c_{12} y_1 y_2 + c_{22} y_2{}^2) \, dt - \int_{\substack{y_1 = y_1' \\ y_2 = y_2'}}^{\substack{y_1 = y_1'' \\ y_2 = y_2''}} (c_{11} \sigma_1{}^2 + 2 c_{12} \sigma_1 \sigma_2$$
$$+ c_{22} \sigma_2{}^2) \, dt = \int_{\substack{y_1 = y_1' \\ y_2 = y_2'}}^{\substack{y_1 = y_1'' \\ y_2 = y_2''}} (\sqrt{c_{11}} \, y_1 + \sqrt{c_{22}} \, y_2)^2 \, dt \geq 0 \quad (20)$$

This proves that whenever it is possible to use the linear feedback control defined by Eq. (10)—i.e., whenever Eqs. (12a) and (16) are satisfied —it is optimal to do so.

The remainder of the optimal policy can now be easily constructed by the backward tracing procedure.   At any point on the singular line values are known for y_1, y_2, λ_1, and λ_2

$$y_1 = y_1' \tag{21a}$$
$$y_2 = y_2' \tag{21b}$$
$$\lambda_1 = -c_{12} y_1' - c_{22} y_2' \tag{21c}$$
$$\lambda_2 = 0 \tag{21d}$$

Equations (1) and (5) can then be integrated in reverse time for $u = u^*$, checking at all times to be sure that $\lambda_2 < 0$.   When λ_2 again returns to zero, we have reached a point on the γ_{-+} switching curve and u is set to u_* and the process continued.   In a similar way, setting u initially to u_* will generate a point on the γ_{+-} switching curve when λ_2 returns to zero. By carrying out this procedure for all points on the singular line the entire switching curve can be generated, just as for the time-optimal problems considered in previous sections.

A final word concerning practical implementation is perhaps in order.   In any real system it will be impossible to maintain the process exactly along the singular line described by Eq. (12a), and careful examination of Eqs. (13) indicates that they are unstable if the system ever deviates from the straight line.   Thus the control law of Eq. (10) must be replaced by one which is completely equivalent when the system is in fact on the singular line but which will be stable if slight excursions do occur. This can be accomplished, for example, by subtracting $3[(c_{11}/c_{22})y_1 + \sqrt{(c_{11}/c_{22})}y_2]$, which vanishes along line (12a), from the right-hand side of Eq. (10), giving

$$u = \left(a_2 - 2\frac{c_{11}}{c_{22}}\right)y_1 + \left(a_1 - 3\sqrt{\frac{c_{11}}{c_{22}}}\right)y_2 \tag{22}$$

with the resulting system equations (1), the stable process

$$\dot{y}_1 = y_2 \tag{23a}$$

$$\dot{y}_2 = -2\frac{c_{11}}{c_{22}}y_1 - 3\sqrt{\frac{c_{11}}{c_{22}}}y_2 \tag{23b}$$

5.10   NONLINEAR MINIMUM–INTEGRAL–SQUARE–ERROR CONTROL

A number of the ideas developed in the previous section can be applied with equivalent ease to nonlinear systems.   In order to demonstrate this fact we shall return again to the now-familiar example of the stirred-tank chemical reactor with control by coolant flow rate, having state equations

$$\dot{Z}_1 = \frac{q}{V}(1 - Z_1) - kZ_1 \tag{1a}$$

$$\dot{Z}_2 = \frac{q}{V}(Z_f - Z_2) - \frac{UKq_c}{VC_p\rho(1 + Kq_c)}(Z_2 - Z_c) + kZ_1 \tag{1b}$$

where $k = k(Z_2)$.   Since $q_c/(1 + Kq_c)$ is a monotonic function of q_c, we may simply define a new decision variable w as the coefficient of $Z_2 - Z_c$ in Eq. (1b) and write

$$\dot{Z}_2 = \frac{q}{V}(Z_f - Z_2) - w(Z_2 - Z_c) + kZ_1 \tag{1c}$$

with constraints derivable from Eq. (3) of Sec. 5.6 as

$$w_* \leq w \leq w^* \tag{2}$$

We shall attempt to maintain the concentration Z_1 and temperature Z_2 near the respective steady-state values Z_{1S}, Z_{2S} by choosing w to minimize the integral

$$\mathcal{E} = \frac{1}{2}\int_0^\theta [(Z_1 - Z_{1S})^2 + c^2(k - k_S)^2]\, dt \tag{3}$$

where $k_S = k(Z_{2S})$. The choice of k, rather than Z_2, in Eq. (3) is one of convenience, though it can be rationalized by the observation that it is deviations in reaction rate, not temperature, which adversely affect the product composition. This is, of course, the objective chosen in Sec. 4.8.

The hamiltonian for this problem is

$$H = \tfrac{1}{2}(Z_1 - Z_{1S})^2 + \tfrac{1}{2}c^2(k - k_S)^2 + \lambda_1 \frac{q}{V}(1 - Z_1) - \lambda_1 k Z_1$$

$$+ \lambda_2 \frac{q}{V}(Z_f - Z_2) - \lambda_2 w(Z_2 - Z_c) + \lambda_2 k Z_1 \quad (4)$$

with multiplier equations

$$\dot{\lambda}_1 = -(Z_1 - Z_{1S}) + \frac{q}{V}\lambda_1 + k\lambda_1 - k\lambda_2 \quad (5a)$$

$$\dot{\lambda}_2 = -c^2(k - k_S)\frac{\partial k}{\partial Z_2} + \lambda_1 Z_1 \frac{\partial k}{\partial Z_2} + \lambda_2 \frac{q}{V} + \lambda_2 w - \lambda_2 Z_1 \frac{\partial k}{\partial Z_2} \quad (5b)$$

and, from the linearity of H in w, the optimum has the form

$$w = \begin{cases} w^* & \lambda_2(Z_2 - Z_c) > 0 \\ w_* & \lambda_2(Z_2 - Z_c) < 0 \end{cases} \quad (6)$$

The possibility of a singular solution, for which λ_2 vanishes over a finite interval, must still be examined. In that case the derivative must also be zero, and since $\partial k/\partial Z_2 \neq 0$, Eq. (5b) reduces to

$$\lambda_1 Z_1 - c^2(k - k_S) = 0 \quad (7a)$$

and Eq. (5a) to

$$\dot{\lambda}_1 = -(Z_1 - Z_{1S}) + \left(\frac{q}{V} + k\right)\lambda_1 \quad (7b)$$

These two equations, together with Eq. (1a), are identical to Eqs. (1), (7), and (8) of Sec. 4.8, and the approximate solution for the singular line is given by Eq. (16) and the corresponding singular control by Eq. (17) of that section. Thus, the singular solution is equivalent to choosing the *temperature* Z_2 which minimizes the objective, Eq. (3), provided that the resulting flow rate is consistant with Eq. (2).

In the previous section we utilized a mathematical argument to prove the optimality of the singular solution. Here, we simply rely on physical reasoning. Clearly, if we are in fact free to specify the temperature directly, we shall choose to do so, since this is our real physical goal, and whenever the choice of flow rate coincides with the optimal choice of temperature, that choice must also lead to the optimal flow rate. Only when the optimal temperature is not accessible by choice of flow rate need we consider the second in this hierarchy of optimization problems, the optimal choice of flow rate itself. Thus, for this reactor problem, the singular solution, when possible, must be optimal.

The construction of the switching curves can now be carried out by the backward tracing technique, since values of Z_1, Z_2, λ_1, and λ_2 are all known along the singular line. We should observe that a closed-form solution for the nonlinear singular control w (but not for the singular line) is in fact available by eliminating λ_1 between Eqs. (7a) and (7b).

5.11 OPTIMAL COOLING RATE IN BATCH AND TUBULAR REACTORS

In Sec. 3.5 we briefly considered the optimal specification of the temperature program in a batch or plug-flow pipeline chemical reactor in order to minimize the time of achieving a given conversion for a single reaction or, equivalently, of obtaining the maximum conversion in a given time. In the batch reactor it is somewhat more realistic to suppose that we can specify the cooling rate as a function of time, rather than the temperature, and in the tubular reactor also the rate of heat removal is somewhat closer to actual design considerations than the temperature at each position. We are now in a position to study this more realistic problem as an example of the methods of this chapter, and we follow a discussion of Siebenthal and Aris in doing so.

Denoting the reactant concentration as a and temperature as T, the equations describing the state of the reactor are

$$\dot{a} = -r(a,T) \tag{1a}$$
$$\dot{T} = Jr(a,T) - u \tag{1b}$$

where J is a constant (heat of reaction divided by the product of density and specific heat) and u, the design or control variable, is the heat removal rate divided by density and specific heat, with bounds

$$u_* = 0 \leq u \leq u^* \tag{2}$$

In a batch system u will be a monotonic function of coolant flow rate. In a tubular reactor the independent variable is residence time, the ratio of axial position to linear velocity. The reaction rate $r(a,T)$ is given by Eq. (13) of Sec. 3.5

$$r(a,T) = p_1 a^{n_1} \exp\left(-\frac{E_1'}{T}\right) - p_2\left(b_0 + \frac{n_2}{n_1}a_0 - \frac{n_2}{n_1}a\right)^{n_2}$$
$$\exp\left(-\frac{E_2'}{T}\right) \tag{3}$$

where a_0 and b_0 are the initial values of reactant and product, respectively. When $E_2' > E_1'$, there is a maximum value of r with respect to

T for fixed a satisfying $\partial r/\partial T = 0$, while r will go to zero for fixed T at some finite value of a (equilibrium). These are shown in Fig. 5.15 as the lines $r = 0$ and $r = r_{\max}$.

We shall first consider the objective of minimizing a (that is, maximizing conversion) while placing some finite cost on operating time (or reactor length), so that the objective is

$$\mathcal{E} = a(\theta) + \rho\theta \tag{4a}$$

or, equivalently, using Eq. (1a),

$$\mathcal{E} = \int_0^\theta [\rho - r(a,T)]\, dt \tag{4b}$$

The hamiltonian is

$$\begin{aligned} H &= \rho - r(a,T) - \lambda_1 r(a,T) + \lambda_2 Jr(a,T) - \lambda_2 u \\ &= \rho - r(a,T)(1 + \lambda_1 - J\lambda_2) - \lambda_2 u \end{aligned} \tag{5}$$

with multiplier equations

$$\dot{\lambda}_1 = -\frac{\partial H}{\partial a} = \frac{\partial r}{\partial a}(1 + \lambda_1 - J\lambda_2) \qquad \lambda_1(\theta) = 0 \tag{6a}$$

$$\dot{\lambda}_2 = -\frac{\partial H}{\partial T} = \frac{\partial r}{\partial T}(1 + \lambda_1 - J\lambda_2) \qquad \lambda_2(\theta) = 0 \tag{6b}$$

Because of the linearity of H in u the optimal decision function is

$$u = \begin{cases} u^* & \lambda_2 > 0 \\ 0 & \lambda_2 < 0 \end{cases} \tag{7}$$

with an intermediate solution possible only if λ_2 vanishes over a finite interval.

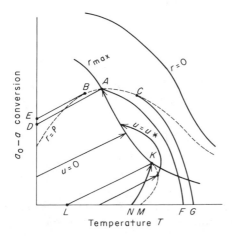

Temperature T

Fig. 5.15 Reaction paths and switching curve for optimal cooling in a batch or tubular reactor. [*After C. D. Siebenthal and R. Aris, Chem. Eng. Sci.,* **19**:747 (1964). *Copyright 1964 by Pergamon Press. Reprinted by permission of the copyright owner.*]

Because θ is not specified, the optimal value of the hamiltonian is zero. At $t = \theta$, using the boundary conditions for λ_1 and λ_2, Eq. (5) reduces to the stopping condition

$$r[a(\theta), T(\theta)] = \rho \tag{8}$$

That is, as common sense dictates, the process should be terminated when the reaction rate falls to the value ρ and the incremental return in the objective changes algebraic sign. The stopping curve $r = \rho$ is shown in Fig. 5.15.

Let us now consider the possibility that the final operation is intermediate. In that case the vanishing of λ_2 implies the vanishing of its derivative, and therefore of $\partial r/\partial T$. That is, if the final operation is singular, it must be along the curve r_{max}, terminating at the highest point of the curve $r = \rho$, shown as point A. This is precisely the policy which we found to give the minimum time and maximum conversion operation in Sec. 3.5, so that we may anticipate that singular operation is optimal when possible for this combined problem as well, and we shall show that this is indeed the case.

We first suppose that an optimal path terminates on the line $r = \rho$ to the right of r_{max}. Here, $\partial r/\partial T < 0$, and it follows from Eq. (6b) that $\lambda_2(\theta) = \partial r/\partial T < 0$, or λ_2 is decreasing to its final value of zero. Thus, from Eq. (7), the final policy must be $u = u^*$, full cooling. There is, however, a point on the curve $r = \rho$, say C, at which a trajectory for $u = u^*$ intersects the line $r = \rho$ from the right; i.e., the slope of the line $-da/dT$ along the trajectory is greater than the slope of the line $r = \rho$. From Eqs. (1),

$$\left(-\frac{da}{dT}\right)_{\substack{u=u^* \\ r=\rho}} = \frac{\rho}{J\rho - u^*} > \left(-\frac{da}{dT}\right)_{r=\rho} = \frac{\partial r/\partial T}{\partial r/\partial a} \tag{9}$$

and, after slight manipulation, the point C is defined by

$$u^* > \rho\left(J - \frac{\partial r/\partial a}{\partial r/\partial T}\right) \tag{10}$$

Since the integrand in Eq. (4b) is positive whenever the system is to the right of $r = \rho$, an optimal policy cannot terminate at $r = \rho$ to the right of point C, for smaller values of $\mathcal{E}$ could be obtained by stopping earlier. Thus, an optimal policy can terminate to the right of r_{max} only under the policy $u = u^*$ on the segment AC of $r = \rho$.

In a similar way we find that the system can terminate to the left of r_{max} only under adiabatic conditions ($u = 0$) on the segment BA of $r = \rho$, where B is defined by $J > (\partial r/\partial a)/(\partial r/\partial T)$. Thus, whenever the system lies to the right of r_{max} and to the left of the full-cooling trajectory AF, the optimal policy is full cooling until the intersection with r_{max}, followed by the intermediate value of u necessary to remain on r_{max}, for the system can never leave the region bounded by DAF, and this is the policy which will minimize the term $\rho\theta$ in the objective, $a(\theta)$ being fixed.　Similarly, to the left of r_{max} and below the adiabatic line DA the policy is adiabatic, followed by the singular operation.　Within the region bounded by $FACG$ the policy is only full cooling, and in $EBAD$ only adiabatic, but even here some trajectories starting in regions $r < \rho$ might give positive values to the objective, indicating a loss, so that some initial states will be completely excluded, as will all initial states outside the envelope $EBACG$.

By combining Eq. (1) and Eq. (14) of Sec. 3.5 it easily follows that the optimal policy on the singular line r_{max} is

$$u = r \left\{ J + \frac{n_1[n_1 b_0 + n_2 a_0 - n_2 a(1 - n_2/n_1)] T^2}{a(n_1 b_0 + n_2 a_0 - n_2 a)} \right\} > 0 \qquad (11)$$

There will be some point on r_{max}, say K, at which this value exceeds u^*, and a switching curve must be constructed by the usual backward tracing method.　This switching curve KM will be to the right of the full-cooling line KN, for we may readily establish that KN cannot form part of an optimal path.　By integrating Eq. (6a) along r_{max} from $t = \theta$ we find that $\lambda_1(t) + 1 > 0$ at K, in which case, for an approach from the left of r_{max}, where $\partial r/\partial T > 0$, $\lambda_2(t) > 0$.　Thus, $\lambda_2 < 0$ just prior to reaching K, and only adiabatic approach is possible.　Furthermore, along an adiabatic path both $\partial r/\partial a$ and λ_2 are negative, so that $\lambda_1 + 1$ is always greater than its value at r_{max}, which is positive.　Hence λ_2 is always positive, so that λ_2 can never reach zero on an adiabatic path, and an adiabatic path cannot be preceded by full cooling.　Thus, for starting points to the right of the adiabatic line KL the optimal policy is adiabatic to the switching line KM, followed by full cooling to r_{max}, then the policy defined by Eq. (11) to the point A.

If we now consider the problem of achieving the maximum conversion in a given duration ($\rho = 0$, θ fixed), clearly the solution is identical, for the hamiltonian now has some nonzero constant value, say $-\mu$, and we may write

$$H + \mu = \mu - r(a,T)(1 + \lambda_1 - J\lambda_2) - \lambda_2 u = 0 \qquad (12)$$

with the multiplier equations and Eq. (7) unchanged.　Furthermore, since $a(t)$ must be a monotone decreasing function, the policy which minimizes $a(\theta)$ for fixed θ must, as noted in Sec. 3.5, also be the policy which reaches a specified value of a in minimum time (or reactor length).

5.12 SOME CONCLUDING COMMENTS

Before concluding this chapter of applications some comments are in order. The bulk of the examples have been formulated as problems in optimal control, largely because such problems admit some discussion without requiring extensive computation. From this point on we shall say less about the construction of feedback control systems and concentrate on computational methods leading to optimal open-loop functions, which are more suitable to process design considerations, although several chapters will contain some significant exceptions.

The reader will have observed that the several optimal-control problems which we have formulated for the second-order system, exemplified by the stirred-tank chemical reactor, have each led to markedly different feedback policies, ranging from relay to linear feedback. The physical objective of each of the controls, however, is the same, the rapid elimination of disturbances, and there will often be situations in which it is not clear that one objective is more meaningful than another, although one "optimal" control may be far easier to implement than another. This arbitrariness in the choice of objective for many process applications will motivate some of our later considerations.

BIBLIOGRAPHICAL NOTES

Section 5.2: The derivation follows

J. M. Douglas and M. M. Denn: *Ind. Eng. Chem.*, **57** (11):18 (1965)

The results are a special case of more general ones derived in Chap. 6, where a complete list of references will be included. A fundamental source is

L. S. Pontryagin, V. G. Boltyanskii, R. V. Gamkrelidze, and E. F. Mishchenko: "Mathematical Theory of Optimal Processes," John Wiley & Sons, Inc., New York, 1962

Sections 5.3 to 5.5: The time-optimal control problem for linear systems is dealt with in the book by Pontryagin and coworkers and in great detail in

M. Athans and P. Falb: "Optimal Control," McGraw-Hill Book Company, New York, 1966
R. Oldenburger: "Optimal Control," Holt, Rinehart & Winston, New York, 1966

Historically, the bang-bang result was obtained by Bushaw and further developed by Bellman and coworkers and LaSalle:

R. Bellman, I. Glicksberg, and O. Gross: *Quart. Appl. Math.*, **14**:11 (1956)
D. W. Bushaw: "Differential Equations with a Discontinuous Forcing Term," *Stevens Inst. Tech. Expt. Towing Tank Rept.* 469, Hoboken, N.J., 1953; Ph.D. thesis, Princeton University, Princeton, N.J., 1952; also in S. Lefschetz (ed.), "Contributions to the Theory of Nonlinear Oscillations," vol. 4, Princeton University Press, Princeton, N.J., 1958

J. P. LaSalle: *Proc. Natl. Acad. Sci. U.S.*, **45**:573 (1959); reprinted in R. Bellman and R. Kalaba (eds.), "Mathematical Trends in Control Theory," Dover Publications, Inc., New York, 1964

The calculations shown here for the reactor problem are from the paper by Douglas and Denn. Further considerations of the problem of nonuniqueness are found in

I. Coward and R. Jackson: *Chem. Eng. Sci.*, **20**:911 (1965)

A detailed discussion of all aspects of relay control is the subject of

I. Flügge-Lotz: "Discontinuous and Optimal Control," McGraw-Hill Book Company, New York, 1968

Section 5.6: The section is based on

J. M. Douglas: *Chem. Eng. Sci.*, **21**:519 (1966)

Other nonlinear time-optimal problems are treated in Athans and Falb and

I. Coward: *Chem. Eng. Sci.*, **22**:503 (1966)
E. B. Lee and L. Markus: "Foundations of Optimal Control Theory," John Wiley & Sons, Inc., New York, 1967
C. D. Siebenthal and R. Aris: *Chem. Eng. Sci.*, **19**:729 (1964)

A simulation and experimental implementation of a nonlinear time-optimal control is described in

M. A. Javinsky and R. H. Kadlec: Optimal Control of a Continuous Flow Stirred Tank Chemical Reactor, *preprint 9C, 63d Natl. Meeting, AIChE, St. Louis, 1968*

Optimal start-up of an autothermic reactor is considered in

R. Jackson: *Chem. Eng. Sci.*, **21**:241 (1966)

Section 5.7: See the books by Pontryagin and coworkers and Athans and Falb.

Section 5.8: The book by Athans and Falb contains an extensive discussion of linear and nonlinear fuel-optimal problems.

Section 5.9: The basic paper is

W. M. Wonham and C. D. Johnson: *J. Basic Eng.*, **86D**:107 (1964)

where the problem is treated for any number of dimensions; see also

Z. V. Rekazius and T. C. Hsia: *IEEE Trans. Autom. Contr.*, **AC9**:370 (1964)
R. F. Webber and R. W. Bass: *Preprints 1967 Joint Autom. Contr. Confer., Philadelphia*, p. 465

In Chap. 6 we establish some further necessary conditions for the optimality of singular solutions and list pertinent references. One useful source is

C. D. Johnson: in C. T. Leondes (ed.), "Advances in Control Systems," vol. 2, Academic Press, Inc., New York, 1965

Section 5.10: The use of a hierarchy of optimization problems to deduce an optimal policy is nicely demonstrated in

N. Blakemore and R. Aris: *Chem. Eng. Sci.*, **17**:591 (1962)

Section 5.11: This section follows

C. D. Siebenthal and R. Aris: *Chem. Eng. Sci.*, **19**:747 (1964)

Section 5.12: The literature contains many examples of further applications of the prin-
ciples developed here, particularly such periodicals as AIAA Journal, Automatica,
Automation and Remote Control, Chemical Engineering Science, Industrial and
Engineering Chemistry Fundamentals Quarterly, IEEE Transactions on Auto-
matic Control, International Journal of Control, Journal of Basic Engineering
(Trans. ASME, Ser. D), and Journal of Optimization Theory and Applications.
The annual reviews of applied mathematics, control, and reactor analysis in Indus-
trial and Engineering Chemistry (monthly) list engineering applications. Some
other recent reviews are

M. Athans: *IEEE Trans. Autom. Contr.*, **AC11**:580 (1966)
A. T. Fuller: *J. Electron. Contr.*, **13**:589 (1962); **15**:513 (1963)
B. Paiewonsky: *AIAA J.*, **3**:1985 (1965)

Applications outside the area of optimal design and control are growing, particularly in
economics. A bibliography on recent applications of variational methods to economic
and business systems, management science, and operations research is

G. S. Tracz: *Operations Res.*, **16**:174 (1968)

PROBLEMS

5.1. Extend the minimum principle of Sec. 5.2 to nth-order systems

$$\dot{x}_i = f_i(x_1, x_2, \ldots x_n, u_1, u_2, \ldots u_m) \qquad i = 1, 2, \ldots, n$$

Show that the hamiltonian is defined by

$$H = \mathfrak{F} + \sum_{i=1}^{n} \lambda_i f_i$$

and the multipliers by

$$\dot{\lambda}_i = -\frac{\partial H}{\partial x_i} = -\frac{\partial \mathfrak{F}}{\partial x_i} - \sum_{j=1}^{n} \lambda_j \frac{\partial f_j}{\partial x_i}$$

5.2. Solve the optimal-temperature-profile problem of Sec. 4.12 for the case of an
upper bound on the temperature. By considering the ratio of multipliers show how
the problem can be solved by a one-dimensional search over initial values of the ratio
of multipliers. For consecutive first-order reactions

$$F(x_1) = x_1 \qquad G(x_2) = x_2$$

obtain an algebraic expression for the time of switching from the upper bound to an
intermediate temperature in terms of the initial ratio of the multipliers. Discuss the
computational effect of including a lower bound on temperature as well.
5.3. Solve the optimal-pressure-profile problem of Sec. 4.13 for the case of an upper
bound on the pressure. By considering the ratio of multipliers show how the problem

can be solved by a one-dimensional search over initial values of the ratio of multipliers. Discuss the computational effect of including a lower bound on the pressure as well.

5.4. Consider the system

$$\dot{x}_1 = a_{11}x_1 + a_{12}x_2 + b_1u$$
$$\dot{x}_2 = a_{21}x_1 + a_{22}x_2 + b_2u$$
$$|u| \leq 1$$

Establish the number of switches possible in the minimum time control to the origin for all values of the parameters and construct the switching curve for the case of complex roots to the characteristic equation.

5.5. A system is termed controllable to the origin ($x_1 = 0$, $x_2 = 0$) if, for each initial state, there exists a piecewise-continuous control $u(t)$ such that the origin can be attained in some finite time θ. We have assumed throughout that the systems with which we are dealing are controllable. For the linear system

$$\dot{x}_1 = a_{11}x_1 + a_{12}x_2 + b_1u$$
$$\dot{x}_2 = a_{21}x_1 + a_{22}x_2 + b_2u$$

show that a necessary and sufficient condition for controllability is

$$b_1(a_{21}b_1 + a_{22}b_2) - b_2(a_{11}b_1 + a_{12}b_2) \neq 0$$

This is a special case of results of Kalman on controllability and the related concept of observability. *Hint:* Solve for x_1 and x_2 in terms of a convolution integral involving $u(t)$ by Laplace transform, variation of parameters, or any other convenient method. "Only if" is most easily demonstrated by counterexample, "if" by construction of a function which satisfies the requirements.

5.6. For the nonlinear system

$$\ddot{x} + g(\dot{x}) = u \qquad |u| \leq 1$$

where g is a differentiable function satisfying $\dot{x}g(\dot{x}) > 0$, develop the feedback control for minimum time to the origin. Show that the optimum is bang-bang and no more than one switch is possible and indicate the procedure for constructing the switching curve. (The problem is due to Lee and Markus.)

5.7. A simplified traffic-control problem is as follows.

Let q_1 and q_2 represent arrival rates at a traffic light in each of two directions during a rush period of length θ, s_1 and s_2 the discharge rates ($s_1 > s_2$), L the time for acceleration and clearing, and c the length of a cycle. If x_1 and x_2 represent the lengths of queues and u the service rate to direction 1, the queues satisfy the equations

$$\dot{x}_1 = q_1 - u$$
$$\dot{x}_2 = q_2 - s_2\left(1 - \frac{L}{c}\right) + \frac{s_2}{s_1}u$$

with initial and final conditions $x_1(0) = x_2(0) = x_1(\theta) = x_2(\theta) = 0$.

The service rate is to be chosen within bounds

$$u_* \leq u \leq u^*$$

to minimize the total holdup

$$\varepsilon = \int_0^\theta (x_1 + x_2)\, dt$$

It is helpful to approximate arrival rates as constant during the latter phases of the rush period in obtaining a completely analytical solution.   (The problem is due to Gazis.)

5.8. The chemical reaction $X \rightleftharpoons Y \rightarrow Z$ is to be carried out isothermally in a catalytic reactor with a blend of two catalysts, one specific to each of the reactions.   The describing equations are

$$\dot{x} = u(k_2 y - k_1 x)$$
$$\dot{y} = u(k_1 x - k_2 y) - (1 - u)k_3 y$$
$$x + y + z = \text{const}$$

Here u is the fraction of catalyst specific to the reaction between X and Y and is bounded between zero and 1.   Initial conditions are 1, 0, and 0 for x, y, and z, respectively.   The catalyst blend is to be specified along the reactor, $0 \le t \le \theta$, to maximize conversion to Z,

$$\max \mathcal{P} = z(\theta) = 1 - x(\theta) - y(\theta)$$

(a) When $k_2 = 0$, show that the optimum is a section with $u = 1$ followed by the remainder of the reactor with $u = 0$ and obtain an equation for the switch point.

(b) When $k_2 \ne 0$, show that the reactor consists of at most three compartments, the two described above, possibly separated by a section of constant intermediate blend.   Obtain the value of the intermediate blend and show that the time interval (t_1, t_2) for intermediate operation is defined by

$$t_1 = \frac{1}{k_1 + k_2} \log \left(1 + \frac{k_2}{k_3} + \frac{k_1}{\sqrt{k_2 k_3}} \right)$$

$$t_2 = \theta - \frac{1}{k_3} \log \left(1 + \sqrt{\frac{k_3}{k_2}} \right)$$

(The problem is due to Gunn, Thomas and Wood, and Jackson.)

5.9. For the physiological system described in Prob. 4.10 find the optimal control when $c = 0$ with the presumption that u is bounded.   Examine possible singular solutions and describe the process of construction of any switching curves.

5.10. Using the device of Secs. 4.4 and 4.9 for treating explicit time dependence, extend the minimum principle of Sec. 5.2 to include problems of the form

$$\dot{x}_1 = f_1(x_1, x_2, u, t)$$
$$\dot{x}_2 = f_2(x_1, x_2, u, t)$$

$$\min \mathcal{E} = \int_0^\theta \mathcal{F}(x_1, x_2, u, t) \, dt$$

5.11. The system

$$\ddot{x} + a\dot{x} + bx = u$$

is to be controlled to minimize the ITES (integral of time times error squared) criterion,

$$\mathcal{E} = \frac{1}{2} \int_0^\theta t(c_1 x^2 + c_2 \dot{x}^2) \, dt$$

Develop the optimal feedback control.   Consider whether intermediate control is ever possible and show how switching curves can be constructed.

5.12. The chemical reactor described by Eqs. (7) of Sec. 4.6 is to be started up from initial conditions Z_{10}, Z_{20} and controlled to steady state Z_{1s}, Z_{2s} by manipulation of coolant flow rate q_c and feed temperature Z_f with bounds

$$0 \leq q_c \leq q_c^* \qquad Z_{f*} \leq Z_f \leq Z_f^*$$

If heating costs are proportional to feed temperature, the loss in profit during start-up is proportional to

$$\varepsilon = \int_0^\theta [(Z_{1s} - Z_1) + c(Z_f - Z_{fs})] \, dt$$

where θ is unspecified. Discuss the optimal start-up procedure. (A more complex version of this problem has been studied by Jackson.)

6

The Minimum Principle

6.1 INTRODUCTION

In the preceding chapters we have obtained and applied necessary conditions for optimality in a wide variety of optimal-design and optimal-control problems. Greater generality is required, and that is one aim of this chapter. A more serious deficiency of the preceding work, however, is that while the results are certainly correct, the motivation for several important operations in the derivations is not at all obvious. Thus we have little direction in attacking new problems by these methods, and, indeed, we sense that our ability to devise efficient computational algorithms is dependent upon our understanding of the logical steps in a proof of necessary conditions.

There is, in fact, an underlying logic which can be applied to both the theoretical and computational aspects of variational problems. The logic is firmly grounded in the elementary theory of linear ordinary differential equations, and the first considerations in this chapter will of necessity be directed to a discussion of this theory. The resulting

relations, which we shall apply to modern variational problems, were
used by Bliss in his analysis of problems of exterior ballistics in 1919.

6.2 INTEGRATING FACTORS AND GREEN'S FUNCTIONS

The principle we are seeking is best introduced by recalling the method
of integrating the linear first-order equation. Consider

$$\dot{x} = a(t)x + b(t) \tag{1}$$

In order to integrate this equation we multiply both sides by an *integrating factor* $\Gamma(t)$, an arbitrary differentiable function. We are inten-
tionally introducing an extra degree of freedom into the problem. Thus,

$$\Gamma(t)\dot{x}(t) = a(t)\Gamma(t)x(t) + \Gamma(t)b(t) \tag{2}$$

or, integrating from $t = 0$ to any time t,

$$\int_0^t \Gamma(\tau)\dot{x}(\tau)\, d\tau = \int_0^t a(\tau)\Gamma(\tau)x(\tau)\, d\tau + \int_0^t \Gamma(\tau)b(\tau)\, d\tau \tag{3}$$

The term on the left can be integrated by parts to give

$$\Gamma(t)x(t) - \Gamma(0)x(0) - \int_0^t \dot{\Gamma}(\tau)x(\tau)\, d\tau$$
$$= \int_0^t a(\tau)\Gamma(\tau)x(\tau)\, d\tau + \int_0^t \Gamma(\tau)b(\tau)\, d\tau \tag{4}$$

If we now remove most of the arbitrariness from Γ by defining it to
be the solution of

$$\dot{\Gamma} = -a\Gamma \tag{5}$$

or

$$\Gamma(t) = \Gamma(0) \exp\left[-\int_0^t a(\xi)\, d\xi\right] \tag{6}$$

then Eq. (4) simplifies to

$$\Gamma(t)x(t) \exp\left[-\int_0^t a(\xi)\, d\xi\right] - \Gamma(0)x(0)$$
$$= \Gamma(0) \int_0^t \exp\left[-\int_0^\tau a(\xi)\, d\xi\right] b(\tau)\, d\tau \tag{7}$$

Assuming that $\Gamma(0)$ is different from zero, we can solve explicitly for
$x(t)$

$$x(t) = x(0) \exp\left[\int_0^t a(\xi)\, d\xi\right] + \int_0^t \exp\left[\int_\tau^t a(\xi)\, d\xi\right] b(\tau)\, d\tau \tag{8}$$

The integrating factor $\Gamma(t)$ is generally known as a weighting function
or, as we prefer, *Green's function.*

We can generalize to a system of n linear equations in a straightforward manner. We consider the system

$$\begin{aligned}
\dot{x}_1 &= a_{11}(t)x_1 + a_{12}(t)x_2 + \cdots + a_{1n}(t)x_n + b_1(t) \\
\dot{x}_2 &= a_{21}(t)x_1 + a_{22}(t)x_2 + \cdots + a_{2n}(t)x_n + b_2(t) \\
&\cdots\cdots\cdots\cdots\cdots\cdots\cdots\cdots\cdots\cdots\cdots\cdots \\
\dot{x}_n &= a_{n1}(t)x_1 + a_{n2}(t)x_2 + \cdots + a_{nn}(t)x_n + b_n(t)
\end{aligned} \tag{9a}$$

or, equivalently,

$$\dot{x}_i = \sum_{j=1}^{n} a_{ij}(t)x_j + b_i(t) \tag{9b}$$

It is convenient to introduce n^2 functions $\Gamma_{ki}(t)$, k, $i = 1, 2, \ldots, n$, to multiply Eq. (9) by Γ_{ki} and sum over all i. Thus

$$\sum_{i=1}^{n} \Gamma_{ki}\dot{x}_i = \sum_{i=1}^{n}\sum_{j=1}^{n} \Gamma_{ki}a_{ij}x_j + \sum_{i=1}^{n} \Gamma_{ki}b_i \tag{10}$$

Integrating from $t = 0$ to any time and integrating the left-hand side by parts, we obtain

$$\int_0^t \sum_{i=1}^{n} \Gamma_{ki}(\tau)\dot{x}_i(\tau)\, d\tau = \sum_{i=1}^{n} \Gamma_{ki}(t)x_i(t) - \sum_{i=1}^{n} \Gamma_{ki}(0)x_i(0)$$

$$- \int_0^t \sum_{i=1}^{n} \dot{\Gamma}_{ki}(\tau)x_i(\tau)\, d\tau = \int_0^t \sum_{i=1}^{n}\sum_{j=1}^{n} \Gamma_{ki}(\tau)a_{ij}(\tau)x_j(\tau)\, d\tau$$

$$+ \int_0^t \sum_{i=1}^{n} \Gamma_{ki}(\tau)b_i(\tau)\, d\tau \quad (11)$$

As for the first-order system, we remove most of the arbitrariness from $\boldsymbol{\Gamma}$ by partially defining $\Gamma_{ki}(t)$ as a solution of

$$\dot{\Gamma}_{ki}(t) = - \sum_{j=1}^{n} \Gamma_{kj}(t)a_{ji} \tag{12}$$

equation (12) is called the *adjoint* of Eq. (9b). Thus,

$$\sum_{i=1}^{n} \Gamma_{ki}(t)x_i(t) - \sum_{i=1}^{n} \Gamma_{ki}(0)x_i(0) = \int_0^t \sum_{i=1}^{n} \Gamma_{ki}(\tau)b_i(\tau)\, d\tau \tag{13}$$

Equation (13) is known as *Green's identity*. It is the one-dimensional equivalent of the familiar relation between volume and surface integrals. The matrix of Green's functions $\boldsymbol{\Gamma}$ is known as *Green's matrix*, the *fundamental matrix*, or sometimes the *adjoint variables*. Unlike the first-order case, it is necessary to specify the value of Γ_{ik} at some time. It is convenient to specify Γ at the time of interest θ as

$$\Gamma_{ik}(\theta) = \delta_{ik} = \begin{cases} 1 & i = k \\ 0 & i \neq k \end{cases} \tag{14}$$

so that Eq. (13) reduces to

$$x_i(\theta) = \sum_{j=1}^{n} \Gamma_{ij}(0)x_j(0) + \int_0^\theta \sum_{j=1}^{n} \Gamma_{ij}(t)b_j(t)\, dt \tag{15}$$

Indeed, for later use it is convenient to note explicitly the moment at which condition (14) is specified, and we do this by specifying two arguments, $\Gamma_{ik}(\theta,t)$, where θ is the time of interest and t any other time. Thus,

$$x_i(\theta) = \sum_{j=1}^{n} \Gamma_{ij}(\theta,0)x_j(0) + \int_0^\theta \sum_{j=1}^{n} \Gamma_{ij}(\theta,t)b_j(t)\, dt \tag{16}$$

where $\Gamma_{ij}(\theta,t)$ satisfies Eq. (12) with respect to t and the condition

$$\Gamma_{ik}(\theta,\theta) = \delta_{ik} \tag{17}$$

We shall sometimes be interested in a linear combination of the components x_i of $\mathbf{x}$ at time θ. We define a vector $\boldsymbol{\gamma}(t)$ with components $\gamma_i(t)$ by

$$\gamma_i(t) = \sum_{j=1}^{n} \gamma_j(\theta)\Gamma_{ji}(\theta,t) \tag{18}$$

where $\gamma_j(\theta)$ are specified numbers; then by multiplying Eqs. (12) and (16) by appropriate components of $\boldsymbol{\gamma}(\theta)$ and summing we obtain

$$\sum_{i=1}^{n} \gamma_i(\theta)x_i(\theta) = \sum_{i=1}^{n} \gamma_i(0)x_i(0) + \int_0^\theta \sum_{i=1}^{n} \gamma_i(t)b_i(t)\, dt \tag{19}$$

and

$$\dot{\gamma}_i = -\sum_{j=1}^{n} \gamma_j(t)a_{ji}(t) \tag{20}$$

As a simple illustrative example consider the linear second-order system with constant coefficients

$$\ddot{x} + \alpha_1\dot{x} + \alpha_2 x = b(t) \tag{21}$$

or, equivalently,

$$\dot{x}_1 = x_2 \tag{22a}$$
$$\dot{x}_2 = -\alpha_2 x_1 - \alpha_1 x_2 + b_2(t) \tag{22b}$$

The adjoint system, defined by Eq. (12), with boundary conditions from Eq. (14), is

$$\dot{\Gamma}_{11} = \alpha_2 \Gamma_{12} \qquad\qquad \Gamma_{11} = 1 \qquad \text{at } t = \theta \qquad\qquad (23a)$$
$$\dot{\Gamma}_{12} = -\Gamma_{11} + \alpha_1 \Gamma_{12} \qquad \Gamma_{12} = 0 \qquad \text{at } t = \theta \qquad\qquad (23b)$$
$$\dot{\Gamma}_{21} = \alpha_2 \Gamma_{22} \qquad\qquad \Gamma_{21} = 0 \qquad \text{at } t = \theta \qquad\qquad (23c)$$
$$\dot{\Gamma}_{22} = -\Gamma_{21} + \alpha_1 \Gamma_{22} \qquad \Gamma_{22} = 1 \qquad \text{at } t = \theta \qquad\qquad (23d)$$

The solution is easily found to be

$$\Gamma_{11}(\theta,t) = \frac{e^{\frac{1}{2}\alpha_1(t-\theta)}}{2\sqrt{\alpha_1{}^2 - 4\alpha_2}} \{(\alpha_1 - \sqrt{\alpha_1{}^2 - 4\alpha_2})$$
$$\exp[\tfrac{1}{2}\sqrt{\alpha_1{}^2 - 4\alpha_2}\,(t-\theta)] - (\alpha_1 + \sqrt{\alpha_1{}^2 - 4\alpha_2})$$
$$\exp[-\tfrac{1}{2}\sqrt{\alpha_1{}^2 - 4\alpha_2}\,(t-\theta)]\} \qquad (24a)$$

$$\Gamma_{12}(\theta,t) = -\frac{e^{\frac{1}{2}\alpha_1(t-\theta)}}{\sqrt{\alpha_1{}^2 - 4\alpha_2}} \{\exp[\tfrac{1}{2}\sqrt{\alpha_1{}^2 - 4\alpha_2}\,(t-\theta)]$$
$$- \exp[-\tfrac{1}{2}\sqrt{\alpha_1{}^2 - 4\alpha_2}\,(t-\theta)]\} \qquad (24b)$$

$$\Gamma_{21}(\theta,t) = \frac{\alpha_2 e^{\frac{1}{2}\alpha_1(t-\theta)}}{\sqrt{\alpha_1{}^2 - 4\alpha_2}} \{\exp[\tfrac{1}{2}\sqrt{\alpha_1{}^2 - 4\alpha_2}\,(t-\theta)]$$
$$- \exp[-\tfrac{1}{2}\sqrt{\alpha_1{}^2 - 4\alpha_2}\,(t-\theta)]\} \qquad (24c)$$

$$\Gamma_{22}(\theta,t) = -\frac{e^{\frac{1}{2}\alpha_1(t-\theta)}}{2\sqrt{\alpha_1{}^2 - 4\alpha_2}} \{(\alpha_1 + \sqrt{\alpha_1{}^2 - 4\alpha_2})$$
$$\exp[\tfrac{1}{2}\sqrt{\alpha_1{}^2 - 4\alpha_2}\,(t-\theta)] - (\alpha_1 - \sqrt{\alpha_1{}^2 - 4\alpha_2})$$
$$\exp[-\tfrac{1}{2}\sqrt{\alpha_1{}^2 - 4\alpha_2}\,(t-\theta)]\} \qquad (24d)$$

Then, for any time θ,

$$x(\theta) = x_1(\theta) = \Gamma_{11}(\theta,0)x(0) + \Gamma_{12}(\theta,0)\dot{x}(0)$$
$$+ \int_0^\theta \Gamma_{12}(\theta,t)b_2(t)\,dt \qquad (25a)$$

$$\dot{x}(\theta) = x_2(\theta) = \Gamma_{21}(\theta,0)x(0) + \Gamma_{22}(\theta,0)\dot{x}(0)$$
$$+ \int_0^\theta \Gamma_{22}(\theta,t)b_2(t)\,dt \qquad (25b)$$

or, substituting just into the equation for $x(t)$,

$$x(\theta) = \frac{e^{-\alpha_1\theta/2}}{\sqrt{\alpha_1{}^2 - 4\alpha_2}} \left\{ \left[\frac{\alpha_1 - \sqrt{\alpha_1{}^2 - 4\alpha_2}}{2} x(0) - \dot{x}(0) \right] \right.$$
$$\exp(-\tfrac{1}{2}\sqrt{\alpha_1{}^2 - 4\alpha_2}\,\theta) - \left[\frac{\alpha_1 + \sqrt{\alpha_1{}^2 - 4\alpha_2}}{2} - \dot{x}(0) \right]$$
$$\left. \exp(\tfrac{1}{2}\sqrt{\alpha_1{}^2 - 4\alpha_2}\,\theta) \right\} - \int_0^\theta \frac{e^{\frac{1}{2}\alpha_1(t-\theta)}}{\sqrt{\alpha_1{}^2 - 4\alpha_2}}$$
$$\{\exp[\tfrac{1}{2}\sqrt{\alpha_1{}^2 - 4\alpha_2}\,(t-\theta)]$$
$$- \exp[-\tfrac{1}{2}\sqrt{\alpha_1{}^2 - 4\alpha_2}\,(t-\theta)]\}b(t)\,dt \qquad (26)$$

6.3 FIRST-ORDER VARIATIONAL EQUATIONS

We now consider a physical system described by the (usually nonlinear) equations

$$\dot{x}_i = f_i(\mathbf{x},\mathbf{u}) \qquad \begin{matrix} 0 < t \leq \theta \\ i = 1, 2, \ldots, n \end{matrix} \tag{1}$$

If we specify a set of initial conditions $\bar{\mathbf{x}}_0$ and specify the components of $\mathbf{u}(t)$ to be particular functions $\bar{u}_k(t)$, $k = 1, 2, \ldots, R$, Eqs. (1) can be integrated to obtain functions which we shall denote by $\bar{x}_i(t)$, $i = 1$, $2, \ldots, n$. That is, we define $\bar{\mathbf{x}}(t)$ as the vector whose components are solutions of the equations

$$\dot{\bar{x}}_i = f_i(\bar{\mathbf{x}},\bar{\mathbf{u}}) \qquad \begin{matrix} 0 < t \leq \theta \\ i = 1, 2, \ldots, n \end{matrix}$$
$$\bar{\mathbf{x}}(0) = \bar{\mathbf{x}}_0 \tag{2}$$

Let us now suppose that we wish to specify new functions $u_k(t)$ and initial values $x_i(0)$ as

$$u_k(t) = \bar{u}_k(t) + \delta u_k(t) \qquad k = 1, 2, \ldots, R \tag{3a}$$
$$x_i(0) = \bar{x}_{i0} + \delta x_{i0} \qquad i = 1, 2, \ldots, n \tag{3b}$$

where, for some predetermined $\epsilon > 0$,

$$|\delta u_k(t)| < \epsilon \qquad \begin{matrix} k = 1, 2, \ldots, R \\ 0 \leq t \leq \theta \end{matrix} \tag{4a}$$
$$|\delta x_{i0}| < \epsilon \qquad i = 1, 2, \ldots, n \tag{4b}$$

If we now solve Eqs. (1) and write the solution as

$$x_i(t) = \bar{x}_i(t) + \delta x_i(t) \tag{5}$$

it follows (see Appendix 6.1) that

$$|\delta x_i(t)| < K\epsilon \tag{6}$$

where K depends on θ but is independent of the variations $\boldsymbol{\delta u}$ and $\boldsymbol{\delta x}_0$. Thus,

$$\dot{x}_i = \dot{\bar{x}}_i + \delta \dot{x}_i = f_i(\bar{\mathbf{x}} + \boldsymbol{\delta x}, \bar{\mathbf{u}} + \boldsymbol{\delta u}) \qquad \begin{matrix} 0 < t \leq \theta \\ i = 1, 2, \ldots, n \end{matrix}$$
$$\mathbf{x}(0) = \bar{\mathbf{x}}_0 + \boldsymbol{\delta x}_0 \tag{7}$$

and, subtracting Eq. (2) from (7),

$$\delta \dot{x}_i = f_i(\bar{\mathbf{x}} + \boldsymbol{\delta x}, \bar{\mathbf{u}} + \boldsymbol{\delta u}) - f_i(\bar{\mathbf{x}},\bar{\mathbf{u}}) \qquad \begin{matrix} 0 < t \leq \theta \\ i = 1, 2, \ldots, n \end{matrix} \tag{8}$$

If f_i has piecewise continuous first and bounded second derivatives with respect to components of $\mathbf{x}$ and $\mathbf{u}$, then everywhere except at isolated discontinuities of $\mathbf{u}$ we may expand the right-hand side of Eq. (8) in a Taylor series to write

$$\delta \dot{x}_i = \sum_{j=1}^{n} \frac{\partial f_i}{\partial x_j} \delta x_j + \sum_{k=1}^{R} \frac{\partial f_i}{\partial u_k} \delta u_k + o(\epsilon) \qquad \begin{matrix} 0 < t \le \theta \\ i = 1, 2, \ldots, n \end{matrix} \qquad (9)$$

where the partial derivatives are evaluated for $\bar{\mathbf{x}}$ and $\bar{\mathbf{u}}$ and, hence, are known functions of t. Comparing the *linear* equation (9) with Eq. (9) of the preceding section, we identify the terms $\partial f_i/\partial x_j$ with a_{ij} and introduce Green's vector $\boldsymbol{\gamma}$ from Eq. (20) of Sec. 6.2

$$\dot{\gamma}_i = - \sum_{j=1}^{n} \gamma_j \frac{\partial f_j}{\partial x_i} \qquad \begin{matrix} 0 \le t < \theta \\ i = 1, 2, \ldots, n \end{matrix} \qquad (10)$$

Green's identity, Eq. (19) of that section, is then

$$\sum_{i=1}^{n} \gamma_i(\theta) \, \delta x_i(\theta) = \sum_{i=1}^{n} \gamma_i(0) \, \delta x_{i0} + \int_0^{\theta} \sum_{i=1}^{n} \sum_{k=1}^{R} \gamma_i \frac{\partial f_i}{\partial u_k} \delta u_k \, dt + o(\epsilon) \qquad (11)$$

To first order, this is the effect on system output of small changes in initial conditions and decision functions.

6.4 THE MINIMIZATION PROBLEM AND FIRST VARIATION OF THE OBJECTIVE

It is convenient to formulate the optimization problem in somewhat different form from the previous chapters. Instead of seeking to minimize an integral we shall suppose that our concern is with some function $\mathcal{E}$ which depends only upon the state of the process at time θ, $\mathbf{x}(\theta)$. We assume that the system is described by the differential equations

$$\dot{x}_i = f_i(\mathbf{x}, \mathbf{u}) \qquad \begin{matrix} 0 < t \le \theta \\ i = 1, 2, \ldots, n \end{matrix} \qquad (1)$$

that the decision variables $u_k(t)$ satisfy inequality constraints

$$U_p(\mathbf{u}) \ge 0 \qquad p = 1, 2, \ldots, P \qquad (2)$$

that the initial state $\mathbf{x}_0$ is subject to equality constraints

$$q_m(\mathbf{x}_0) = 0 \qquad m = 1, 2, \ldots, M \qquad (3)$$

and the final state to equality constraints

$$g_s[\mathbf{x}(\theta)] = 0 \qquad s = 1, 2, \ldots, S \qquad (4)$$

The constraint equations (2) to (4) are more general than those considered previously. We shall see later that the choice of the objective $\mathcal{E}[\mathbf{x}(\theta)]$ also includes the form of previous chapters as a special case.

We shall first assume that θ is specified. If we specify $\bar{\mathbf{u}}(t)$ and $\boldsymbol{\delta}\mathbf{u}(t)$ subject to the constraint equation (2) and $\bar{\mathbf{x}}_0$ and $\boldsymbol{\delta}\mathbf{x}_0$ subject to Eq. (3), we may write Eq. (11) of Sec. 6.3

$$\sum_{i=1}^{n} \gamma_i(\theta) \, \delta x_i(\theta) = \sum_{i=1}^{n} \gamma_i(0) \, \delta x_i(0) + \int_0^\theta \sum_{i=1}^{n} \sum_{k=1}^{R} \gamma_i \frac{\partial f_i}{\partial u_k} \, \delta u_k \, dt + o(\epsilon)$$

(5)

Furthermore, $\mathbf{q}(\bar{\mathbf{x}}_0 + \boldsymbol{\delta}\mathbf{x}_0)$ and $\mathbf{q}(\bar{\mathbf{x}}_0)$ must both equal zero, as must $\mathbf{g}[\bar{\mathbf{x}}(\theta) + \boldsymbol{\delta}\mathbf{x}(\theta)]$ and $\mathbf{g}[\bar{\mathbf{x}}(\theta)]$, so that

$$q_m(\bar{\mathbf{x}}_0 + \boldsymbol{\delta}\mathbf{x}_0) - q_m(\bar{\mathbf{x}}_0) = \sum_{i=1}^{n} \frac{\partial q_m}{\partial x_{i0}} \, \delta x_{i0} + o(\epsilon) = 0$$

$$m = 1, 2, \ldots, M \quad (6)$$

$$g_s[\bar{\mathbf{x}}(\theta) + \boldsymbol{\delta}\mathbf{x}(\theta)] - g_s[\bar{\mathbf{x}}(\theta)] = \sum_{i=1}^{n} \frac{\partial g_s}{\partial x_i} \, \delta x_i(\theta) + o(\epsilon) = 0$$

$$s = 1, 2, \ldots, S \quad (7)$$

The change in $\mathcal{E}$ as a result of decision changes is reflected through a change in $\mathbf{x}(\theta)$

$$\delta\mathcal{E} = \mathcal{E}[\bar{\mathbf{x}}(\theta) + \boldsymbol{\delta}\mathbf{x}(\theta)] - \mathcal{E}[\bar{\mathbf{x}}(\theta)] = \sum_{i=1}^{n} \frac{\partial \mathcal{E}}{\partial x_i} \, \delta x_i(\theta) + o(\epsilon)$$

(8)

We have not yet specified the value of the vector $\boldsymbol{\gamma}(\theta)$. Let us write this as

$$\gamma_i(\theta) = \frac{\partial \mathcal{E}}{\partial x_i} + \tilde{\gamma}_i$$

(9)

From Eq. (8) we may then write

$$\delta\mathcal{E} = \sum_{i=1}^{n} [\gamma_i(\theta) - \tilde{\gamma}_i] \, \delta x_i(\theta) + o(\epsilon)$$

(10)

and, from Eq. (5),

$$\delta\mathcal{E} = -\sum_{i=1}^{n} \tilde{\gamma}_i \, \delta x_i(\theta) + \sum_{i=1}^{n} \gamma_i(0) \, \delta x_{i0} + \int_0^\theta \sum_{i=1}^{n} \sum_{k=1}^{R} \gamma_i \frac{\partial f_i}{\partial u_k} \, \delta u_k \, dt$$

$$+ o(\epsilon) \quad (11)$$

We can obtain an expression for $\delta\mathcal{E}$ entirely in terms of the variations in decision $\delta \mathbf{u}(t)$ by choosing particular boundary conditions for $\boldsymbol{\gamma}(t)$

$$\tilde{\gamma}_i = \sum_{s=1}^{S} \nu_s \frac{\partial g_s}{\partial x_i} \tag{12a}$$

or

$$\gamma_i(\theta) = \frac{\partial \mathcal{E}}{\partial x_i} + \sum_{s=1}^{S} \nu_s \frac{\partial g_s}{\partial x_i} \tag{12b}$$

and

$$\gamma_i(0) = \sum_{m=1}^{M} \eta_m \frac{\partial g_m}{\partial x_i} \tag{13}$$

Here, $\mathbf{n}$ and $\mathbf{v}$ are undetermined multipliers which must be found as part of the solution. Using Eqs. (6) and (7), Eq. (11) becomes, upon substitution of the boundary conditions,

$$\delta \mathcal{E} = \int_0^\theta \sum_{i=1}^{n} \sum_{k=1}^{R} \gamma_i \frac{\partial f_i}{\partial u_k} \delta u_k \, dt + o(\epsilon) \tag{14}$$

The $2n$ differential equations for the x_i and γ_i require a total of $2n$ specified values. Equations (3) and (4) give a total of $M + S$ conditions. Equation (12b) specifies $n - S$ values and Eq. (13) $n - M$, so that the total is indeed the required number. The special case in which $x_i(\theta)$ or x_{i0} is directly specified is included by setting

$$g_s[\mathbf{x}(\theta)] = x_s(\theta) - x_s^*(\theta) = 0 \tag{15a}$$

and

$$q_m(\mathbf{x}_0) = x_{m0} - x_{m0}^* = 0 \tag{15b}$$

where $x_s^*(\theta)$ and x_{m0}^* are fixed numbers. In the former case

$$\gamma_s(\theta) = \frac{\partial \mathcal{E}}{\partial x_s} + \nu_s \tag{16a}$$

where ν_s is unspecified, so that $\gamma_s(\theta)$ is unspecified, while in the latter

$$\gamma_m(0) = \eta_m \tag{16b}$$

where η_m is an unspecified number.

When θ is not fixed, we must allow for the possibility that the changes $\delta \mathbf{u}(t)$, $\delta \mathbf{x}_0$ will require a change $\delta \theta$ in order to satisfy all the conditions on $\mathbf{x}(\theta)$. Let $\bar{\theta}$ refer to the interval associated with $\bar{\mathbf{u}}(t)$, $\bar{\mathbf{x}}_0$. Then

$$\delta \mathcal{E} = \mathcal{E}[\mathbf{x}(\bar{\theta} + \delta \theta)] - \mathcal{E}[\bar{\mathbf{x}}(\bar{\theta})] \tag{17a}$$

or, writing $\mathcal{E}[\mathbf{x}(\bar{\theta} + \delta\theta)]$ in terms of $\bar{\theta}$,

$$\delta\mathcal{E} = \mathcal{E}[\mathbf{x}(\bar{\theta})] + \sum_{i=1}^{n} \frac{\partial\mathcal{E}}{\partial x_i} f_i[\mathbf{x}(\bar{\theta}),\mathbf{u}(\bar{\theta})] \, \delta\theta + o(\delta\theta) - \mathcal{E}[\bar{\mathbf{x}}(\bar{\theta})]$$

$$= \mathcal{E}[\mathbf{x}(\bar{\theta})] - \mathcal{E}[\bar{\mathbf{x}}(\bar{\theta})] + \sum_{i=1}^{n} \frac{\partial\mathcal{E}}{\partial x_i} f_i[\bar{\mathbf{x}}(\bar{\theta}),\bar{\mathbf{u}}(\bar{\theta})] \, \delta\theta + o(\delta\theta) + o(\epsilon)$$

$$(17b)$$

and, finally,

$$\delta\mathcal{E} = \sum_{i=1}^{n} \frac{\partial\mathcal{E}}{\partial x_i} [\delta x_i(\bar{\theta}) + f_i \, \delta\theta] + o(\epsilon) \tag{17c}$$

where we have included $o(\delta\theta)$ as $o(\epsilon)$.   Similarly,

$$\delta g_s = \sum_{i=1}^{n} \frac{\partial g_s}{\partial x_i} [\delta x_i(\bar{\theta}) + f_i \, \delta\theta] + o(\epsilon) = 0 \tag{18}$$

Defining $\gamma_i(\bar{\theta})$ by Eq. (12b), $\gamma_i(0)$ by Eq. (13), and substituting Eqs. (5) and (18) into (17c), we obtain

$$\delta\mathcal{E} = \sum_{i=1}^{n} \gamma_i(\bar{\theta})f_i[\bar{\mathbf{x}}(\bar{\theta}),\bar{\mathbf{u}}(\bar{\theta})] \, \delta\theta + \int_{0}^{\theta} \sum_{i=1}^{n} \sum_{k=1}^{R} \gamma_i \frac{\partial f_i}{\partial u_k} \, \delta u_k \, dt + o(\epsilon) \quad (19)$$

Finally, because of the extra degree of freedom resulting from the non-specification of θ we may set

$$\theta \text{ unspecified:} \qquad \sum_{i=1}^{n} \gamma_i(\bar{\theta})f_i[\bar{\mathbf{x}}(\bar{\theta}),\bar{\mathbf{u}}(\bar{\theta})] = 0 \tag{20}$$

and $\delta\mathcal{E}$ is again represented by Eq. (14).   In other words, the unspecified stopping time θ is defined by the additional Eq. (20).

6.5 THE WEAK MINIMUM PRINCIPLE

It is convenient to introduce again the hamiltonian notation

$$H = \sum_{i=1}^{n} \gamma_i f_i \tag{1}$$

As previously, the system and multiplier equations may be written in the canonical form

$$\dot{x}_i = \frac{\partial H}{\partial \gamma_i} \tag{2a}$$

$$\dot{\gamma}_i = -\frac{\partial H}{\partial x_i} \tag{2b}$$

with the boundary conditions Eqs. (3), (4), (12), and (13) of Sec. 6.4 and the additional condition

θ unspecified: $H = 0$ at $t = \theta$ $\hspace{3cm}$ (3)

If we now assume that $\bar{\mathbf{u}}(t)$ is in fact the set of allowable functions which minimize $\mathcal{E}$, then, from Eq. (14) of the preceding section, we may write

$$\delta\mathcal{E} = \int_0^\theta \sum_{k=1}^R \frac{\partial H}{\partial u_k} \delta u_k\, dt + o(\epsilon) \geq 0 \hspace{3cm} (4)$$

The functions $u_k(t)$ must be chosen subject to the restriction

$$U_p(\mathbf{u}) \geq 0 \qquad p = 1, 2, \ldots, P \hspace{3cm} (5)$$

If, for a particular value of k, all constraints are at inequality, we may choose δu_k (sufficiently small) as we wish, and it follows (as in Sec. 5.2) that

$$\frac{\partial H}{\partial u_k} = 0 \qquad u_k \text{ unconstrained} \hspace{3cm} (6)$$

If some constraint U_p is at equality for the optimum, it must be true that

$$\sum_{k=1}^R \frac{\partial U_p}{\partial u_k} \delta u_k + o(\epsilon) \geq 0 \hspace{3cm} (7)$$

In the spirit of Sec. 5.2, we choose all variations but one to be zero and that one as

$$\delta u_k = \sigma \frac{\partial U_p}{\partial u_k} \frac{\partial H}{\partial u_k} \hspace{3cm} (8)$$

From Eq. (7) δu_k and $\partial U_p/\partial u_k$ must have the same algebraic sign. Substituting into Eq. (4), we obtain

$$\sigma \frac{\partial U_p}{\partial u_k} \left(\frac{\partial H}{\partial u_k}\right)^2 \geq 0 \hspace{3cm} (9a)$$

or

$$\sigma \frac{\partial U_p}{\partial u_k} \geq 0 \hspace{3cm} (9b)$$

Then, multiplying Eq. (8) by $\partial H/\partial u_k$,

$$\delta u_k \frac{\partial H}{\partial u_k} = \sigma \frac{\partial U_p}{\partial u_k} \left(\frac{\partial H}{\partial u_k}\right)^2 \geq 0 \hspace{3cm} (10a)$$

or, equivalently,

$$\delta u_k \frac{\partial H}{\partial u_k} \geq 0 \hspace{3cm} (10b)$$

Thus, when a constraint is in force, the hamiltonian is a minimum with respect to decisions affected by the constraint. (In exceptional cases the hamiltonian may only be stationary at a constraint.) An equivalent result holds when the hamiltonian is not differentiable for some $\bar{u}_k$.

For the special case considered in Sec. 5.2 we were able to establish that the hamiltonian is a constant along the optimal path. This result, too, carries over to the more general situation considered here. We calculate

$$\dot{H} = \sum_{i=1}^{n} \frac{\partial H}{\partial x_i} \dot{x}_i + \sum_{i=1}^{n} \frac{\partial H}{\partial \gamma_i} \dot{\gamma}_i + \sum_{k=1}^{R} \frac{\partial H}{\partial u_k} \dot{u}_k \tag{11}$$

The first two terms are equal with opposite signs as a result of the canonical equations (2). When the optimal u_k is unconstrained, $\partial H/\partial u_k$ vanishes. When at a constraint for a finite interval, however, we might move along the constraint surface with changing u_k if more than one decision enters into the constraint. Thus, we cannot simply set $\dot{u}_k$ to zero at a constraint. However, when the constraint U_p is equal to zero for a finite time, its time derivative must vanish, so that

$$\dot{U}_p = \sum_{k=1}^{R} \frac{\partial U_p}{\partial u_k} \dot{u}_k = 0 \tag{12}$$

Furthermore, if H is minimized subject to the constraint $U_p = 0$, the Lagrange multiplier rule requires that

$$\frac{\partial H}{\partial u_k} = -\lambda \frac{\partial U_p}{\partial u_k} \tag{13}$$

or, substituting into Eq. (11),

$$\dot{H} = -\lambda \sum_{k=1}^{R} \frac{\partial U_p}{\partial u_k} \dot{u}_k \tag{14}$$

But from Eq. (12) this is equal to zero. Thus,

$$H = \text{const along optimal path} \tag{15a}$$

Furthermore, from Eq. (3),

$$\theta \text{ unspecified:} \quad H = 0 \qquad 0 < t \le \theta \tag{15b}$$

We may summarize the necessary conditions as the weak minimum principle:

Given $\mathcal{E}[\mathbf{x}(\theta)]$ *and*

$$H = \sum_{i=1}^{n} \gamma_i f_i$$

where

$$\dot{x}_i = \frac{\partial H}{\partial \gamma_i} \qquad \dot{\gamma}_i = -\frac{\partial H}{\partial x_i}$$

with 2n boundary conditions

$$q_m(\mathbf{x}_0) = 0 \qquad m = 1, 2, \ldots, M$$
$$g_s[\mathbf{x}(\theta)] = 0 \qquad s = 1, 2, \ldots, S$$

$$\gamma_i(0) = \sum_{m=1}^{M} \eta_m \frac{\partial q_m}{\partial x_i} \qquad i = 1, 2, \ldots, n$$

$$\gamma_i(\theta) = \frac{\partial \mathcal{E}}{\partial x_i} + \sum_{s=1}^{S} \nu_s \frac{\partial g_s}{\partial x_i} \qquad i = 1, 2, \ldots, n$$

The decisions $u_k(t)$ which minimize $\mathcal{E}[\mathbf{x}(\theta)]$ subject to the restrictions

$$U_p(\mathbf{u}) \geq 0 \qquad p = 1, 2, \ldots, P$$

make the hamiltonian H stationary when a constraint is not at equality and minimize the hamiltonian (or make it stationary) with respect to constraints at equality or at nondifferentiable points. Along the optimal path the hamiltonian is a constant, and if θ is unspecified, that constant value is zero.

One final remark should be made. Since an integral is unaffected by values of the integrand at discrete points, the arguments made here and in Sec. 5.2 may fail to hold at a set of discrete points (in fact, over a set of measure zero). This in no way affects the application of the results.

6.6 EQUIVALENT FORMULATIONS

In assuming that the objective function to be minimized depends only upon the state $\mathbf{x}(\theta)$ we have written the *Mayer form* of the optimization problem. In previous chapters we have been concerned with the *Lagrange form*, in which an integral of a function of state and decision is minimized. It is useful to establish that the latter problem can be treated as a special case of the results of Sec. 6.5.

We consider the system

$$\dot{x}_i = f_i(\mathbf{x},\mathbf{u}) \qquad i = 1, 2, \ldots, n \tag{1}$$

with appropriately bounded decisions and end points, and we assume that $\mathbf{u}$ is to be chosen in order to minimize

$$\mathcal{E} = \int_0^\theta \mathcal{F}(\mathbf{x},\mathbf{u}) \, dt \tag{2}$$

By defining a new variable, say x_0, such that

$$\dot{x}_0 = \mathcal{F}(\mathbf{x},\mathbf{u}) \qquad x_0(0) = 0 \tag{3}$$

we may write

$$\mathcal{E} = \int_0^\theta \dot{x}_0 \, dt = x_0(\theta) \tag{4}$$

Thus, numbering from $k = 0$ to n, the hamiltonian, Eq. (1) of Sec. 6.5, becomes

$$H = \gamma_0 \mathcal{F}(\mathbf{x},\mathbf{u}) + \sum_{k=1}^{n} \gamma_k f_k(\mathbf{x},\mathbf{u}) \tag{5}$$

with

$$\dot{x}_i = \frac{\partial H}{\partial \gamma_i} = f_i \qquad i = 1, 2, \ldots, n \tag{6a}$$

$$\dot{\gamma}_i = -\frac{\partial H}{\partial x_i} = -\gamma_0 \frac{\partial \mathcal{F}}{\partial x_i} - \sum_{k=1}^{n} \gamma_k \frac{\partial f_k}{\partial x_i} \tag{6b}$$

and

$$\dot{\gamma}_0 = -\frac{\partial H}{\partial x_0} = 0 \tag{6c}$$

Furthermore, since $x_0(\theta)$ is completely free, the boundary condition at $t = \theta$ for γ_0 is

$$\gamma_0(\theta) = \frac{\partial \mathcal{E}}{\partial x_0} = 1 \tag{7}$$

which, together with Eq. (6c), establishes that γ_0 is identically unity.†
The hamiltonian is therefore

$$H = \mathcal{F} + \sum_{i=1}^{n} \gamma_i f_i \tag{8}$$

as in Chaps. 4 and 5.

It is also useful to consider cases in which the independent variable t appears explicitly in the system equations or in the objective. Here we consider systems

$$\dot{x}_i = f_i(\mathbf{x},\mathbf{u},t) \qquad i = 1, 2, \ldots, n \tag{9}$$

and objectives

$$\mathcal{E} = \mathcal{E}[\mathbf{x}(\theta),\theta] \tag{10}$$

---

† Any positive constant multiple of an integral of $\mathcal{F}$ could be minimized, so that γ_0 will be any positive constant. An unstated *regularity assumption*, similar to the one used for the Lagrange multiplier rule in Sec. 1.8, prevents γ_0 from becoming zero, and hence there is no loss of generality in taking it to be unity.

The device which we use here is identical to that above. We define a new variable x_0 such that

$$\dot{x}_0 = 1 \qquad x_0(0) = 0 \tag{11}$$

in which case x_0 and t are identical. Equations (9) and (10) now depend explicitly on $x_0, x_1, \ldots, x_n$ but (formally) not on t. Thus, the results of Sec. 6.5 must apply, and we write

$$H = \gamma_0 + \sum_{i=1}^{n} \gamma_i f_i \tag{12}$$

with

$$\dot{\gamma}_0 = -\frac{\partial H}{\partial x_0} = -\sum_{i=1}^{n} \gamma_i \frac{\partial f_i}{\partial t} \tag{13a}$$

$$\gamma_0(\theta) = \begin{cases} \dfrac{\partial \mathcal{E}}{\partial x_0} = \dfrac{\partial \mathcal{E}}{\partial \theta} & \theta \text{ unspecified} \\ \text{unspecified} & x_0(\theta) = \theta \text{ specified} \end{cases} \tag{13b}$$

Clearly nothing is changed except that the sum $\sum_{i=1}^{n} \gamma_i f_i$ is not equal to a constant when some f_i depend on t, since $H = $ const and γ_0 varies with t according to Eq. (13a).

6.7 AN APPLICATION WITH TRANSVERSALITY CONDITIONS

The essential improvement in the form of the weak minimum principle developed in the two preceding sections over that used in Chap. 5 is the generalization of allowable boundary conditions for the state variable and the corresponding boundary conditions for the multipliers, or, as we now prefer to think of them, Green's functions. These latter conditions are usually referred to as *transversality conditions*. We can demonstrate the application of the transversality conditions by again considering the simple second-order system of Sec. 5.3

$$\dot{x}_1 = x_2 \tag{1a}$$
$$\dot{x}_2 = u \tag{1b}$$
$$|u| \le 1 \tag{1c}$$

where we now seek the minimum time control not to the origin but to a circle about the origin of radius R; that is, we impose the final condition

$$g[\mathbf{x}(\theta)] = x_1^2(\theta) + x_2^2(\theta) - R^2 = 0 \qquad t = \theta$$

Using the equivalent formulation of Sec. 6.6, we have $\mathcal{F} = 1$ for minimum time control, and the hamiltonian is

$$H = 1 + \gamma_1 x_2 + \gamma_2 u \tag{2}$$

where γ_1 and γ_2 satisfy the canonical equations

$$\dot{\gamma}_1 = -\frac{\partial H}{\partial x_1} = 0 \tag{3a}$$

$$\dot{\gamma}_2 = -\frac{\partial H}{\partial x_2} = -\gamma_1 \tag{3b}$$

Thus, as previously,

$$\gamma_1 = c_1 = \text{const} \tag{4a}$$
$$\gamma_2 = -c_1 t - c_2 \tag{4b}$$

and the optimal control is

$$u = -\,\text{sgn}\,\gamma_2 = \text{sgn}\,(c_1 t + c_2) \tag{5}$$

which can switch between extremes at most once. The allowable trajectories are again the parabolas

$$x_1 = \pm x_2^2 + \text{const} \tag{6}$$

Since the objective does not depend explicitly on $x_1(\theta)$ or $x_2(\theta)$, Eq. (12b) of Sec. 6.4 for the boundary conditions of the Green's functions becomes

$$\gamma_1(\theta) = \frac{\partial \mathscr{E}^{0}}{\partial x_1} + \nu \frac{\partial g}{\partial x_1} = 2\nu x_1(\theta) \tag{7a}$$

$$\gamma_2(\theta) = \frac{\partial \mathscr{E}^{0}}{\partial x_2} + \nu \frac{\partial g}{\partial x_2} = 2\nu x_2(\theta) \tag{7b}$$

or, eliminating the unknown constant ν,

$$\frac{\gamma_1(\theta)}{\gamma_2(\theta)} = \frac{x_1(\theta)}{x_2(\theta)} \tag{8}$$

Thus, evaluating Eqs. (4) at time θ, Eq. (8) becomes

$$\frac{c_1}{-c_1 \theta - c_2} = \frac{x_1(\theta)}{x_2(\theta)} \tag{9}$$

or

$$c_1 = \frac{-c_2}{\theta + x_2(\theta)/x_1(\theta)} \tag{10}$$

The optimal control, Eq. (5), then becomes

$$u = \text{sgn}\left\{ c_2 \left[1 - \frac{t}{\theta + x_2(\theta)/x_1(\theta)} \right] \right\} \tag{11}$$

When $x_2(\theta)$ and $x_1(\theta)$ have the same algebraic sign (the first and third quadrants), the algebraic sign of the argument in Eq. (11) never changes ($t \leq \theta$ and therefore $t/[\theta + x_2(\theta)/x_1(\theta)] < 1$). Thus, all trajectories ending on the circle of radius R in the first and third quadrant must do so without switching between control extremes and must consist of one of the parabolas defined by Eq. (6). Inspection of Fig. 5.4 clearly indicates that for approaches from outside the circle the optimum is

$$u = -1 \qquad \begin{matrix} x_1(\theta) > 0 \\ x_2(\theta) > 0 \end{matrix} \qquad\qquad (12a)$$

$$u = +1 \qquad \begin{matrix} x_1(\theta) < 0 \\ x_2(\theta) < 0 \end{matrix} \qquad\qquad (12b)$$

For trajectories ending in the fourth quadrant, where $x_1(\theta) < 0$, $x_2(\theta) > 0$, there is a switch in the optimal control at a value

$$t_s = \theta + \frac{x_2(\theta)}{x_1(\theta)} < \theta \qquad\qquad (13)$$

where the argument of the signum function in Eq. (11) passes through zero. Here the optimal trajectories, which must finish with control $u = -1$ in order to intersect the circle (we assume $R < 1$ to avoid the possibility of intersection along a line $u = +1$), can have utilized $u = -1$ for a time equal at most to

$$\theta - t_s = -\frac{x_2(\theta)}{x_1(\theta)} > 0 \qquad\qquad (14)$$

prior to which the control must have been $u = +1$. The point on the parabolic trajectory ending at $x_1(\theta)$, $x_2(\theta)$ corresponding to a time $\theta - t_s$ units earlier is

$$x_1(t_s) = \frac{R^2}{x_1(\theta)} - \frac{1}{2}\frac{x_2{}^2(\theta)}{x_1{}^2(\theta)}$$

$$x_2(t_s) = \frac{x_2(\theta)}{x_1(\theta)}[x_1(\theta) - 1]$$

which then defines the switching curve as $x_1(\theta)$ runs from zero to $-R$, $x_2(\theta)$ from R to zero. The switching curve in the second quadrant is similarly constructed, giving the entire feedback policy.

6.8 THE STRONG MINIMUM PRINCIPLE

We have called the necessary conditions derived thus far the weak minimum principle because, with the experience we have now developed and little more effort, a stronger result can be obtained. Specifically, it is possible to prove that the optimal function $\mathbf{u}(t)$ not only makes the hamiltonian stationary but that it makes the hamiltonian an absolute minimum. This latter result is of importance in some applications.

The essential part of the derivation of the weak minimum principle was in obtaining an infinitesimal change in $\mathbf{x}(\theta)$ and relating this change to the hamiltonian and to the infinitesimal change in $\mathbf{u}$ for all t. It is possible to obtain an infinitesimal change in $\mathbf{x}(\theta)$ (and thus $\mathcal{E}$) by making a change in $\mathbf{u}$ of finite magnitude if that change is made over a sufficiently small time interval, and the hamiltonian enters in a different way.

Let us suppose that the optimal decision function $\bar{\mathbf{u}}(t)$ and optimal state vector $\bar{\mathbf{x}}(t)$ are available and that at time t we have effected a change in the state such that

$$\mathbf{x}(t_1) = \bar{\mathbf{x}}(t_1) + \delta\mathbf{x}(t_1) \qquad |\delta\mathbf{x}(t_1)| \le \epsilon \tag{1}$$

If in the interval $t_1 < t \le \theta$ we employ the optimal decision $\bar{\mathbf{u}}(t)$, we may write

$$\dot{x}_i = f_i(\mathbf{x},\mathbf{u}) = f_i(\bar{\mathbf{x}},\bar{\mathbf{u}}) + \sum_{j=1}^{n} \frac{\partial f_i}{\partial x_j} \delta x_j + o(\epsilon) \tag{2a}$$

or

$$\delta\dot{x}_i = \sum_{j=1}^{n} \frac{\partial f_i}{\partial x_j} \delta x_j + o(\epsilon) \qquad t_1 < t \le \theta \tag{2b}$$

The equation is of this form because $\delta\mathbf{u} = \mathbf{0}$, $t_1 < t \le \theta$. Green's identity then becomes

$$\sum_{i=1}^{n} \gamma_i(\theta)\, \delta x_i(\theta) = \sum_{i=1}^{n} \gamma_i(t_1)\, \delta x_i(t_1) + o(\epsilon) \tag{3}$$

where $\boldsymbol{\gamma}$ satisfies the equations

$$\dot{\gamma}_i = -\sum_{j=1}^{n} \gamma_j \frac{\partial f_j}{\partial x_i} \tag{4}$$

Next we presume that for all time earlier than $t_1 - \Delta$ we make no changes at all. Thus, $\mathbf{x}(t_1 - \Delta) = \bar{\mathbf{x}}(t_1 - \Delta)$, or

$$\delta\mathbf{x}(t_1 - \Delta) = 0 \tag{5}$$

During the interval $t_1 - \Delta < t < t_1$ we set

$$\mathbf{u}(t) = \bar{\mathbf{u}}(t) + \delta\mathbf{u} \qquad t_1 - \Delta < t < t_1 \tag{6}$$

where $|\delta\mathbf{u}|$ is finite. It follows then by direct integration that

$$x_i(t) = \bar{x}_i(t_1 - \Delta) + \int_{t_1-\Delta}^{t} f_i(\mathbf{x}, \bar{\mathbf{u}} + \delta\mathbf{u})\, dt \qquad t_1 - \Delta \le t \le t_1 \tag{7}$$

while the optimal values may be written

$$\bar{x}_i(t) = \bar{x}_i(t_1 - \Delta) + \int_{t_1 - \Delta}^{t} f_i(\bar{\mathbf{x}}, \bar{\mathbf{u}}) \, dt \qquad t_1 - \Delta \le t \le t_1 \tag{8}$$

The variation in the state is then

$$\delta x_i(t) = x_i(t) - \bar{x}_i(t) = \int_{t_1 - \Delta}^{t} [f_i(\mathbf{x}, \bar{\mathbf{u}} + \delta\mathbf{u}) - f_i(\bar{\mathbf{x}}, \bar{\mathbf{u}}) \, dt$$
$$t_1 - \Delta \le t \le t_1 \tag{9}$$

and we may note a useful bound on its magnitude,

$$|\delta x_i(t)| \le \max_{t_1 - \Delta < t < t_1} |f_i(\mathbf{x}, \bar{\mathbf{u}} + \delta\mathbf{u}) - f_i(\bar{\mathbf{x}}, \bar{\mathbf{u}})|(t - t_1 + \Delta)$$
$$t_1 - \Delta \le t \le t_1 \tag{10}$$

We are now able to relate the change in the objective to the finite change $\delta\mathbf{u}$. Truncating a Taylor series after one term, we write

$$f_i(\mathbf{x}, \bar{\mathbf{u}} + \delta\mathbf{u}) = f_i(\bar{\mathbf{x}}, \bar{\mathbf{u}} + \delta\mathbf{u}) + \sum_{j=1}^{n} \frac{\partial f_i}{\partial x_j} \delta x_j \tag{11}$$

where the partial derivatives are evaluated somewhere between $\mathbf{x}$ and $\bar{\mathbf{x}}$. Because Eq. (10) establishes $\delta\mathbf{x}$ as being of the same order as Δ, any integral of $\delta\mathbf{x}$ over an interval Δ must be of order Δ^2. Thus

$$\int_{t_1 - \Delta}^{t_1} \sum_{j=1}^{n} \frac{\partial f_i}{\partial x_j} \delta x_j \, dt = o(\Delta) \tag{12}$$

and Eq. (9) becomes, at $t = t_1$,

$$\delta x_i(t_1) = \int_{t_1 - \Delta}^{t_1} [f_i(\bar{\mathbf{x}}, \bar{\mathbf{u}} + \delta\mathbf{u}) - f_i(\bar{\mathbf{x}}, \bar{\mathbf{u}})] \, dt + o(\Delta) \tag{13}$$

Furthermore, if the region $t_1 - \Delta < t < t_1$ includes no points of discontinuity of $\mathbf{u}$, the integrand is continuous and we can write the exact relationship as

$$\delta x_i(t_1) = [f_i(\bar{\mathbf{x}}, \bar{\mathbf{u}} + \delta\mathbf{u}) - f_i(\bar{\mathbf{x}}, \bar{\mathbf{u}})] \Delta + o(\Delta) \tag{14}$$

and Eq. (3) as

$$\sum_{i=1}^{n} \gamma_i(\theta) \, \delta x_i(\theta) = \sum_{i=1}^{n} [\gamma_i f_i(\bar{\mathbf{x}}, \bar{\mathbf{u}} + \delta\mathbf{x}) - \gamma_i f_i(\bar{\mathbf{x}}, \bar{\mathbf{u}})]\Big|_{t=t_1} \Delta + o(\Delta) \tag{15}$$

By imposing on γ the boundary conditions of Eq. (12) of Sec. 6.4, Eq. (15) becomes

$$\delta\mathcal{E} = \sum_{i=1}^{n} [\gamma_i f_i(\bar{\mathbf{x}}, \bar{\mathbf{u}} + \delta\mathbf{u}) - \gamma_i f_i(\bar{\mathbf{x}}, \bar{\mathbf{u}})]\Big|_{t=t_1} \Delta + o(\Delta) \tag{16}$$

relating the finite change in **u** to the corresponding infinitesimal change in ε.

The condition of optimality, $\delta\varepsilon \geq 0$, requires

$$\Delta \sum_{i=1}^{n} \gamma_i f_i(\bar{\mathbf{x}}, \bar{\mathbf{u}} + \boldsymbol{\delta}\mathbf{u}) + o(\Delta) \geq \Delta \sum_{i=1}^{n} \gamma_i f_i(\bar{\mathbf{x}},\bar{\mathbf{u}}) \qquad t = t_1 \qquad (17)$$

or, dividing by the positive number Δ and taking the limit as $\Delta \to 0$,

$$\sum_{i=1}^{n} \gamma_i f_i(\bar{\mathbf{x}}, \bar{\mathbf{u}} + \boldsymbol{\delta}\mathbf{u}) \geq \sum_{i=1}^{n} \gamma_i f_i(\bar{\mathbf{x}},\bar{\mathbf{u}}) \tag{18}$$

where t_1 is any point of continuity of $\bar{\mathbf{u}}$. The sum in Eq. (18) is the hamiltonian

$$H = \sum_{i=1}^{n} \gamma_i f_i \tag{19}$$

and so we have established that:

> *The hamiltonian takes on its absolute minimum value when evaluated for the optimal decision function* $\bar{\mathbf{u}}(t)$.

This result is significantly stronger than the condition that the hamiltonian be a minimum at boundaries and simply stationary for interior values of the optimal decision, and we call this, together with all the conditions of Sec. 6.5, the *strong minimum principle*.

6.9 THE STRONG MINIMUM PRINCIPLE: A SECOND DERIVATION

Before applying the strong minimum principle to some practical problems, it is useful for future comparisons to consider a second derivation of the strong minimum principle based on Picard's method of the solution of ordinary differential equations. We return to the idea of a continuous infinitesimal variation $\boldsymbol{\delta}\mathbf{u}(t)$, $|\boldsymbol{\delta}\mathbf{u}| \leq \epsilon$, $0 < t \leq \theta$, and, as in Sec. 6.3, we write the equations describing the corresponding infinitesimal variation $\boldsymbol{\delta}\mathbf{x}$. Now, however, we retain terms to the second order in ϵ

$$\delta\dot{x}_i = \sum_{j=1}^{n} \frac{\partial f_i}{\partial x_j} \delta x_j + \sum_{k=1}^{R} \frac{\partial f_i}{\partial u_k} \delta u_k + \frac{1}{2} \sum_{j,k=1}^{n} \frac{\partial^2 f_i}{\partial x_j \, \delta x_k} \delta x_j \, \delta x_k$$

$$+ \sum_{j=1}^{n} \sum_{k=1}^{R} \frac{\partial^2 f_i}{\partial x_j \, \partial u_k} \delta x_j \, \delta u_k + \frac{1}{2} \sum_{j,k=1}^{R} \frac{\partial^2 f_i}{\partial u_j \, \partial u_k} \delta u_j \, \delta u_k + o(\epsilon^2) \quad (1)$$

The essence of Picard's method is first to integrate Eq. (1) by considering the second-order terms as nonhomogeneous forcing terms in a linear equation. In this light we introduce the appropriate Green's functions $\Gamma_{ik}(t,\tau)$ as defined by Eq. (12) of Sec. 6.2

$$\dot{\Gamma}_{ij}(t,\tau) = -\sum_{k=1}^{n} \Gamma_{ik}(t,\tau) \frac{\partial f_k}{\partial x_j} \tag{2a}$$

$$\Gamma_{ij}(t,t) = \delta_{ij} = \begin{cases} 1 & i = j \\ 0 & i \neq j \end{cases} \tag{2b}$$

Then Green's identity becomes

$$\delta x_i(t) = \sum_{j=1}^{n} \Gamma_{ij}(t,0)\, \delta x_j(0) + \int_0^t \sum_{j=1}^{n} \sum_{k=1}^{R} \Gamma_{ij}(t,s) \frac{\partial f_j}{\partial u_k} \delta u_k\, ds$$

$$+ \frac{1}{2} \int_0^t \sum_{j=1}^{n} \sum_{k,l=1}^{R} \Gamma_{ij}(t,s) \frac{\partial^2 f_j}{\partial u_k\, \partial u_l} \delta u_k\, \delta u_l\, ds$$

$$+ \frac{1}{2} \int_0^t \sum_{j,k,l=1}^{n} \Gamma_{ij}(t,s) \frac{\partial^2 f_j}{\partial x_k\, \partial x_l} \delta x_k\, \delta x_l\, ds$$

$$+ \int_0^t \sum_{j,k=1}^{n} \sum_{l=1}^{R} \Gamma_{ij}(t,s) \frac{\partial^2 f_j}{\partial x_k\, \partial u_l} \delta x_k\, \delta u_l\, ds + o(\epsilon^2) \qquad 0 \leq t \leq \theta \tag{3}$$

which is an integral equation for $\boldsymbol{\delta x}$. We next substitute $\boldsymbol{\delta x}(t)$ from Eq. (3) into the last two terms of Eq. (3). For simplicity we set $\boldsymbol{\delta x}(0)$ to zero, either as a special variation or because of fixed initial values, and we obtain the explicit representation at $t = \theta$,

$$\delta x_i(\theta) = \int_0^\theta \sum_{j=1}^{n} \sum_{k=1}^{R} \Gamma_{ij}(\theta,s) \frac{\partial f_j}{\partial u_k} \delta u_k\, ds$$

$$+ \frac{1}{2} \int_0^\theta \sum_{j=1}^{n} \sum_{k,l=1}^{R} \Gamma_{ij}(\theta,s) \frac{\partial^2 f_j}{\partial u_k\, \partial u_l} \delta u_k\, \delta u_l\, ds$$

$$+ \int_0^\theta \sum_{j,l,m=1}^{n} \sum_{k,r=1}^{R} \Gamma_{ij}(\theta,s) \frac{\partial^2 f_j}{\partial u_k\, \partial x_l} \delta u_k \left[\int_0^s \Gamma_{lm}(s,\sigma) \frac{\partial f_m}{\partial u_r} \delta u_r\, d\sigma \right] ds$$

$$+ \frac{1}{2} \int_0^\theta \sum_{j,k,l,m,v=1}^{n} \sum_{r,w=1}^{R} \Gamma_{ij}(\theta,s) \frac{\partial^2 f_j}{\partial x_k\, \partial x_l}$$

$$\left[\int_0^s \Gamma_{km}(s,\sigma) \frac{\partial f_m}{\partial u_r} \delta u_r\, d\sigma \right] \left[\int_0^s \Gamma_{lv}(s,\tau) \frac{\partial f_v}{\partial u_w} \delta u_w\, d\tau \right] ds + o(\epsilon^2) \tag{4}$$

where all products of variations higher than the second have been included in $o(\epsilon^2)$.

We now make a special choice of the variation $\boldsymbol{\delta u}(t)$. As we have already established that the hamiltonian is a minimum for noninterior values of the optimal decision, we set δu_i to zero whenever the corresponding decision u_i does not correspond to a stationary value of the hamiltonian. Furthermore, we allow $\boldsymbol{\delta u}$ to be nonzero only over an infinitesimal interval $t_1 - \Delta \leq t \leq t_1$

$$
\delta u_i(t) = \begin{cases} 0 & \dfrac{\partial H}{\partial u_i} \neq 0 \\ y_i(t) \neq 0 & t_1 - \Delta \leq t \leq t_1 \\ 0 & \text{otherwise} \end{cases} \tag{5}
$$

In that case each of the first two terms in Eq. (4) is a single integral and hence of order Δ, while the third and fourth terms involve products of integrals containing $\boldsymbol{\delta u}$ and are of order Δ^2. Thus,

$$
\delta x_i(\theta) = \int_{t_1-\Delta}^{t_1} \sum_{j=1}^{n} \sum_{k=1}^{R} \Gamma_{ij}(\theta,s) \frac{\partial f_j}{\partial u_k} y_k \, ds
$$

$$
+ \frac{1}{2} \int_{t_1-\Delta}^{t_1} \sum_{j=1}^{n} \sum_{k,l=1}^{R} \Gamma_{ij}(\theta,s) \frac{\partial^2 f_j}{\partial u_k \, \partial u_l} y_k y_l \, ds + o(\epsilon^2) + o(\Delta) \tag{6}
$$

If we multiply Eqs. (2) and (6) by a constant, $\gamma_i(\theta)$, and sum over $i = 1$ to n, then, as in Sec. 6.2, we define a vector $\boldsymbol{\gamma}(t)$ satisfying

$$
\dot{\gamma}_i = - \sum_{j=1}^{n} \frac{\partial f_j}{\partial x_i} \gamma_j \qquad 0 \leq t < \theta
$$

and

$$
\sum_{i=1}^{n} \gamma_i(\theta) \, \delta x_i(\theta) = \int_{t_1-\Delta}^{t_1} \sum_{j=1}^{n} \sum_{k=1}^{R} \gamma_j \frac{\partial f_j}{\partial u_k} y_k \, ds
$$

$$
+ \frac{1}{2} \int_{t_1-\Delta}^{t_1} \sum_{j=1}^{n} \sum_{k,l=1}^{R} \gamma_j \frac{\partial^2 f_j}{\partial u_k \, \partial u_l} y_k y_l \, ds + o(\epsilon^2) + o(\Delta) \tag{7}
$$

We shall assume θ to be fixed for simplicity. The variation in $\mathcal{E}$ caused by the change $\boldsymbol{\delta x}(\theta)$ is then, to second order,

$$
\delta\mathcal{E} = \sum_{i=1}^{n} \frac{\partial \mathcal{E}}{\partial x_i} \delta x_i(\theta) + \frac{1}{2} \sum_{i,j=1}^{n} \frac{\partial^2 \mathcal{E}}{\partial x_i \, \partial x_j} \delta x_i(\theta) \, \delta x_j(\theta) + o(\epsilon^2) \tag{8}
$$

But, from Eq. (3),

$$
\delta x_i(\theta) \, \delta x_j(\theta) = o(\Delta) + o(\epsilon^2) \tag{9}
$$

so that

$$
\delta\mathcal{E} = \sum_{i=1}^{n} \frac{\partial \mathcal{E}}{\partial x_i} \delta x_i(\theta) + o(\Delta) + o(\epsilon^2) \tag{10}
$$

Thus, if for simplicity we assume $\mathbf{x}(\theta)$ to be unspecified and write

$$\gamma_i(\theta) = \frac{\partial \mathcal{E}}{\partial x_i} \tag{11}$$

and introduce the hamiltonian, the combination of Eqs. (10), (11), and (7) becomes

$$\delta \mathcal{E} = \int_{t_1-\Delta}^{t_1} \sum_{i=1}^{n} \frac{\partial H}{\partial u_i} y_i \, ds + \frac{1}{2} \int_{t_1-\Delta}^{t_1} \sum_{i,j=1}^{n} \frac{\partial^2 H}{\partial u_i \, \partial u_j} y_i y_j \, ds + o(\epsilon^2) + o(\Delta) \tag{12}$$

The first integral vanishes by virtue of the arguments used in the derivation of the weak minimum principle. The nonnegativity of the variation $\delta \mathcal{E}$ requires that

$$\sum_{i,j=1}^{n} \frac{\partial^2 H}{\partial u_i \, \partial u_j} y_i y_j \geq 0 \tag{13}$$

for arbitrary (small) $\mathbf{y}$ and, in fact, equality in Eq. (13) is an exceptional case which cannot occur when the minimum is taken on at discrete points, so that Eq. (13), taken together with the weak minimum principle, implies that the hamiltonian is a (local) minimum. If equality should obtain in Eq. (13), a consideration of higher-order terms would lead to the same conclusion.

This derivation leads to weaker results than that of the previous section. Rather than establishing, as in Sec. 6.8, that the minimizing function $\mathbf{u}(t)$ causes the hamiltonian to take on its absolute minimum, we have shown here only that the hamiltonian is a local minimum. The extension to constrained end points and variable θ is straightforward. The particular usefulness of the somewhat weaker result of this section will become apparent in considering sufficiency and in the discussion of discrete systems in the next chapter.

6.10 OPTIMAL TEMPERATURES FOR CONSECUTIVE REACTIONS

In Sec. 4.12 we applied the weak minimum principle to the problem of determining the optimal temperature program for the consecutive-reaction scheme

$$X_1 \rightarrow X_2 \rightarrow \text{decomposition products}$$

We return to that problem now to demonstrate the usefulness of the strong principle and to impose a limitation on the solution we have obtained.

The system is described by equations

$$\dot{x}_1 = -k_1(u)F(x_1) \tag{1a}$$
$$\dot{x}_2 = \nu k_1(u)F(x_1) - k_2(u)G(x_2) \tag{1b}$$

where F and G must be positive functions for physical reasons and k_1 and k_2 have the form

$$k_i(u) = k_{i0}e^{-E_i'/u} \qquad i = 1, 2 \tag{2}$$

The goal is to maximize $x_2(\theta)$ or, equivalently, to minimize $-x_2(\theta)$, with $x_1(\theta)$ unspecified. Thus, the hamiltonian is

$$H = -\gamma_1 k_1 F + \gamma_2 \nu k_1 F - \gamma_2 k_2 G \tag{3}$$

with multiplier equations and boundary conditions

$$\dot{\gamma}_1 = -\frac{\partial H}{\partial x_1} = (\gamma_1 - \nu\gamma_2)k_1 F' \qquad \gamma_1(\theta) = 0 \tag{4a}$$

$$\dot{\gamma}_2 = -\frac{\partial H}{\partial x_2} = \gamma_2 k_2 G' \qquad \gamma_2(\theta) = -1 \tag{4b}$$

Equation (4b) may be integrated to give the important result

$$\gamma_2(t) = -\exp\left(\int_\theta^t k_2 G' \, ds\right) < 0 \tag{5}$$

If we assume that the optimal temperature function $u(t)$ is unconstrained, the conditions for a minimum are

$$\frac{\partial H}{\partial u} = -\gamma_1 k_1' F + \gamma_2 \nu k_1' F - \gamma_2 k_2' G = 0 \tag{6}$$

$$\frac{\partial^2 H}{\partial u^2} = -\gamma_1 k_1'' F + \gamma_2 \nu k_1'' F - \gamma_2 k_2'' G \geq 0 \tag{7}$$

Equation (6) can be solved for γ_1 as

$$\gamma_1 = \nu\gamma_2 - \gamma_2 \frac{k_2' G}{k_1' F} \tag{8}$$

in which case Eq. (7) becomes

$$\gamma_2 G\left(\frac{k_1'' k_2'}{k_1'} - k_2''\right) \geq 0 \tag{9}$$

or, making use of the negativity of γ_2 and the positivity of G, a necessary condition for optimality is

$$\frac{k_1'' k_2'}{k_1'} - k_2'' \leq 0 \tag{10}$$

Now, from the defining equation (2) it follows that

$$k_i' = \frac{k_{i0}E_i'}{u^2} e^{-E_i'/u} \tag{11a}$$

$$k_{ij}'' = \frac{k_{i0}E_i'^2}{u^4} e^{-E_i'/u} - \frac{2k_{i0}E_i'}{u^3} e^{-E_i'/u} \tag{11b}$$

so that Eq. (10) becomes, after some simplification,

$$\frac{k_{20}E_2'}{u^4} e^{-E_2'/u}(E_1 - E_2) \leq 0 \tag{12}$$

or, eliminating the positive coefficient,

$$E_1 \leq E_2 \tag{13}$$

That is, the solution derived in Sec. 4.12 is optimal only when $E_1 \leq E_2$. When $E_1 > E_2$, the optimum is the highest possible temperature.

6.11 OPTIMALITY OF THE STEADY STATE

It has been found recently that certain reaction and separation processes which are normally operated in the steady state give improved performance when deliberately forced to operate in a time-varying manner, where the measure of performance is taken to be a time-averaged quantity such as conversion. Horn has shown a straightforward procedure for determining under certain conditions when improved performance may be expected in the unsteady state by the obvious but profound observation that if the problem is posed as one of finding the time-varying operating conditions to optimize a time-averaged quantity, the best steady state can be optimal only if it satisfies the necessary conditions for the time-varying problem. As we shall see, the application of this principle requires the strong minimum principle.

We shall consider an example of Horn and Lin of parallel chemical reactions carried out in a continuous-flow stirred-tank reactor. The reactions are

where the reaction $X_1 \rightarrow X_2$ is of order n and the reaction $X_1 \rightarrow X_3$ is of first order. X_2 is the desired product, and the goal is to choose the operating temperature which will maximize the amount of X_2. The reactor equations are

$$\dot{x}_1 = -ux_1{}^n - au^rx_1 - x_1 + 1 \tag{1a}$$
$$\dot{x}_2 = ux_1{}^n - x_2 \tag{1b}$$

where x_1 and x_2 are dimensionless concentrations, r is the ratio of activation energies of the second reaction to the first, u is the decision variable, the temperature-dependent rate coefficient of the first reaction, and time is dimensionless.

For steady-state operation the time derivatives are zero, and we obtain a single equation in x_2 and u by solving Eq. (1b) and substituting into (1a)

$$0 = -x_2 - x_2{}^{1/n}(au^{r-1/n} + u^{-1/n}) + 1 \tag{2}$$

As we wish to maximize x_2 by choice of u, we differentiate Eq. (2) with respect to u to obtain

$$0 = -\frac{\partial x_2}{\partial u} - \frac{1}{n} x_2{}^{1/n-1} \frac{\partial x_2}{\partial u} (au^{r-1/n} + u^{-1/n})$$
$$- x_2{}^{1/n}\left(a \frac{nr-1}{n} u^{(nr-1-n)/n} - \frac{1}{n} u^{-(1+n)/n}\right) \tag{3}$$

and since $\partial x_2/\partial u$ must be zero at an internal maximum, solving for u leads to the optimum steady-state value

$$u = \left[\frac{1}{a(nr-1)}\right]^{1/r} \tag{4}$$

Clearly this requires

$$nr - 1 > 0 \tag{5}$$

To establish that Eq. (4) does indeed lead to a maximum we take the second derivative of Eq. (2) with respect to u to obtain, after using Eq. (4) and the vanishing of $\partial x_2/\partial u$,

$$0 = -\frac{\partial^2 x_2}{\partial u^2}\left(1 + x_2{}^{1/n-1}u^{-1/n} \frac{r}{nr-1}\right) - x_2{}^{1/n} \frac{r}{n} u^{-(1+2n)/n} \tag{6}$$

Since u and x_2 are both positive, it follows at once that $\partial^2 x_2/\partial u^2$ is negative and that a maximum is obtained.

If we formulate the problem as a dynamical one, our goal is to choose $u(t)$ in the interval $0 \le t \le \theta$ in order to maximize the time-average value of $x_2(t)$ or, equivalently, minimize the negative of the time-average value. Thus,

$$\varepsilon = -\frac{1}{\theta} \int_0^\theta x_2(t)\ dt \tag{7}$$

or, using the equivalent formulation which we have developed,

$$\mathcal{F} = -\frac{x_2}{\theta} \tag{8}$$

Using Eqs. (1a) and (1b), the hamiltonian is then

$$H = -\frac{x_2}{\theta} + \gamma_1(-ux_1{}^n - au^r x_1 - x_1 + 1) + \gamma_2(ux_1{}^n - x_2) \tag{9}$$

The multiplier equations are

$$\dot{\gamma}_1 = -\frac{\partial H}{\partial x_1} = \gamma_1(nux_1{}^{n-1} + au^r + 1) - n\gamma_2 ux_1{}^{n-1} \tag{10a}$$

$$\dot{\gamma}_2 = -\frac{\partial H}{\partial x_2} = \frac{1}{\theta} + \gamma_2 \tag{10b}$$

and the partial derivatives of H with respect to u are

$$\frac{\partial H}{\partial u} = -\gamma_1 x_1{}^n - a\gamma_1 r u^{r-1} x_1 + \gamma_2 x_1{}^n \tag{11}$$

$$\frac{\partial^2 H}{\partial u^2} = -a\gamma_1 r(r-1)u^{r-2} x_1 \tag{12}$$

Since we are testing the steady state for optimality, it follows that all time derivatives must vanish and the Green's functions γ_1 and γ_2 must also be constant. Thus, from Eqs. (10),

$$\gamma_1 = -\frac{nux_1{}^{n-1}}{\theta(nux_1{}^{n-1} + au^r + 1)} \tag{13a}$$

$$\gamma_2 = -\frac{1}{\theta} \tag{13b}$$

(Clearly we are excluding small initial and final transients here.) Setting $\partial H/\partial u$ to zero in Eq. (11) and using Eq. (13) leads immediately to the solution

$$u = \left[\frac{1}{a(nr-1)}\right]^{1/r} \tag{14}$$

the optimal steady state, so that the optimal steady state does satisfy the first necessary condition for dynamic optimality. However, it follows from Eq. (13a) that γ_1 is always negative, so that $\partial^2 H/\partial u^2$ in Eq. (12) has the algebraic sign of $r - 1$. Thus, when $r > 1$, the second derivative is positive and the hamiltonian is a minimum, as required. When $r < 1$, however, the second derivative is negative and the hamiltonian is a local maximum. Thus, for $r < 1$ the best steady-state operation can always be improved upon by dynamic operation.

We shall return to this problem in a later chapter with regard to the computational details of obtaining a dynamic operating policy.   We leave as a suggestion to the interested reader the fruitfulness of investigating the relation between steady-state optimality and minimization of the steady-state lagrangian, with emphasis on the meaning of steady-state Lagrange multipliers.

6.12 OPTIMAL OPERATION OF A CATALYTIC REFORMER

One of the interesting applications of the results of this chapter to an industrial process has been Pollock's recent preliminary study of the operation of a catalytic reformer.   The reforming process takes a feed-stock of hydrocarbons of low octane number and carries out a dehydro-genation over a platinum-halide catalyst to higher octane product.   A typical reaction is the conversion of cyclohexane, with octane number of 77, to benzene, with an octane number of over 100.

A simple diagram of the reformer is shown in Fig. 6.1.   The feed is combined with a hydrogen recycle gas stream and heated to about 900°F, then passed into the first reactor.   The dehydrogenation reaction is endo-thermic, and the effluent stream from the first reactor is at about 800°F. This is reheated and passed to the second reactor, where the temperature drop is typically half that in the first reactor.   The stream is again heated and passed through the final reactor, where the temperature varies only slightly and may even rise.   Finally, the stream is passed to a separator for recovery of hydrogen gas, and the liquid is debutanized to make the

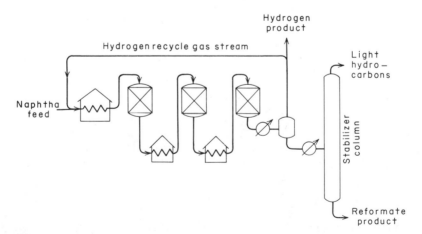

Fig. 6.1 Schematic of a catalytic reforming process.   (*Courtesy of A. W. Pollock.*)

reformate product, which may have an octane number as much as 60 greater than the feed.

Numerous side reactions deposit coke on the catalyst, reducing its efficiency, and it becomes necessary eventually to shut down the reactor and regenerate the catalyst. The problem posed by Pollock, then, was to determine how to operate the reformer and when to shut down in order to maximize the profit over the total period, including both operating time and downtime.

The decision variable is taken by means of a simplified model to be the desired octane number of the product, which is adjusted by an operator. For convenience we define u as the octane number less 60. Pollock has used estimates from plant and literature data to write the equation for coke accumulation, the single state variable, as

$$\dot{x}_1 = b_2 u^\beta \qquad x_1(0) = 0 \tag{1}$$

and the time-average profit as proportional to

$$\mathcal{P} = \frac{1}{\theta + \tau} \int_0^\theta (B + u)[1 - (b_0 + b_1 x_1)u^2] \, dt - \frac{Q - A\theta}{\theta + \tau} \tag{2}$$

Here τ is the time needed for regeneration, $B + u$ the difference in value between product and feed divided by the (constant) marginal return for increased octane, $1 - (b_0 + b_1 x_1)u^2$ the fractional yield of product, Q the fixed cost of regeneration divided by flow rate and marginal return, and A the difference in value between other reaction products and feed divided by marginal return. We refer the reader to the original paper for the construction of the model, which is at best very approximate. The following values of fixed parameters were used for numerical studies:

$$b_0 = 10^{-4} \qquad b_1 = 2 \times 10^{-8}$$
$$b_2 = 35^{-\beta+1} \qquad \tau = 2$$

The objective equation (2) is not in the form which we have used, and to convert it we require two additional state variables, defined as

$$\dot{x}_2 = (B + u)[1 - (b_0 + b_1 x_1)u^2] \qquad x_2(0) = 0 \tag{3a}$$
$$\dot{x}_3 = 1 \qquad x_3(0) = 0 \tag{3b}$$

In that case we can express the objective as one of minimizing

$$\mathcal{E} = \frac{Q - A x_3(\theta) - x_2(\theta)}{x_3(\theta) + \tau} \tag{4}$$

The hamiltonian is then

$$H = \gamma_1 b_2 u^\beta + \gamma_2 (B + u)[1 - (b_0 + b_1 x_1)u^2] + \gamma_3 \tag{5}$$

with multiplier equations

$$\dot{\gamma}_1 = -\frac{\partial H}{\partial x_1} = -b_1 u^2 (B + u)\gamma_2 \qquad \gamma_1(\theta) = \frac{\partial \mathcal{E}}{\partial x_1} = 0 \tag{6a}$$

$$\dot{\gamma}_2 = -\frac{\partial H}{\partial x_2} = 0 \qquad \gamma_2(\theta) = \frac{\partial \mathcal{E}}{\partial x_2} = -\frac{1}{x_3(\theta) + \tau} \tag{6b}$$

$$\gamma_3 = -\frac{\partial H}{\partial x_3} = 0 \qquad \gamma_3(\theta) = \frac{\partial \mathcal{E}}{\partial x_3} = -\frac{Q - A\tau - x_2(\theta)}{[x_3(\theta) + \tau]^2} \tag{6c}$$

It is convenient to define

$$Z = \gamma_1(\theta + \tau) \tag{7}$$

which, together with Eqs. (3), (5), and (6), leads to

$$H = -\frac{1}{\theta + \tau}\left\{ Z b_2 u^\beta + (B + u)[1 - (b_0 + b_1 x_1)u^2 \right.$$
$$\left. + \frac{Q - A\tau - x_2(\theta)}{\theta + \tau}\right\} \tag{8}$$

$$\dot{Z} = b_1 u^2 (B + u) \qquad Z(\theta) = 0 \tag{9}$$

Eliminating constant terms, minimizing the hamiltonian is equivalent to *maximizing*

$$H^* = Z b_2 u^\beta + (B + u)[1 - (b_0 + b_1 x_1)u^2] \tag{10}$$

The optimal θ is determined by the vanishing of the hamiltonian

$$(\theta + \tau)(B + u)[1 - (b_0 + b_1 x_1)u^2]$$
$$= \int_0^\theta (B + u)[1 - (b_0 + b_1 x_1)u^2]\, dt - (Q - A\tau) \qquad t = \theta \tag{11}$$

Because of the preliminary nature of the model and the uncertainty of the parameters, particularly A, B, Q, and β, a full study was not deemed warranted. Instead, in order to obtain some idea of the nature of the optimal policy, the value of Z at $t = 0$ was arbitrarily set at -0.1. Equations (1) and (9) were then integrated numerically from $t = 0$, choosing u at each time instant by maximizing Eq. (10), subject to constraints

$$30 \leq u \leq 40 \tag{12}$$

(octane number between 90 and 100). θ was determined by the point at which Z reached zero. Equation (11) then determined the values of Q and A for which the policy was optimal. A complete study would require a one-parameter search over all values $Z(0)$ to obtain optimal policies for all Q and A. Figures 6.2, 6.3, and 6.4 show the optimal policies for $\beta = 1$, 2, and 4, respectively, with values of B from 30 to

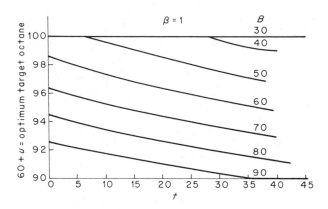

Fig. 6.2 Optimum target octane schedule for $\beta = 1$. (*Courtesy of A. W. Pollock.*)

90.   The near constancy of the policies for $\beta = 2$ is of particular interest, for conventional industrial practice is to operate at constant target octane.

The remarkable agreement for some values of the parameters between Pollock's preliminary investigation of optimal policies and industrial practice suggests further study. We consider, for example, the possibility that constant octane might sometimes represent a rigorously optimal policy. Differentiating H^* in Eq. (10) with respect to u and setting the derivative to zero for an internal maximum yields, after some rearrangement,

$$x_1 = \frac{\beta b_2 u^{\beta - 1}}{b_1[(B + 2)u^2 + 2Bu]} Z + \frac{B - b_0[(B + 2)u^2 + 2Bu]}{b_1[(B + 2)u^2 + 2Bu]} \qquad (13)$$

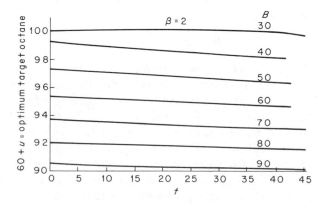

Fig. 6.3 Optimum target octane schedule for $\beta = 2$. (*Courtesy of A. W. Pollock.*)

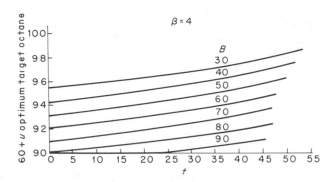

Fig. 6.4 Optimum target octane schedule for $\beta = 4$. (*Courtesy of A. W. Pollock.*)

Now, if u is a constant, then differentiating Eq. (13) with respect to time

$$\dot{x}_1 = \frac{\beta b_2 u^{\beta-1}}{b_1[(B + 2)u^2 + 2Bu]} \dot{Z} \tag{14}$$

and combining this result with Eqs. (1) and (9), we obtain a value for u

$$u = \frac{B(\beta - 2)}{B + 2 - \beta} \tag{15}$$

We test for a maximum of H^* by calculating the second derivative

$$\frac{\partial^2 H^*}{\partial u^2} = -2(b_0 + b_1 x_1)[(B + 2)u + B] + Z b_2 \beta(\beta - 1)u^{\beta-2} \tag{16}$$

Since $Z < 0$, it follows that we always have a maximum for $u \geq 0$, or, from Eq. (15),

$$2 \leq \beta < B + 2 \tag{17}$$

Equation (11) is easily shown to be satisfied for positive θ.

For β near 2 and large B (small marginal return), the rigorously constant policy corresponds to very small u, which is verified by examination of Fig. 6.3. For small values of B (B near $\beta - 2$) constant values of u of the order of 35 to 40 can be accommodated. For very large β ($\beta > 10$) and large B corresponding large values of u are also obtained. It is clear from Figs. 6.2 to 6.4 that only minor changes in shape occur over wide ranges of B, so that essentially constant policies will generally be expected. Thus, we conclude that within the accuracy of Pollock's model the industrial practice of constant-octane operation is close to optimal for a significant range of parameters.

6.13 THE WEIERSTRASS CONDITION

The minimum principle is a generalization of a well-known necessary condition of Weierstrass in the calculus of variations. In Sec. 4.4 we considered the problem of finding the function $x(t)$ which minimized the integral

$$\mathcal{E} = \int_0^\theta \mathcal{F}(x,\dot{x},t)\, dt \tag{1}$$

where the hamiltonian was found to be

$$H = \mathcal{F} + \lambda_1 \dot{x} + \lambda_2 \tag{2}$$

and $\dot{x}$ is the decision variable. The minimum condition is then

$$\mathcal{F}(\bar{x},\dot{\bar{x}},t) + \lambda_1 \dot{\bar{x}} \leq \mathcal{F}(\bar{x},\dot{x},t) + \lambda_1 \dot{x} \tag{3}$$

where $\bar{x}$ denotes the optimal function. From the stationary condition Eq. (7) of Sec. 4.4,

$$\lambda_1 = -\frac{\partial \mathcal{F}}{\partial \dot{x}} \tag{4}$$

or

$$\mathcal{F}(\bar{x},\dot{\bar{x}},t) - \mathcal{F}(\bar{x},\dot{x},t) - \frac{\partial \mathcal{F}}{\partial \dot{x}}(\dot{\bar{x}} - \dot{x}) \leq 0 \tag{5}$$

which is the *Weierstrass inequality*.

The somewhat weaker condition that the second derivative of the hamiltonian be positive for a minimum becomes, here,

$$\frac{\partial^2 \mathcal{F}}{\partial \dot{x}^2} \geq 0 \tag{6}$$

which is known as the *Legendre condition*. Both these results are generalized to situations with differential-equation side conditions.

6.14 NECESSARY CONDITION FOR SINGULAR SOLUTIONS

In Chap. 5 we considered several applications in which a portion of the optimal solution lay along a singular curve. These were internal optima in problems in which the decision appeared linearly, so that first derivatives of the hamiltonian were independent of the decision. Consequently the second-derivative test for a minimum cannot be applied, and different conditions are needed. To this end we return to the analysis in Sec. 6.9.

Our starting point is Eq. (3) of Sec. 6.9, which we multiply by $\gamma_i(\theta)$

and sum over i to obtain

$$\delta\mathcal{E} = \frac{1}{2} \int_0^\theta \sum_{k,l} \frac{\partial^2 H}{\partial x_k\,\partial x_l} \delta x_k\,\delta x_l\,dt + \int_0^\theta \sum_k \frac{\partial^2 H}{\partial x_k\,\partial u} \delta x_k\,\delta u\,dt$$
$$+ o(\epsilon^2) \geq 0 \quad (1)$$

Here we have assumed a single decision variable and made use of the fact that $\partial H/\partial u$ is zero for an optimum and $\partial^2 H/\partial u^2$ vanishes for singular solutions. **δx**(0) was taken to be zero, and we have deleted the term

$$\sum_{j,k} \frac{\partial^2 \mathcal{E}}{\partial x_j\,\partial x_k} \delta x_j(\theta)\,\delta x_k(\theta)$$

leaving to the reader the task of demonstrating that for the special variation which we shall choose this term is of negligible magnitude compared to the terms retained. We now assume the symmetric special variation

$$\delta u = \begin{cases} +\epsilon & t_0 < t < t_0 + \Delta \\ -\epsilon & t_0 - \Delta < t < t_0 \end{cases} \quad (2)$$

in which case both integrals in Eq. (1) need be evaluated only over $t_0 - \Delta < t < t_0 + \Delta$ to within the specified order. Now we expand the integrands in Taylor series about their values at t_0 as follows:

$$\frac{\partial^2 H}{\partial x_k\,\partial x_l} \delta x_k\,\delta x_l = \left[\frac{\partial^2 H}{\partial x_k\,\partial x_l} + \frac{d}{dt}\frac{\partial^2 H}{\partial x_k\,\partial x_l} (t - t_0) + \cdots \right] \left[\epsilon\Delta \frac{\partial f_k}{\partial u} \right.$$
$$+ \sum_j \left(\epsilon\Delta \frac{\partial f_k}{\partial x_j}\frac{\partial f_j}{\partial u} \pm \epsilon \frac{\partial f_k}{\partial u} \right)(t - t_0) + \cdots \right]\left[\epsilon\Delta \frac{\partial f_l}{\partial u} \right.$$
$$\left. + \sum_j \left(\epsilon\Delta \frac{\partial f_l}{\partial x_j}\frac{\partial f_j}{\partial u} \pm \epsilon \frac{\partial f_k}{\partial u} \right)(t - t_0) + \cdots \right] \quad (3a)$$

$$\frac{\partial^2 H}{\partial x_k\,\partial u} \delta x_k\,\delta u = \pm\,\epsilon \left[\frac{\partial^2 H}{\partial x_k\,\partial u} + \frac{d}{dt}\frac{\partial^2 H}{\partial x_k\,\partial u}(t - t_0) + \cdots \right]\left[\epsilon\Delta \frac{\partial f_k}{\partial u} \right.$$
$$+ \left(\sum_j \epsilon\Delta \frac{\partial f_k}{\partial x_j}\frac{\partial f_j}{\partial u} \pm \epsilon \frac{\partial f_k}{\partial u} \right)(t - t_0) + \frac{d}{dt}\left(\sum_j \epsilon\Delta \frac{\partial f_k}{\partial x_j}\frac{\partial f_j}{\partial u} \right.$$
$$\left. \pm\,\epsilon \frac{\partial f_k}{\partial u} \right)\frac{(t - t_0)^2}{2!} + \cdots \right] \quad (3b)$$

Here all derivatives are evaluated at t_0, and $\delta\dot{x}_k$ has been calculated from the equation

$$\delta\dot{x}_k = \sum_j \frac{\partial f_k}{\partial x_j} \delta x_j + \frac{\partial f_k}{\partial u} \delta u \quad (4)$$

The symbol $\pm$ denotes $+$ for $t > t_0$, $-$ for $t < t_0$.

Substituting Eqs. (3) into Eq. (1) and integrating, we obtain

$$
\delta \mathcal{E} = \frac{\epsilon^2 \Delta^3}{3} \left[\sum_{i,j} \frac{\partial^2 H}{\partial x_i \partial x_j} \frac{\partial f_i}{\partial u} \frac{\partial f_j}{\partial u} + \sum_i \frac{\partial^2 H}{\partial x_i \partial u} \left(\frac{d}{dt} \frac{\partial f_i}{\partial u} - 2 \sum_j \frac{\partial f_i}{\partial x_j} \frac{\partial f_j}{\partial u} \right) \right.
$$
$$
\left. - \sum_i \left(\frac{d}{dt} \frac{\partial^2 H}{\partial x_i \partial u} \right) \frac{\partial f_i}{\partial u} \right] + o(\epsilon^2 \Delta^3) \geq 0 \quad (5)
$$

in which case the term in square brackets must be nonnegative. This may be shown to be equivalent to the statement

$$
\frac{\partial}{\partial u} \left(\frac{d^2}{dt^2} \frac{\partial H}{\partial u} \right) \leq 0 \tag{6}
$$

which is, in fact, a special case of the more general relation which we shall not derive

$$
(-1)^k \frac{\partial}{\partial u} \left(\frac{d^{2k}}{dt^{2k}} \frac{\partial H}{\partial u} \right) \geq 0 \qquad k = 0, 1, 2, \ldots \tag{7}
$$

Consider, for example, the batch- and tubular-reactor temperature-profile problem of Sec. 5.11. The hamiltonian is

$$
H = \rho - r(a,T)(1 + \lambda_1 - J\lambda_2) - \lambda_2 u \tag{8}
$$

where a and T are the state variables, and the singular arc corresponds to $\lambda_2 = 0$, $\partial r / \partial T = 0$, $\partial^2 r / \partial T^2 < 0$, the line of maximum reaction rate. The state equations are

$$
\dot{a} = -r(a,T) \tag{9a}
$$
$$
\dot{T} = Jr(a,T) - u \tag{9b}
$$

so that the bracketed term in Eq. (5) becomes

$$
- \frac{\partial^2 r}{\partial T^2} (1 + \lambda_1 - J\lambda_2) \geq 0 \tag{10}
$$

But $\lambda_2 = 0$, $\partial^2 r / \partial T^2 < 0$, and since the singular arc has been shown to be a final arc with $\lambda_1(\theta) = 0$, it easily follows from Eq. (6a) of Sec. 5.11 that $\lambda_1 > 0$. Thus Eq. (10) is satisfied, and the singular arc does satisfy the further necessary condition.

6.15 MIXED CONSTRAINTS

Our considerations thus far have been limited to processes in which only the decision variables are constrained during operation. In fact, it may be necessary to include limitations on state variables or combinations of state and decision variables. The latter problem is easier, and we shall consider it first.

We suppose that the minimization problem is as described in

Sec. 6.4 but with added restrictions of the form

$$Q_k(\mathbf{x},\mathbf{u}) \geq 0 \qquad k = 1, 2, \ldots, K \tag{1}$$

When every constraint is at strict inequality, Eq. (1) places no restrictions on variations in the optimal decision, and the minimum principle as derived in Sec. 6.8 clearly applies. We need only ask, then, what changes are required when one or more of the constraints is at equality for a finite interval.

It is instructive to consider first the case in which there is a single decision variable u and only one constraint is at equality. Dropping the subscript on the constraint, we then have

$$Q(\mathbf{x},u) = 0 \tag{2}$$

If the state is specified, Eq. (2) determines the decision. Any change in $\mathbf{x}$ or u must satisfy, to first order,

$$\delta Q = \sum_{i=1}^{n} \frac{\partial Q}{\partial x_i} \delta x_i + \frac{\partial Q}{\partial u} \delta u = 0 \tag{3}$$

while, since the differential equations must be satisfied,

$$\delta \dot{x}_i = \sum_{j=1}^{n} \frac{\partial f_i}{\partial x_j} \delta x_j + \frac{\partial f_i}{\partial u} \delta u \tag{4}$$

Substituting for δu from Eq. (3),

$$\delta \dot{x}_i = \sum_{j=1}^{n} \left[\frac{\partial f_i}{\partial x_j} - \frac{\partial f_i}{\partial u} \left(\frac{\partial Q}{\partial u} \right)^{-1} \frac{\partial Q}{\partial x_j} \right] \delta x_j \tag{5}$$

Thus, using the results of Sec. 6.2, along this section of the optimal path the Green's vector must satisfy the equation

$$\dot{\gamma}_i = - \sum_{j=1}^{n} \left[\frac{\partial f_j}{\partial x_i} - \frac{\partial f_j}{\partial u} \left(\frac{\partial Q}{\partial u} \right)^{-1} \frac{\partial Q}{\partial x_i} \right] \gamma_j \tag{6}$$

Although the canonical equations are not valid over the constrained part of the trajectory we define the hamiltonian as before, and it is easily shown that H is constant over the constrained section. If continuity of H is required over the entrance to the constrained section, the multipliers are continuous and H retains the same constant value throughout and, in particular, $H = 0$ when θ is unspecified.

In the general case of K_1 independent constraints at equality involving $R_1 \geq K_1$ components of the decision vector we solve the K_1

equations

$$\delta Q_k = \sum_{i=1}^{R_1} \frac{\partial Q_k}{\partial u_i} \delta u_i + \sum_{i=1}^{n} \frac{\partial Q_k}{\partial x_i} \delta x_i = 0 \qquad k = 1, 2, \ldots, K_1 \qquad (7)$$

for the first K_1 components δu_i. (It is assumed that $\partial Q_k/\partial u_i$ is of rank K_1.) The equation for the Green's vector is then

$$\dot{\gamma}_i = -\sum_{j=1}^{n} \left(\frac{\partial f_j}{\partial x_i} - \sum_{p,k=1}^{K_1} \frac{\partial f_j}{\partial u_p} S_{pk} \frac{\partial Q_k}{\partial x_i} \right) \gamma_j \qquad (8)$$

Here S_{pk} is the inverse of the matrix consisting of the first K_1 elements $\partial Q_k/\partial u_i$; that is,

$$\sum_{k=1}^{K_1} S_{jk} \frac{\partial Q_k}{\partial u_i} = \delta_{ji} \qquad i, j = 1, 2, \ldots, K_1 \qquad (9)$$

Upon substitution into the variational equations it follows that the $R_1 - K_1$ independent components of $\mathbf{u}$ must be chosen to satisfy the weak minimum principle and that H is again constant.

6.16 STATE-VARIABLE CONSTRAINTS

We are now in a position to consider constraints on state variables only. Clearly we need only consider the time intervals when such constraints are actually at equality. For simplicity we consider here only a single active constraint

$$Q(\mathbf{x}) = 0 \qquad (1)$$

The generalization is straightforward.

Let us consider an interval $t_1 \leq t \leq t_2$ during which the optimal trajectory lies along the constraint. Since Eq. (1) holds as an identity, we can differentiate it with respect to t as often as we wish to obtain

$$Q = \dot{Q} = \ddot{Q} = \cdots = Q^{(m)} = 0 \qquad t_1 \leq t \leq t_2 \qquad (2)$$

In particular, we assume that the decision vector $\mathbf{u}$ first enters explicitly in the mth derivative. If we then let

$$Q^{(m)}(\mathbf{x}) = \tilde{Q}(\mathbf{x},\mathbf{u}) = 0 \qquad t_1 \leq t \leq t_2 \qquad (3)$$

we can apply the results of the previous section to the mixed constraint $\tilde{Q}$ in order to obtain the multiplier equations. In addition, however, we now have m additional constraints

$$Q^{(j)} = 0 \qquad j = 0, 1, 2, \ldots, m - 1 \qquad (4)$$

which must be satisfied at $t = t_1$.

If we apply Green's identity across the vanishingly small interval $t_1^- \leq t \leq t_1^+$, we obtain

$$\sum_{i=1}^{n} \gamma_i \, \delta x_i \Big|_{t_1^-} = \sum_{i=1}^{n} \gamma_i \, \delta x_i \Big|_{t_1^+} \tag{5}$$

The Green's vector $\boldsymbol{\gamma}(t_1^-)$ may be expressed as the sum of $m + 1$ vectors, the first m being normal to the surfaces defined by Eq. (4). Thus

$$\gamma_i(t_1^-) = \sum_{j=0}^{m-1} \mu_j \frac{\partial Q^{(j)}}{\partial x_i} + \sigma_i \tag{6}$$

or, from Eq. (5),

$$\sum_{j=0}^{m-1} \sum_{i=1}^{n} \mu_j \frac{\partial Q^{(j)}}{\partial x_i} \, \delta x_i \Big|_{t_1^-} + \sum_{i=1}^{n} \sigma_i \, \delta x_i \Big|_{t_1^-} = \sum_{i=1}^{n} \gamma_i \, \delta x_i \Big|_{t_1^+} \tag{7}$$

But the constraints imply that

$$\delta Q^{(j)} = \sum_{i=1}^{n} \frac{\partial Q^{(j)}}{\partial x_i} \, \delta x_i = 0 \qquad j = 0, 1, \ldots, m - 1 \tag{8}$$

so that Eq. (7) becomes

$$\sum_{i=1}^{n} \sigma_i \, \delta x_i \Big|_{t_1^-} = \sum_{i=1}^{n} \gamma_i \, \delta x_i \Big|_{t_1^+} \tag{9}$$

Since $\delta \mathbf{x}$ is continuous, Eq. (9) has a solution

$$\boldsymbol{\sigma} = \boldsymbol{\gamma}(t_1^+) \tag{10}$$

and the Green's vector may then have a discontinuity of the form

$$\gamma_i(t_1^-) = \gamma_i(t_1^+) + \sum_{j=0}^{m-1} \mu_j \frac{\partial Q^{(j)}}{\partial x_i} \tag{11}$$

This *jump condition* is unique if we require that $\boldsymbol{\gamma}(t_2)$ be continuous, and the m components of $\boldsymbol{\mu}$ are found from the m extra conditions of Eq. (4). As before, the hamiltonian has a continuous constant value along the optimal path.

6.17 CONTROL WITH INERTIA

In all the control problems we have studied we have assumed that instantaneous switching is possible between extremes. A more realistic approximation might be an upper limit on the rate of change of a control setting. Consider again, for example, the optimal-cooling problem of Sec. 5.11.

The state equations are

$$\dot{a} = -r(a,T) \tag{1a}$$
$$\dot{T} = Jr(a,T) - q \tag{1b}$$

where we have written q for the cooling rate. Now we wish to bound the rate of change of q, and so we write a third equation

$$\dot{q} = u \tag{1c}$$

with constraints

$$u_* \leq u \leq u^* \tag{2a}$$
$$0 \leq q \leq q^* \tag{2b}$$

and perhaps a temperature constraint

$$T_* \leq T \leq T^* \tag{2c}$$

The objective is again to minimize

$$\mathcal{E} = \int_0^\theta [\rho - r(a,T)] \, dt \tag{3}$$

We may consider u to be the decision variable, in which case Eqs. (2b) and (2c) denote constraints on the state. The hamiltonian is

$$H = -\gamma_1 r + \gamma_2 (Jr - q) + \gamma_3 u + \rho - r \tag{4}$$

and if neither constraints (2b) nor (2c) are in effect, the multiplier equations are

$$\dot{\gamma}_1 = -\frac{\partial H}{\partial a} = \frac{\partial r}{\partial a}(1 + \gamma_1 - J\gamma_2) \tag{5a}$$

$$\dot{\gamma}_2 = -\frac{\partial H}{\partial T} = \frac{\partial r}{\partial T}(1 + \gamma_1 - J\gamma_2) \tag{5b}$$

$$\dot{\gamma}_3 = -\frac{\partial H}{\partial q} = \gamma_2 \tag{5c}$$

If control is not intermediate, u lies at one of its extremes, so that the optimum is to change controls as fast as possible. If the constraint surface $q = 0$ or $q = q^*$ is reached, the control appears in the first derivative and $m = 1$:

$$Q^{(1)} = \frac{d}{dt}(q^* - q) = \tilde{Q} = -u = 0 \tag{6}$$

Since $\tilde{Q}$ is independent of $\mathbf{x}$, the multiplier equations are unchanged and we simply reduce to the problem of Sec. 5.11.

Intermediate control is possible only if $\gamma_3 = 0$, which, from Eq. (5c), implies $\gamma_2 = 0$. But γ_2 vanishes only along the line of maximum reaction rate, so that the same intermediate policy is optimal. If a temperature constraint is reached, say $T = T^*$, we have

$$Q^{(1)} = \frac{d}{dt}(T^* - T) = -\dot{T} = q - Jr(a,T^*) = 0 \tag{7a}$$

$$Q^{(2)} = \frac{d^2}{dt^2}(T^* - T) = -\ddot{T} = u - J\frac{\partial r}{\partial a}\dot{a} = \tilde{Q} = 0 \tag{7b}$$

so that $m = 2$ and the coolant is varied at a rate

$$u = Jr(a,T^*)\frac{\partial r(a,T^*)}{\partial a} \tag{8}$$

It will be impossible to satisfy equality constraints on both temperature and rate of cooling, as seen by comparing Eqs. (6) and (8), and so some initial states will be excluded.

We wish to emphasize here the manner in which we have developed the optimal policy for this problem. In Sec. 3.5 we studied the optimal temperature policy. We found in Sec. 5.11 that, when possible, the optimal coolant policy reduces to one of choosing the optimal temperature policy. Here we have found that, except when switching, the optimal coolant rate of change is one which gives the optimal coolant policy. This hierarchy of optimization problems is typical of applications, and the physical understanding eliminates some of the need for rigorous establishment of optimality of intermediate policies.

6.18 DISCONTINUOUS MULTIPLIERS

One of the most striking features of the presence of state-variable constraints is the possibility of a discontinuity in the Green's functions. We shall demonstrate this by a problem of the utmost simplicity, a first-order process,

$$\dot{x}_1 = u \tag{1}$$

We assume that x_1 must lie within the region bounded by the straight lines

$$Q_1 = x_1 - (\alpha_1 - \beta_1 t) \geq 0 \tag{2a}$$
$$Q_2 = -x_1 + (\alpha_2 - \beta_2 t) \geq 0 \tag{2b}$$

and that we choose u to minimize

$$\mathcal{E} = \frac{1}{2}\int_0^\theta (x_1{}^2 + c^2 u^2)\, dt \tag{3}$$

while obtaining $x_1(\theta) = 0$. The reader seeking a physical motivation might think of this as a power-limited mixing process in which the constraints are blending restrictions.

It is convenient to define a new state variable

$$\dot{x}_2 = \tfrac{1}{2}(x_1{}^2 + c^2 u^2) \qquad x_2(0) = 0 \tag{4}$$

and, because time enters explicitly into the constraints

$$\dot{x}_3 = 1 \qquad x_3(0) = 0 \qquad x_3(\theta) = \theta \tag{5}$$

We then have

$$\mathcal{E} = x_2(\theta) \tag{6}$$
$$Q_1 = x_1 + \beta_1 x_3 - \alpha_1 \geq 0 \tag{7a}$$
$$Q_2 = -x_1 - \beta_2 x_3 + \alpha_2 \geq 0 \tag{7b}$$

The hamiltonian is

$$H = \gamma_1 u + \tfrac{1}{2}\gamma_2(x_1{}^2 + c^2 u^2) + \gamma_3 \tag{8}$$

and, along unconstrained sections of the trajectory,

$$\dot{\gamma}_1 = -\gamma_2 x_1 \tag{9}$$

$$\dot{\gamma}_2 = 0 \qquad \gamma_2(\theta) = 1 \tag{10}$$

$$\dot{\gamma}_3 = 0 \tag{11}$$

Furthermore, along unconstrained sections the optimal path is defined by

$$c^2 \ddot{x}_1 - x_1 = 0 \tag{12}$$

From the time t_1, when x_1 intersects the constraint $Q_1 = 0$, until leaving at t_2, we have

$$\dot{Q}_1 = 0 = u + \beta_1 \qquad t_1 \leq t \leq t_2 \tag{13}$$

and since this is independent of $\mathbf{x}$, the multiplier equations are unchanged. In particular, γ_2 and γ_3 are constants with possible jumps at intersections of the free curve and the constraint.

The jump condition, Eq. (11) of Sec. 6.16, requires that

$$\gamma_1(t_1{}^-) = \gamma_1(t_1{}^+) + \mu \tag{14a}$$
$$\gamma_2(t_1{}^-) = \gamma_2(t_1{}^+) \tag{14b}$$
$$\gamma_3(t_1{}^-) = \gamma_3(t_1{}^+) + \mu\beta_1 \tag{14c}$$

and together with the continuity of the hamiltonian it follows that the control is continuous

$$u(t_1{}^-) = -\beta_1 = u(t_1{}^+) \tag{15}$$

In the interval $0 \leq t \leq t_1$ Eq. (12) must be satisfied, and the solution with

$$x_1(t_1) = \alpha_1 - \beta_1 t_1 \tag{16}$$

is

$$x_1 = [\alpha_1 - \beta_1 t_1 - x_1(0)e^{-t_1/c}] \frac{\sinh (t/c)}{\sinh (t_1/c)} + x_1(0)e^{-t/c} \tag{17a}$$

$$u = [\alpha_1 - \beta_1 t_1 - x_1(0)e^{-t_1/c}] \frac{\cosh (t/c)}{c \sinh (t_1/c)} - \frac{x_1(0)}{c} e^{-t/c} \tag{17b}$$

Because of the continuous concave curvature of $x_1(t)$ the point t_1 is the first at which $u = -\beta_1$. Thus t_1 is the solution of

$$(\alpha_1 - \beta_1 t_1) \cosh \frac{t_1}{c} + \beta_1 c \sinh \frac{t_1}{c} - x_1(0) = 0 \tag{18}$$

The solution curve $x_1(t)$ leaves the constraint $Q_1 = 0$ at $t = t_2$. In the interval between constraints Eq. (12) must again be satisfied, and it is a simple consequence of the curvature of the solution of that equation and the continuity of $u(t_2)$ that the solution curve $x_1(t)$ can never intersect the line $Q_2 = 0$ for any $t < \theta$. The solution of Eq. (12) satisfying $x_1(t_2) = \alpha_1 - \beta_1 t_2$ and $x_1(\theta) = 0$ is

$$x_1(t) = (\alpha_1 - \beta_1 t_2) \frac{\sinh [(\theta - t)/c]}{\sinh [(\theta - t_2)/c]} \tag{19a}$$

$$u = -(\alpha_1 - \beta_1 t_2) \frac{\cosh [(\theta - t)/c]}{c \sinh [(\theta - t_2)/c]} \tag{19b}$$

The contribution to the objective from this final section is made as small as possible by choosing θ as large as allowable, $\theta = \alpha_2/\beta_2$. Evaluating Eq. (19b) at t_2, the continuity of $u(t_2)$ requires that t_2 be the solution of

$$(\alpha_1 - \beta_1 t_2) \coth \frac{\alpha_2 - \beta_2 t_2}{c\beta_2} - c\beta_1 = 0 \tag{20}$$

Finally, we return to the multipliers. Since γ_2 is continuous, it follows from Eq. (10) that $\gamma_2 = 1$. Thus, setting $\partial H/\partial u$ to zero, we obtain for unconstrained sections

$$\gamma_1 = -c^2 u \tag{21}$$

In particular, at $t = t_1$ and, from Eq. (13), $t = t_2$

$$\gamma_1(t_1^-) = c^2 \beta_1 = \gamma_1(t_2) \tag{22}$$

But since Eq. (9) applies along the constraint,

$$\dot{\gamma}_1 = -x_1 = \beta_1 t - \alpha_1 \qquad t_1 < t \leq t_2 \tag{23}$$

and integrating with the boundary condition at t_2,

$$\gamma_1(t_1{}^+) = c^2\beta_1 + \tfrac{1}{2}\beta_1(t_1{}^2 - t_2{}^2) - \alpha_1(t_1 - t_2) \tag{24}$$

Thus, from Eqs. (14a), (22), and (24),

$$\mu = \alpha_1(t_1 - t_2) - \tfrac{1}{2}\beta_1(t_1{}^2 - t_2{}^2) \tag{25}$$

and $\boldsymbol{\gamma}$ is discontinuous at t_1.

6.19 BOTTLENECK PROBLEMS

Bottleneck problems are a class of control problem typified by constraints of the form

$$u \leq \phi(\mathbf{x}) \tag{1}$$

When there is little of the state variable and ϕ is small, a bottleneck exists and little effort can be exerted. When ϕ has increased and the bottleneck has been removed, large effort can be used. Many economic problems are of this type, and a bottleneck model has recently been employed to describe the growth of a bacterial culture, suggesting an optimum-seeking mechanism in the growth pattern.

The example of a bottleneck problem we shall examine has been used by Bellman to demonstrate the application of dynamic programming. We have

$$\dot{x}_1 = a_1u_1 \tag{1a}$$
$$\dot{x}_2 = a_2u_2 - u_1 \tag{1b}$$

with

$$u_1,\ u_2 \geq 0 \tag{2}$$
$$Q_1 = x_2 - u_1 - u_2 \geq 0 \tag{3a}$$
$$Q_2 = x_1 - u_2 \geq 0 \tag{3b}$$

and we seek to maximize $x_2(\theta)$. In this case the multiplier boundary conditions are

$$\gamma_1(\theta) = 0 \qquad \gamma_2(\theta) = -1 \tag{4}$$

and the hamiltonian is

$$H = \gamma_1a_1u_1 + \gamma_2a_2u_2 - \gamma_2u_1 \tag{5}$$

At $t = \theta$ the hamiltonian becomes

$$H = -a_2u_2 + u_1 \qquad t = \theta \tag{6}$$

in which case we require u_2 to be a maximum, u_1 a minimum. Thus,

$$u_1(\theta) = 0 \tag{7a}$$

and, from Eqs. (3),

$$u_2(\theta) = \min \begin{cases} x_1(\theta) \\ x_2(\theta) \end{cases} \tag{7b}$$

We shall carry out our analysis for $x_2(\theta) > x_1(\theta)$, in which case

$$u_2(\theta) = x_1(\theta) \tag{8}$$

Then $Q_2 = 0$ is in force, and just prior to $t = \theta$ the state satisfies the equations

$$\dot{x}_1 = 0 \tag{9a}$$
$$\dot{x}_2 = a_2 x_1 \tag{9b}$$

or

$$x_1(t) = x_1(\theta) \tag{10a}$$
$$x_2(t) = a_2 x_1(\theta)(t - \theta) + x_2(\theta) \tag{10b}$$

The Green's functions satisfy Eq. (5) of Sec. 6.15 when $Q_2 = 0$ is in effect, which leads to

$$\dot{\gamma}_1 = -\left[\frac{\partial f_1}{\partial x_1} - \frac{\partial f_1}{\partial u_2}\left(\frac{\partial Q_2}{\partial u_2}\right)^{-1}\frac{\partial Q_2}{\partial x_1}\right]\gamma_1$$
$$-\left[\frac{\partial f_2}{\partial x_1} - \frac{\partial f_2}{\partial u_2}\left(\frac{\partial Q_2}{\partial u_2}\right)^{-1}\frac{\partial Q_2}{\partial x_1}\right]\gamma_2 = -a_2\gamma_2 \tag{11a}$$

$$\dot{\gamma}_2 = -\left[\frac{\partial f_1}{\partial x_2} - \frac{\partial f_1}{\partial u_2}\left(\frac{\partial Q_2}{\partial u_2}\right)^{-1}\frac{\partial Q_2}{\partial x_2}\right]\gamma_1$$
$$-\left[\frac{\partial f_2}{\partial x_2} - \frac{\partial f_2}{\partial u_2}\left(\frac{\partial Q_2}{\partial u_2}\right)^{-1}\frac{\partial Q_2}{\partial x_2}\right]\gamma_2 = 0 \tag{11b}$$

or

$$\gamma_2 = 1 \tag{12a}$$
$$\gamma_1 = a_2(\theta - t) \tag{12b}$$

This final interval is preceded by one in which either

$$\gamma_2(t) - a\gamma_1(t) \geq 0 \qquad x_2(t) > x_1(t) \tag{13a}$$

or

$$\gamma_2(t) - a_1\gamma_1(t) < 0 \qquad x_1(t) > x_2(t) \tag{13b}$$

The former condition ends at time t_1, when γ_2 equals $a_1\gamma_1$, which is, from Eq. (12),

$$t_1 = \theta - \frac{1}{a_1 a_2} \tag{14a}$$

The latter ends at t_2, when $x_1 = x_2$, from Eq. (10)

$$t_2 = \theta - \left[\frac{x_2(\theta)}{x_1(\theta)} - 1 \right] \frac{1}{a_2} \tag{14b}$$

The simpler case encompasses final states for which $t_2 > t_1$. For times just preceding t_2, then, the optimal policy is

$$u_1 = 0 \tag{15a}$$

$$u_2 = \min \begin{Bmatrix} x_1(t) \\ x_2(t) \end{Bmatrix} = x_2 \tag{15b}$$

The state equations are

$$\dot{x}_1 = 0 \tag{16a}$$

$$\dot{x}_2 = a_2 x_2 \tag{16b}$$

in which case

$$x_1(t < t_2) = x_1(t_2) \tag{17a}$$

$$x_2(t < t_2) < x_2(t_2) \tag{17b}$$

so that $x_2 < x_1$ for all $t < t_2$ during which this policy is in effect. But now $Q_1 = 0$ is in effect, leading to multiplier equations

$$\dot{\gamma}_1 = 0 \qquad \gamma_1 = \gamma_1(t_2) = a_2(\theta - t_2) \tag{18a}$$

$$\dot{\gamma}_2 = -a_2\gamma_2 \qquad \gamma_2 = \gamma_2(t_2)e^{-a_2(t-t_2)} = e^{-a_2(t-t_2)} \tag{18b}$$

and

$$\gamma_2(t) - a_1\gamma_1(t) > \gamma_2(t_2) - a_1\gamma_1(t_2) > 0 \qquad t < t_2 \tag{19}$$

so that the optimal policy is unchanged for all $t < t_2$.

The second possibility is $t_1 > t_2$. For times just prior to t_1 the coefficient of u_2 in the hamiltonian is negative, while $x_2 > x_1$. If the coefficient of u_1 is zero and u_1 is intermediate while u_2 is at its maximum, the multiplier equations are unchanged and $\gamma_2 - a\gamma_1 < 0$, which is a contradiction. Thus, the coefficient of u_1 is also negative, requiring u_1 to be at a maximum. The only possibilities compatable with $x_2 > x_1$ are

$$u_2 = x_1 \qquad u_1 = x_2 - x_1 \tag{20a}$$

$$u_2 = 0 \qquad u_1 = x_1 \tag{20b}$$

The second case corresponds to $Q_1 = 0$, Q_2 not in effect, in which case it follows from the previous discussion that the coefficient of u_1 is positive, resulting in a contradiction. Thus, for $t < t_1$ the policy is Eq. (20a).

It follows directly by substitution into the state and multiplier equations that this policy remains unchanged for all $t < t_1$. Thus, the optimal policy can be summarized as

$$0 \le t \le \theta - \frac{1}{a_1 a_2} \quad \begin{matrix} x_2 \le x_1, \ u_1 = 0, \ u_2 = x_2 \\ x_2 \ge x_1, \ u_1 = x_2 - x_1, \ u_2 = x_1 \end{matrix}$$

$$\theta - \frac{1}{a_1 a_2} < t \le \theta \quad \begin{matrix} u_1 = 0 \\ u_2 = \min (x_1, x_2) \end{matrix} \tag{21}$$

Bellman has established the optimality of this policy by an interesting duality relation which establishes strong connections between this type of linear bottleneck problem and a continuous analog of linear programming.

6.20 SUFFICIENCY

In Sec. 1.3 we found that only a slight strengthening of the necessary conditions for optimality was required in order to obtain sufficient conditions for a policy that was at least locally optimal. Sufficient conditions in the calculus of variations are far more difficult to establish, and, in general, only limited results can be obtained.

Our starting point is Eq. (4) of Sec. 6.9, which, after multiplication by $\gamma_i(\theta)$ and summation from $i = 1$ to n yields

$$\sum_{i=1}^{n} \gamma_i(\theta) \, \delta x_i(\theta) = \int_0^{\theta} \left(\sum_{k=1}^{R} \frac{\partial H}{\partial u_k} \delta u_k + \frac{1}{2} \sum_{j,k=1}^{R} \frac{\partial^2 H}{\partial u_j \, \partial u_k} \delta u_j \, \delta u_k \right.$$

$$\left. + \sum_{k=1}^{R} \sum_{j=1}^{n} \frac{\partial^2 H}{\partial u_k \, \partial x_j} \delta u_k y_j + \frac{1}{2} \sum_{j,k=1}^{n} \frac{\partial^2 H}{\partial x_j \, \partial x_k} y_j y_k \right) ds + o(\epsilon^2) \tag{1}$$

where $\mathbf{y}$ is defined as

$$y_j(s) = \int_0^s \sum_{i=1}^{n} \sum_{k=1}^{R} \Gamma_{ji}(s,t) \frac{\partial f_i}{\partial u_k} \delta u_k(t) \, dt \tag{2}$$

If we restrict attention to situations in which $\mathcal{E}$ is a linear function of components of $\mathbf{x}(\theta)$ so that the second derivative vanishes identically, Eq. (1) is an expression of $\delta \mathcal{E}$ for an arbitrary (small) variation in $\mathbf{u}$. We require conditions, then, for which $\delta \mathcal{E}$ is always strictly positive for nonzero $\delta \mathbf{u}$.

When a function satisfying the minimum principle lies along a boundary, the first-order terms dominate. Thus we need only consider interior values at which $\partial H / \partial u_k$ vanishes and the hessian is positive definite. The two remaining terms in Eq. (1) prevent us from stating that $\delta \mathcal{E}$ is positive for *any* change in $\mathbf{u}$, rather than the very special changes which we have considered. A special case of importance is that of minimizing

$$\mathcal{E} = \int_0^\theta \mathcal{F}(\mathbf{x}, \mathbf{u})\, dt \tag{3}$$

with linear state equations

$$\dot{x}_i = \sum_{j=1}^n A_{ij} x_j + \sum_{k=1}^R B_{ik} u_k \tag{4}$$

By introduction of a new state variable this is equivalent to a form in which $\mathcal{E}$ is a linear function of the final state. The hamiltonian is then

$$H = \mathcal{F} + \sum_{i,j=1}^n \gamma_i A_{ij} x_j + \sum_{i=1}^n \sum_{k=1}^R \gamma_i B_{ik} u_k \tag{5}$$

and for a function satisfying the minimum principle Eq. (1) becomes

$$\delta\mathcal{E} = \frac{1}{2} \int_0^\theta \left(\sum_{j,k=1}^R \frac{\partial^2 \mathcal{F}}{\partial u_j\, \partial u_k} \delta u_j\, \delta u_k + 2 \sum_{j=1}^n \sum_{k=1}^R \frac{\partial^2 \mathcal{F}}{\partial x_j\, \partial u_k} \delta u_k\, y_j \right.$$
$$\left. + \sum_{j,k=1}^n \frac{\partial^2 \mathcal{F}}{\partial x_j\, \partial x_k} y_j y_k \right) ds + o(\epsilon^2) \tag{6}$$

If $\mathcal{F}$ is strictly convex, that is,

$$\sum_{j,k=1}^R \frac{\partial^2 \mathcal{F}}{\partial u_j\, \partial u_k} \alpha_j \alpha_k + 2 \sum_{j=1}^n \sum_{k=1}^R \frac{\partial^2 \mathcal{F}}{\partial x_j\, \partial u_k} \beta_j \alpha_k$$
$$+ \sum_{j,k=1}^n \frac{\partial^2 \mathcal{F}}{\partial x_j\, \partial x_k} \beta_j \beta_k > 0 \tag{7}$$

at all differentiable points for *arbitrary* nonzero α, β, then $\delta\mathcal{E}$ is positive. Thus, the minimum principle is sufficient for local optimality in a linear system if the objective is the integral of a strictly convex function. An important case is the positive definite quadratic objective, which we have considered several times.

APPENDIX 6.1 CONTINUOUS DEPENDENCE OF SOLUTIONS

The linearizations in this chapter are all justified by a fundamental result about the behavior of the solutions of differential equations when small changes are made in initial conditions or arguments. We establish that result here.

Consider

$$\dot{\mathbf{x}} = \mathbf{f}(\mathbf{x}, \mathbf{u}) \qquad 0 < t \le \theta \tag{1}$$

where $\mathbf{u}(t)$ is a piecewise-continuous vector function. We assume that $\mathbf{f}$ satisfies a Lipschitz condition with respect to $\mathbf{x}$ and $\mathbf{u}$

$$|\mathbf{f}(\mathbf{x_1}, \mathbf{u_1}) - \mathbf{f}(\mathbf{x_2}, \mathbf{u_2})| \le Lp(|\mathbf{x_1} - \mathbf{x_1}| + |\mathbf{u_1} - \mathbf{u_2}|) \tag{2}$$

where Lp is a constant and any definition of the magnitude of a vector may be used. If we specify a function $\bar{\mathbf{u}}(t)$ and initial condition $\bar{\mathbf{x}}_0$, Eq. (2) is sufficient to ensure a unique solution $\bar{\mathbf{x}}(t)$ to Eq. (1). Similarly, if we specify $\mathbf{u} = \bar{\mathbf{u}} + \delta\mathbf{u}$ and $\mathbf{x}_0 = \bar{\mathbf{x}}_0 + \delta\mathbf{x}_0$, we obtain a unique solution $\bar{\mathbf{x}} + \delta\mathbf{x}$. We seek the magnitude of $\delta\mathbf{x}$ at some finite terminal time θ for which both solutions remain finite

Let

$$|\delta\mathbf{u}(t)|, |\delta\mathbf{x}_0| \le \epsilon \tag{3}$$

If we integrate Eq. (1) for the two choices of $\mathbf{x}_0$ and $\mathbf{u}$ and subtract, we obtain

$$\delta\mathbf{x}(t) = \delta\mathbf{x}_0 + \int_0^t [\mathbf{f}(\bar{\mathbf{x}} + \delta\mathbf{x}, \bar{\mathbf{u}} + \delta\bar{\mathbf{u}}) - \mathbf{f}(\bar{\mathbf{x}}, \bar{\mathbf{u}})]\, ds \qquad 0 \le t \le \theta \tag{4}$$

Making use of Eqs. (2) and (3), this leads to

$$|\delta\mathbf{x}(t)| \le \epsilon + Lp \int_0^t (|\delta\mathbf{x}| + |\delta\mathbf{u}|)\, ds \qquad 0 \le t \le \theta \tag{5}$$

If $|\delta\mathbf{x}|_m$ is equal to the least upper bound of $|\delta\mathbf{x}|$ in $0 \le t \le \theta$, then

$$|\delta\mathbf{x}(t)| \le \epsilon(1 + Lpt) + Lp|\delta\mathbf{x}|_m t \tag{6}$$

Substituting Eq. (6) back into Eq. (5), we obtain

$$|\delta\mathbf{x}(t)| \le \epsilon(1 + Lpt) + \epsilon t + \tfrac{1}{2} Lpt^2 + \tfrac{1}{2} Lp|\delta\mathbf{x}|_m t^2 \tag{7}$$

and by continued substitution we obtain, finally,

$$|\delta\mathbf{x}(t)| \le \epsilon(e^{2Lpt} - 1) \tag{8}$$

or

$$|\delta\mathbf{x}(\theta)| \le K\epsilon \tag{9}$$

where

$$K = 2e^{Lp\theta} - 1 \tag{10}$$

and depends on θ but not on $\delta\mathbf{x}_0$ or $\delta\mathbf{u}(t)$.

BIBLIOGRAPHICAL NOTES

Section 6.2: Solution procedures for linear ordinary differential equations are discussed in any good text, such as

E. A. Coddington and N. Levinson: "Theory of Ordinary Differential Equations," McGraw-Hill Book Company, New York, 1955

The linear analysis used here is covered in

M. M. Denn and R. Aris: *Ind. Eng. Chem. Fundamentals,* **4**:7 (1965)
L. A. Zadeh and C. A. Desoer: "Linear System Theory," McGraw-Hill Book Company, New York, 1963

Green's identity introduced here is a one-dimensional version of the familiar surface integral–volume integral relation of the same name, which is established in any book on advanced calculus, such as

R. C. Buck: "Advanced Calculus," 2d ed., McGraw-Hill Book Company, New York, 1965

Section 6.3: The properties of the adjoint system of the variational equations were exploited in pioneering studies of Bliss in 1919 on problems of exterior ballistics and subsequently in control studies of Laning and Battin; see

G. A. Bliss: "Mathematics for Exterior Ballistics," John Wiley & Sons, Inc., New York, 1944
J. H. Laning, Jr., and R. H. Battin: "Random Processes in Automatic Control," McGraw-Hill Book Company, New York, 1956

Sections 6.4 and 6.5: The development is similar to

M. M. Denn and R. Aris: *AIChE J.,* **11**:367 (1965)

The derivation in that paper is correct only for the case of fixed θ.

Section 6.8: The earliest derivation of a result equivalent to the strong minimum principle was by Valentine, using the Weierstrass condition of the classical calculus of variations and slack variables to account for inequality constraints; see

F. A. Valentine: in "Contributions to the Theory of Calculus of Variations, 1933–37," The University of Chicago Press, Chicago, 1937

The result was later obtained independently by Pontryagin and coworkers under somewhat weaker assumptions and is generally known as the Pontryagin maximum (or minimum) principle; see

V. G. Boltyanskii, R. V. Gamkrelidze, and L. S. Pontryagin: *Rept. Acad. Sci. USSR,* **110**:7 (1956); reprinted in translation in R. Bellman and R. Kalaba (eds.), "Mathematical Trends in Control Theory," Dover Publications, Inc., New York, 1964
L. S. Pontryagin, V. A. Boltyanskii, R. V. Gamkrelidze, and E. F. Mishchenko: "The Mathematical Theory of Optimal Processes," John Wiley & Sons, Inc., New York, 1962

A number of different approaches can be taken to the derivation of the minimum principle. These are typified in the following references, some of which duplicate each other in approach:

M. Athans and P. L. Falb: "Optimal Control," McGraw-Hill Book Company, New York, 1966
L. D. Berkovitz: *J. Math. Anal. Appl.,* **3**:145 (1961)
A. Blaquiere and G. Leitmann: in G. Leitmann (ed.), "Topics in Optimization," Academic Press, Inc., New York, 1967

S. Dreyfus: "Dynamic Programming and the Calculus of Variations," Academic Press, Inc., New York, 1965

H. Halkin: in G. Leitmann (ed.), "Topics in Optimization," Academic Press, Inc., New York, 1967

M. Hestenes: "Calculus of Variations and Optimal Control Theory," John Wiley & Sons, Inc., New York, 1966

R. E. Kalman: in R. Bellman (ed.), "Mathematical Optimization Techniques," University of California Press, Berkeley, 1963

E. B. Lee and L. Markus: "Foundations of Optimal Control Theory," John Wiley & Sons, Inc., New York, 1967

G. Leitmann: "An Introduction to Optimal Control," McGraw-Hill Book Company, New York, 1966

L. Neustadt: *SIAM J. Contr.*, **4**:505 (1966); **5**:90 (1967)

J. Warga: *J. Math. Anal. Appl.*, **4**:129 (1962)

The derivation used here is not rigorous for the constrained final condition, but the transversality conditions can be obtained by using penalty functions for end-point constraints and taking limits as the penalty constant becomes infinite. We have assumed here and throughout this book the existence of an optimum. This important question is discussed in several of the above references.

Section 6.9: We follow here

M. M. Denn and R. Aris: *Chem. Eng. Sci.*, **20**:373 (1965)

The result is equivalent to the Legendre-Clebsch condition of classical calculus of variations.

Section 6.10: This result was obtained by Aris; see

R. Aris: "Optimal Design of Chemical Reactors," Academic Press, Inc., New York, 1961

Section 6.11: The example and mathematical development follow

F. Horn and R. C. Lin: *Ind. Eng. Chem. Process Design Develop.*, **6**:21 (1967)

A good introduction to the notion of process improvement by unsteady operation is

J. M. Douglas and D. W. T. Rippin: *Chem. Eng. Sci.*, **21**:305 (1966)

Section 6.12: The example is from

A. W. Pollock: Applying Pontryagin's Maximum Principle to the Operation of a Catalytic Reformer, *16th Chem. Eng. Conf. Chem. Inst. Can., Windsor, Ontario, 1966*

Applications to other problems with decaying catalysts have been carried out by

A. Chou, W. H. Ray, and R. Aris: *Trans. Inst. Chem. Engrs. (London)*, **45**:T153 (1967)

S. Szepe and O. Levenspiel: unpublished research, Illinois Institute of Technology, Chicago

Section 6.13: The classical Weierstrass and Legendre conditions are treated in any of the texts on calculus of variations cited for Sec. 3.2.

Section 6.14: See

B. S. Goh: *SIAM J. Cont.*, **4**:309 (1966)
C. D. Johnson: in C. T. Leondes (ed.), "Advances in Control Systems," vol. 2, Academic Press, Inc., New York, 1965
H. J. Kelley, R. E. Kopp, and H. G. Moyer: in G. Leitmann (ed.), "Topics in Optimization," Academic Press, Inc., New York, 1967

These papers develop the necessary conditions in some detail and deal with several nontrivial applications. A very different approach applicable to certain linear problems is taken in

A. Miele: in G. Leitmann (ed.), "Optimization Techniques with Applications to Aerospace Systems," Academic Press, Inc., New York, 1962

Sections 6.15 and 6.16: The approach is motivated by

A. E. Bryson, Jr., W. F. Denham, and S. E. Dreyfus: *AIAA J.*, **1**:2544 (1963)

State-variable constraints are included in Valentine's formulation cited above and are treated in most of the other references for Sec. 6.8. See also the review paper

J. McIntyre and B. Paiewonsky: in C. T. Leondes (ed.), "Advances in Control Systems," vol. 5, Academic Press, Inc., New York, 1967

Section 6.17: This problem was solved using other methods by

N. Blakemore and R. Aris: *Chem. Eng. Sci.*, **17**:591 (1962)

Other applications of the theory can be found in the references cited for Secs. 6.8 and 6.15 and

J. Y. S. Luh, J. S. Shafran, and C. A. Harvey: *Preprints 1967 Joint Autom. Contr. Conf., Philadelphia*, p. 144
C. L. Partain and R. E. Bailey: *Preprints 1967 Joint Autom. Contr. Conf., Philadelphia*, p. 71
C. D. Siebenthal and R. Aris: *Chem. Eng. Sci.*, **19**:729 (1964)

Section 6.19: The example was solved by other methods in

R. E. Bellman: "Dynamic Programming," Princeton University Press, Princeton, N.J., 1957

An intriguing analysis of batch microbial growth as an optimum-seeking bottleneck process is in

C. H. Swanson, R. Aris, A. G. Fredrickson, and H. M. Tsuchiya: *J. Theoret. Biol.*, **12**:228 (1966)

Section 6.20: Sufficiency for more general situations is considered in several of the references cited for Sec. 6.8.

Appendix 6.1: This is a straightforward extension of a well-known result in differential equations; see, for example, the book by Coddington and Levinson cited above.

PROBLEMS

6.1. Show that the equation

$$\frac{d}{dt} p(t)\dot{x} + h(t)x = u(t)$$

is *self-adjoint* in that the adjoint equation is

$$\frac{d}{dt} p(t)\dot{\Gamma} + h(t)\Gamma = 0$$

6.2. Obtain the control which takes the second-order system

$$\ddot{x} + a\dot{x} + bx = u$$
$$|u| \leq 1$$

from some initial state to a circle about the origin while minimizing

(a) time

(b) $\varepsilon = \dfrac{1}{2} \displaystyle\int_0^\theta (x^2 + cu^2)\, dt$

6.3. Determine the relation between optimality of the steady state and the character of stationary points of the steady-state lagrangian. Interpret this result in terms of the meaning of steady-state Lagrange multipliers.

6.4. Examine the singular solutions of Secs. 5.9 and 5.10 in the context of the necessary condition of Sec. 6.14.

6.5. *Goddard's problem* of maximizing the height of a vertical flight rocket is described by

$$\dot{h} = v$$
$$\dot{v} = \frac{1}{m}(T - D) - g$$
$$\dot{m} = \frac{-T}{c}$$

Here h is the altitude, v the velocity, m the mass, g the acceleration due to gravity, and c the exhaust velocity. D is the drag, a function of v and h. The control variable is the thrust T, to be chosen subject to

$$\theta \leq T \leq T^*$$

to maximize the terminal value of h for fixed m. Show that intermediate thrust is possible only for drag laws satisfying

$$\frac{\partial^2 D}{\partial v^2} + \frac{2}{c}\frac{\partial D}{\partial v} + \frac{D}{c^2} \geq 0$$

6.6. Given

$$\ddot{x} = u$$
$$x(0) = x(1) = 0$$
$$\dot{x}(0) = v_0 > 0 \qquad \dot{x}(1) = -v_1 < 0$$
$$|x(t)| \leq L$$

find $u(t)$, $0 \leq t \leq 1$ to minimize

$$\mathcal{E} = \frac{1}{2} \int_0^\theta u^2(t) \, dt$$

(The problem is due to Bryson, Denham, and Dreyfus.)

6.7. Consider the problem of determining the curve of minimum length between point (x_{10}, x_{20}) and the origin which cannot pass through a closed circular region. The equations are

$$\dot{x}_1 = \sin u$$
$$\dot{x}_2 = \cos u$$
$$(x_1 - a)^2 + x_2{}^2 \geq R^2$$
$$\min \mathcal{E} = \theta = \int_0^\theta dt$$

Find $u(t)$. (The problem is due to Leitmann.)

6.8. In some practical problems state variables may be discontinuous at discrete points (e.g., the mass in a multistage rocket) and the form of the system equations might change. Consider the system

$$\dot{x}_i{}^{(n)} = f_i{}^{(n)}(\mathbf{x}^{(n)}, u) \qquad t_{n-1} < t < t_n$$

with discontinuities of the form

$$\mathbf{x}^{(n+1)}(t_n) = \lim_{\epsilon \to 0} \mathbf{x}(t_n + \epsilon) = \lim_{\epsilon \to 0} \mathbf{x}(t_n - \epsilon) + \xi_n = \mathbf{x}^{(n)}(t_n) + \xi_n$$

where ξ_n is a specified constant and t_n is defined by the condition

$$\psi_n(\mathbf{x}, t) = 0$$

For specified initial conditions $\mathbf{x}_0$ and decision and final state constraints

$$u_* \leq u \leq u^*$$
$$\phi_k[\mathbf{x}^{(N)}(t_N)] = 0$$

find the control which minimizes

$$\mathcal{E} = \sum_{n=1}^{N} \int_{t_{n-1}}^{t_n} \mathfrak{F}[\mathbf{x}^{(n)}(t), u(t)] \, dt$$

(The problem is due to Graham.)

7
Staged Systems

7.1 INTRODUCTION

In Chap. 1 we briefly examined some properties of optimal systems in which the state is described by finite difference equations, while the preceding four chapters have involved the detailed study of continuous systems. We shall now return to the consideration of staged systems in order to generalize some of the results of the first chapter and to place the optimization of discrete systems within the mathematical framework developed in Chap. 6 for continuous systems. We shall find that the conditions we develop are conveniently represented in a hamiltonian formulation and that many analogies to continuous variational problems exist, as well as significant differences.

In the study of continuous systems it has often been possible to suppress the explicit dependence of variables on the independent variable t so that inconvenient notational problems have rarely arisen. This is not the case with staged processes, where it is essential to state the location in space or discretized time precisely. For typographical reasons we shall denote the location at which a particular variable is considered by a *superscript*. Thus, if $\mathbf{x}$ is the vector representing the state of the system, the state at position n is $\mathbf{x}^n$, and the ith component of $\mathbf{x}^n$ is $x_i{}^n$. The systems under consideration are represented by the block diagram in Fig. 7.1, and the relation between the input and output of any stage is

$$\mathbf{x}^n = \mathbf{f}^n(\mathbf{x}^{n-1}, \mathbf{u}^n) \qquad n = 1, 2, \ldots, N \tag{1a}$$

or

$$x_i{}^n = f_i{}^n(\mathbf{x}^{n-1}, \mathbf{u}^n) \qquad \begin{matrix} i = 1, 2, \ldots, S \\ n = 1, 2, \ldots, N \end{matrix} \tag{1b}$$

S is the number of variables needed to represent the state. The functions $\mathbf{f}^n$ need not be the same for each stage in the process.

7.2 GREEN'S FUNCTIONS

Linear difference equations may be summed in a manner identical to the integration of linear differential equations by means of Green's functions. We consider the system

$$x_i{}^n = \sum_{j=1}^{S} A_{ij}{}^n x_j{}^n + b_i{}^n \qquad \begin{matrix} i = 1, 2, \ldots, S \\ n = 1, 2, \ldots, N \end{matrix} \tag{1}$$

We now introduce S^2 functions $\Gamma_{ki}{}^{Nn}$, $i, k = 1, 2, \ldots, S$, at each stage, denoting both the stage number n and the final stage N, with which we multiply Eq. (1) and sum over n from 1 to N and over i from 1 to S. Thus,

$$\sum_{n=1}^{N} \sum_{i=1}^{S} \Gamma_{ki}{}^{Nn} x_i{}^n = \sum_{n=1}^{N} \sum_{i,j=1}^{S} \Gamma_{ki}{}^{Nn} A_{ij}{}^n x_j{}^{n-1} + \sum_{n=1}^{N} \sum_{i=1}^{S} \Gamma_{ki}{}^{Nn} b_i{}^n$$

$$k = 1, 2, \ldots, S \tag{2}$$

The left-hand side of Eq. (2) may be rewritten (summed by parts) as

$$\sum_{n=1}^{N} \sum_{i=1}^{S} \Gamma_{ki}{}^{Nn} x_i{}^n = \sum_{n=1}^{N} \sum_{j=1}^{S} \Gamma_{kj}{}^{N,n-1} x_j{}^{n-1} + \sum_{i=1}^{S} \Gamma_{ki}{}^{NN} x_i{}^N$$

$$- \sum_{i=1}^{S} \Gamma_{ki}{}^{N0} x_i{}^0 \tag{3}$$

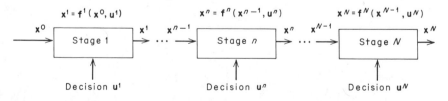

Fig. 7.1 Schematic of a staged system.

which yields, upon substitution into Eq. (2),

$$
\sum_{i=1}^{S} \Gamma_{ki}{}^{NN} x_i{}^{N} - \sum_{i=1}^{S} \Gamma_{ki}{}^{N0} x_i{}^{0}
$$

$$
= \sum_{n=1}^{N} \sum_{j=1}^{S} \left(\sum_{i=1}^{S} \Gamma_{ki}{}^{Nn} A_{ij}{}^{n} - \Gamma_{kj}{}^{N,n-1} \right) x_j{}^{n-1} + \sum_{n=1}^{N} \sum_{i=1}^{S} \Gamma_{ki}{}^{Nn} b_i{}^{n}
$$

$$
k = 1, 2, \ldots, S \quad (4)
$$

Since the goal of summing the equations is to express the output $\mathbf{x}^N$ in terms of the input $\mathbf{x}^0$ and the forcing functions $\mathbf{b}^n$, we define the Green's functions by the adjoint difference equations

$$
\Gamma_{kj}{}^{N,n-1} = \sum_{i=1}^{S} \Gamma_{ki}{}^{Nn} A_{ij}{}^{n} \qquad \begin{array}{l} k, j = 1, 2, \ldots, S \\ n = 1, 2, \ldots, N \end{array} \quad (5)
$$

$$
\Gamma_{kj}{}^{NN} = \delta_{kj} = \begin{cases} 1 & k = j \\ 0 & k \neq j \end{cases} \quad (6)
$$

Thus, Eq. (4) becomes a special form of Green's identity

$$
x_i{}^{N} = \sum_{j=1}^{S} \Gamma_{ij}{}^{N0} x_j{}^{0} + \sum_{n=1}^{N} \sum_{j=1}^{S} \Gamma_{ij}{}^{Nn} b_j{}^{n} \quad (7)
$$

These equations are analogous to Eqs. (12), (14), and (15) of Sec. 6.2, but it should be noted that the difference equation (5) has an algebraic sign opposite that of the differential equation (12) of that section.

We shall generally be interested in a linear combination of the components $x_i{}^{N}$. If we define a vector $\boldsymbol{\gamma}^n$ with components $\gamma_i{}^n$ by

$$
\gamma_i{}^{n} = \sum_{j=1}^{S} \gamma_j{}^{N} \Gamma_{ji}{}^{Nn} \qquad i = 1, 2, \ldots, S \quad (8)
$$

where $\gamma_j{}^{N}$ are specified numbers, then by multiplying Eqs. (5) to (7) by $\gamma_i{}^{N}$ and summing over i we obtain the scalar form of Green's identity

$$
\sum_{i=1}^{S} \gamma_i{}^{N} x_i{}^{N} = \sum_{i=1}^{S} \gamma_i{}^{0} x_i{}^{0} + \sum_{n=1}^{N} \sum_{i=1}^{S} \gamma_i{}^{n} b_i{}^{n} \quad (9)
$$

and

$$
\gamma_i{}^{n-1} = \sum_{j=1}^{S} \gamma_j{}^{n} A_{ji}{}^{n} \qquad \begin{array}{l} i = 1, 2, \ldots, S \\ n = 1, 2, \ldots, N \end{array} \quad (10)
$$

7.3 THE FIRST VARIATION

We now turn to the optimization problem. The state is described by the equations

$$x_i{}^n = f_i{}^n(\mathbf{x}^{n-1}, \mathbf{u}^n) \qquad \begin{aligned} i &= 1, 2, \ldots, S \\ n &= 1, 2, \ldots, N \end{aligned} \qquad (1)$$

with the objective the minimization of a function of the final state $\mathcal{E}(\mathbf{x}^N)$. The decision functions $\mathbf{u}^n$ may be restricted by inequality constraints

$$U_p{}^n(\mathbf{u}^n) \geq 0 \qquad \begin{aligned} p &= 1, 2, \ldots, P \\ n &= 1, 2, \ldots, N \end{aligned} \qquad (2)$$

and the initial and final states by equality constraints,

$$q_k(\mathbf{x}^0) = 0 \qquad k = 1, 2, \ldots, K \qquad (3)$$
$$g_l(\mathbf{x}^N) = 0 \qquad l = 1, 2, \ldots, L \qquad (4)$$

We assume that we have chosen a sequence of decisions $\bar{\mathbf{u}}^n$ and an initial state $\bar{\mathbf{x}}^0$, after which we determine a sequence of states $\bar{\mathbf{x}}^n$ from Eq. (1). We now make small changes consistant with the constraints

$$u_k{}^n = \bar{u}_k{}^n + \delta u_k{}^n \qquad |\delta u_k{}^n| \leq \epsilon \qquad (5a)$$
$$x_i{}^0 = \bar{x}_i{}^0 + \delta x_i{}^0 \qquad |\delta x_i{}^0| \leq \epsilon \qquad (5b)$$

for some predetermined $\epsilon > 0$. If the functions $\mathbf{f}^n$ are continuous, it follows that the outputs from stage N change only to order ϵ. For piecewise continuously differentiable functions f^n the variational equations for $\delta \mathbf{x}^n$ are then

$$\delta x_i{}^n = \sum_{j=1}^{S} \frac{\partial f_i{}^n}{\partial x_j{}^{n-1}} \delta x_j{}^{n-1} + \sum_{k=1}^{R} \frac{\partial f_i{}^n}{\partial u_k{}^n} \delta u_k{}^n + o(\epsilon) \qquad (6)$$

where the partial derivatives are evaluated at $\bar{\mathbf{x}}^n$, $\bar{\mathbf{u}}^n$. Equation (6) is linear and has Green's functions defined by Eq. (10) of the preceding section as

$$\gamma_i{}^{n-1} = \sum_{j=1}^{S} \gamma_j{}^n \frac{\partial f_j{}^n}{\partial x_i{}^{n-1}} \qquad (7)$$

in which case Green's identity, Eq. (9), becomes

$$\sum_{i=1}^{S} \gamma_i{}^N \delta x_i{}^N = \sum_{i=1}^{S} \gamma_i{}^0 \delta x_i{}^0 + \sum_{n=1}^{N} \sum_{i=1}^{S} \sum_{k=1}^{R} \gamma_i{}^n \frac{\partial f_i{}^n}{\partial u_k{}^n} \delta u_k{}^n + o(\epsilon) \qquad (8)$$

For staged systems we need only consider the case where N is fixed, so that the change in $\mathcal{E}$ brought about by the changes $\delta \mathbf{u}^n$, $\delta \mathbf{x}^0$ is

$$\delta \mathcal{E} = \sum_{i=1}^{S} \frac{\partial \mathcal{E}}{\partial x_i{}^N} \delta x_i{}^N + o(\epsilon) \tag{9}$$

If, exactly as in Sec. 6.4 for continuous systems, we define the boundary conditions for the difference equations for the Green's functions as

$$\gamma_i{}^N = \frac{\partial \mathcal{E}}{\partial x_i{}^N} + \sum_{l=1}^{L} \nu_l \frac{\partial g_l}{\partial x_i{}^N} \tag{10a}$$

$$\gamma_i{}^0 = \sum_{k=1}^{K} \eta_k \frac{\partial q_k}{\partial x_i{}^0} \tag{10b}$$

then Eqs. (8) to (10) combine to give

$$\delta \mathcal{E} = \sum_{n=1}^{N} \sum_{i=1}^{S} \sum_{k=1}^{R} \gamma_i{}^n \frac{\partial f_i{}^n}{\partial u_k{}^n} \delta u_k{}^n + o(\epsilon) \tag{11}$$

7.4 THE WEAK MINIMUM PRINCIPLE

In order to reinforce the analogy to continuous systems we introduce the stage hamiltonian H^n as

$$H^n = \sum_{i=1}^{S} \gamma_i{}^n f_i{}^n \tag{1}$$

We then have the canonical equations

$$x_i{}^n = \frac{\partial H^n}{\partial \gamma_i{}^n} \tag{2a}$$

$$\gamma_i{}^{n-1} = \frac{\partial H^n}{\partial x_i{}^{n-1}} \tag{2b}$$

and Eq. (11) of Sec. 7.3 for $\delta \mathcal{E}$ may be written

$$\delta \mathcal{E} = \sum_{n=1}^{N} \sum_{k=1}^{R} \frac{\partial H^n}{\partial u_k{}^n} \delta u_k{}^n + o(\epsilon) \geq 0 \tag{3}$$

where the inequality follows from the fact that the sequence $\bar{\mathbf{u}}^n$ minimizes $\mathcal{E}$. By a proof identical in every respect to that in Sec. 6.5 for continuous systems we obtain the following weak minimum principle:

> *The decisions $u_k{}^n$ which minimize $\mathcal{E}(\mathbf{x}^N)$ subject to the constraints $U_p{}^n(\mathbf{u}^n) \geq 0$ make the stage hamiltonian H^n stationary when a constraint is not at equality and minimize the hamiltonian (or make it*

stationary) with respect to constraints at equality or at nondifferentiable points.

It should be noted that, unlike continuous systems, it is *not* true that H^n is a constant for all n or, as we shall subsequently prove, that H^n is a minimum at stationary points. Equivalent formulations of the objective may be accommodated by defining new state equations as for continuous systems.

7.5 LAGRANGE MULTIPLIERS

The only essential generalization other than notation which we have introduced beyond the discussion of staged systems in Chap. 1 is the restriction on the allowable values of the decision. It follows, then, that in situations in which the optimal decisions are unconstrained the stationary condition on the staged hamiltonian should be derivable from the Lagrange multiplier rule. We show here that this is so.

We shall write the system equations and boundary restrictions as

$$-x_i^n + f_i^n(\mathbf{x}^{n-1}, \mathbf{u}^n) = 0 \qquad \begin{array}{l} i = 1, 2, \ldots, S \\ n = 1, 2, \ldots, N \end{array} \tag{1}$$

$$g_l(\mathbf{x}^N) = 0 \qquad l = 1, 2, \ldots, L \tag{2}$$

$$-q_k(\mathbf{x}^0) = 0 \qquad k = 1, 2, \ldots, K \tag{3}$$

The choice of positive or negative signs is motivated by a desire to relate the results directly to those of the previous section. The minimizing values of $\mathbf{u}^n$ for the objective $\mathcal{E}(\mathbf{x}^N)$ are found from stationary values of the lagrangian

$$\mathcal{L} = \mathcal{E}(\mathbf{x}^N) + \sum_{n=1}^{N} \sum_{i=1}^{S} \lambda_i^n [-x_i^n + f_i^n(\mathbf{x}^{n-1}, \mathbf{u}^n)]$$

$$+ \sum_{l=1}^{L} \nu_l g_l(\mathbf{x}^N) - \sum_{k=1}^{K} \eta_k q_k(\mathbf{x}^0) \tag{4}$$

Here we have introduced a multiplier λ_i^n for each of the constraint Eqs. (1), a multiplier ν_l for each Eq. (2), and η_k for each Eq. (3).

Setting partial derivatives with respect to u_j^n to zero, we obtain

$$\sum_{i=1}^{S} \gamma_i^n \frac{\partial f_i^n}{\partial u_j^n} = 0 \qquad j = 1, 2, \ldots, R \tag{5}$$

For $n \neq 1$ the partial derivatives with respect to x_i^{n-1} yield

$$-\lambda_i^{n-1} + \sum_{j=1}^{S} \lambda_j^n \frac{\partial f_j^n}{\partial x_i^{n-1}} = 0 \qquad \begin{array}{l} i = 1, 2, \ldots, S \\ n = 2, 3, \ldots, N \end{array} \tag{6}$$

Partial derivatives with respect to x_i^N give the equations

$$\frac{\partial \mathcal{E}}{\partial x_i^N} - \lambda_i^N + \sum_{l=1}^{L} \nu_l \frac{\partial g_l}{\partial x_i^N} = 0 \tag{7}$$

while derivatives with respect to x_i^0 give

$$\sum_{j=1}^{S} \lambda_j^1 \frac{\partial f_j^1}{\partial x_i^0} - \sum_{k=1}^{K} \eta_k \frac{\partial q_k}{\partial x_i^0} = 0 \tag{8a}$$

By defining λ_i^0 through Eq. (6) this last relation may be written

$$\lambda_i^0 - \sum_{k=1}^{K} \eta_k \frac{\partial q_k}{\partial x_i^0} = 0 \tag{8b}$$

Equations (5) to (8) are identical to the equations for the weak minimum principle when the decisions are unconstrained if the Lagrange multipliers λ_i^n are identified with the Green's functions γ_i^n. Surprisingly some authors have failed to recognize this simple relation and have devoted considerable space in the literature to the complicated derivation (or rederivation) through a "minimum principle" of results in unconstrained systems more easily (and frequently, previously) obtained through application of lagrangian methods. Indeed, the interpretation of the multipliers in Sec. 1.15 in terms of partial derivatives of the objective leads directly to the weak minimum principle with constraints, and it is only to simplify and unify later considerations of computation that we have adopted the form of presentation in this chapter.

7.6 OPTIMAL TEMPERATURES FOR CONSECUTIVE REACTIONS

In order to illustrate the difficulties which arise in even the simplest situations when constraints are imposed on the decisions in a staged system it is helpful to return to the example of Sec. 1.12. We desire the optimum sequence of operating temperatures in carrying out the chemical reaction

$$X \rightarrow Y \rightarrow \text{products}$$

in N stirred-tank reactors. We shall retain x and y to denote concentrations and u for temperature, but we now use superscripts to signify the stage number.

Equations (6) defining the state in Sec. 1.12 are

$$0 = x^{n-1} - x^n - \theta^n k_1(u^n) F(x^n) \tag{1a}$$

$$0 = y^{n-1} - y^n + \nu \theta^n k_1(u^n) F(x^n) - \theta^n k_2(u^n) G(y^n) \tag{1b}$$

and the goal is to choose $u^1, u^2, \ldots, u^N$ in order to minimize

$$\mathcal{E} = -\rho x^N - y^N \tag{2}$$

In Chap. 1 we found that if u^n is not constrained, the design problem is reduced to the simultaneous solution of two coupled nonlinear difference equations with an iterative search over one initial condition, a rather simple calculation. We now add the restriction that u^n lie within bounds

$$u_* \leq u^n \leq u^* \tag{3}$$

Equations (1) are not of the form for which we have developed the theory, which would be

$$x^n = f^n(x^{n-1}, y^{n-1}, u^n) \tag{4a}$$
$$y^n = g^n(x^{n-1}, y^{n-1}, u^n) \tag{4b}$$

Since we need only partial derivatives of f^n and g^n in order to apply the theory, however, this causes no difficulty. For example, partially differentiating Eq. (1a) with respect to x^{n-1} yields

$$1 - \frac{\partial f^n}{\partial x^{n-1}} - \theta^n k_1(u^n) F'(x^n) \frac{\partial f^n}{\partial x^{n-1}} = 0 \tag{5a}$$

or

$$\frac{\partial f^n}{\partial x^{n-1}} = \frac{1}{1 + \theta^n k_1(u^n) F'(x^n)} \tag{5b}$$

The other partial derivatives are similarly obtained, and we can write the multiplier equations

$$\gamma_1^{n-1} = \gamma_1^n \frac{\partial f^n}{\partial x^{n-1}} + \gamma_2^n \frac{\partial g^n}{\partial x^{n-1}} = \frac{\gamma_1^n}{1 + \theta^n k_1(u^n) F'(x^n)}$$
$$+ \frac{\gamma_2^n \nu \theta^n k_1(u^n) F'(x^n)}{[1 + \theta^n k_1(u^n) F'(x^n)][1 + \theta^n k_2(u^n) G'(y^n)]} \tag{6a}$$
$$\gamma_2^{n-1} = \gamma_1^n \frac{\partial f^n}{\partial y^{n-1}} + \gamma_2^n \frac{\partial g^n}{\partial y^{n-1}} = \frac{\gamma_2^n}{1 + \theta^n k_2(u^n) G'(y^n)} \tag{6b}$$

with boundary conditions

$$\gamma_1^N = \frac{\partial \mathcal{E}}{\partial x^N} = -\rho \tag{7a}$$

$$\gamma_2^N = \frac{\partial \mathcal{E}}{\partial y^N} = -1 \tag{7b}$$

Also,

$$
\frac{\partial H^n}{\partial u^n} = \gamma_1{}^n \frac{\partial f^n}{\partial u^n} + \gamma_2{}^n \frac{\partial g^n}{\partial u^n} = -\frac{\gamma_1{}^n \theta^n k_1'(u^n) F(x^n)}{1 + \theta^n k_1(u^n) F'(x^n)}
$$
$$
+ \frac{\gamma_2{}^n \nu \theta^n k_1'(u^n) F(x^n)}{[1 + \theta^n k_2(u^n) G'(y^n)][1 + \theta^n k_1(u^n) F'(x^n)]}
$$
$$
- \frac{\gamma_2{}^n \theta^n k_2'(u^n) G(y^n)}{1 + \theta^n k_2(u^n) G'(y^n)} \quad (8)
$$

The equation $\partial H^n/\partial u^n = 0$ will have a unique solution for variables of interest, in which case the optimum u^n for solutions which lie outside the bounds established by Eq. (3) is at the nearest bound.

Because Eq. (6b) is independent of $\gamma_1{}^n$ and $\gamma_2{}^n$ cannot go to zero, we may divide Eqs. (6) to (8) by $\gamma_2{}^n$ and by defining

$$
\zeta^n = \frac{\gamma_1{}^n}{\gamma_2{}^n} \quad (9)
$$

we obtain

$$
\zeta^n = \frac{1 + \theta^n k_1(u^n) F'(x^n)}{1 + \theta^n k_2(u^n) G'(y^n)} \zeta^{n-1} - \frac{\nu \theta^n k_1(u^n) F'(x^n)}{1 + \theta^n k_2(u^n) G'(y^n)} \quad (10)
$$
$$
\zeta^N = \rho \quad (11)
$$

with solutions of $\partial H^n/\partial u^n = 0$ at solutions of

$$
-\zeta^n \frac{\theta^n k_1'(u^n) F(x^n)}{1 + \theta^n k_1(u^n) F'(x^n)} + \frac{\nu \theta^n k_1'(u^n) F(x^n)}{[1 + \theta^n k_2(u^n) G'(y^n)][1 + \theta^n k_1(u^n) F'(x^n)]}
$$
$$
- \frac{\theta^n k_2'(u^n) G(y^n)}{1 + \theta^n k_2(u^n) G'(y^n)} = 0 \quad (12)
$$

The required computational procedure is then as follows:

1. Assume ζ^1.
2. Solve Eqs. (1) and (12) simultaneously for x^1, y^1, u^1.
3. If u^1 exceeds a bound, set u^1 to the nearest bound and recompute x^1, y^1 from Eqs. (1).
4. Compute x^2, y^2, u^2, ζ^2 simultaneously from Eqs. (1), (10), and (12).
5. If u^2 exceeds a bound, set u^2 to the nearest bound and recompute x^2, y^2, ζ^2 from Eqs. (1) and (10).
6. Repeat steps 4 and 5 for $n = 3, 4, \ldots, N$.
7. Repeat steps 1 to 6 for changing ζ^1 until $\zeta^N = \rho$.

This is substantially more computation than required for the unconstrained case. Note that we could have chosen x^N and y^N and computed backward, adjusting x^N and y^N until matching the specified x^0, y^0, but that without some rational iteration procedure this requires substantially more calculations than a one-dimensional search on ζ^1.

In the special case that

$$F(x^n) = x^n \tag{13a}$$
$$G(y^n) = y^n \tag{13b}$$

the problem simplifies greatly. Defining

$$\eta^n = \frac{y^n}{x^n} \tag{14}$$

we combine Eqs. (1) to yield

$$\eta^{n-1} = \frac{1 + \theta^n k_2(u^n)}{1 + \theta^n k_1(u^n)} \eta^n - \frac{\nu \theta^n k_1(u^n)}{1 + \theta^n k_1(u^n)} \tag{15}$$

while Eqs. (10) and (12) become, respectively,

$$\zeta^{n-1} = \frac{1 + \theta^n k_2(u^n)}{1 + \theta^n k_1(u^n)} \zeta^n + \frac{\nu \theta^n k_1(u^n)}{1 + \theta^n k_1(u^n)} \tag{16}$$

$$\zeta^n \frac{k_1'(u^n)}{1 + \theta^n k_1(u^n)} + \eta^n \frac{k_2'(u^n)}{1 + \theta^n k_2(u^n)}$$
$$- \frac{\nu k_1'(u^n)}{[1 + \theta^n k_1(u^n)][1 + \theta^n k_2(u^n)]} = 0 \tag{17}$$

We may now either follow the procedure outlined above or choose η^N, compute u^N since we are given ζ^N, then compute η^{N-1}, ζ^{N-1}, u^{N-1}, etc., by means of Eqs. (15) to (17), iterating on η^N until matching the specified feed ratio η^0. The advantage of this latter calculation is that whatever the η^0 corresponding to the chosen η^N, the complete set of necessary conditions has been used and the policy is optimal for that particular η^0. Thus a complete set of optimal policies is mapped out in the course of the computation, which is not the case in calculations which assume ζ^0.

7.7 THE STRONG MINIMUM PRINCIPLE: A COUNTEREXAMPLE

At the end of Sec. 7.4 we stated that it is not generally true for discrete systems that the stage hamiltonian assumes a minimum at stationary points. From the discussion in Sec. 7.5 it may be seen that such a minimization would correspond to minimizing the lagrangian, which we showed to be incorrect in Sec. 1.8, but because of substantial confusion about this point in the engineering literature we shall pursue it somewhat more. In this section we shall demonstrate a counterexample to a strong minimum principle; in the next section we shall construct some of those situations for which a strong minimum principle does exist.

Consider the system

$$x_1^n = x_1^{n-1}(1 + x_2^{n-1}) - \tfrac{1}{2}(u^n)^2 \qquad x_1^0 = -\tfrac{1}{4} \tag{1a}$$
$$x_2^n = 4x_1^{n-1} - 2x_2^{n-1} + u^n \qquad x_2^0 = +1 \tag{1b}$$

where u^1 and u^2 are to be chosen subject to

$$0 < u* \leq u^n \qquad n = 1, 2 \tag{2}$$

in order to minimize

$$\mathcal{E} = -x_1^2 \tag{3}$$

By direct substitution

$$\mathcal{E} = -\tfrac{1}{2}[1 + (u^1)^2](2 - u^1) + \tfrac{1}{2}(u^2)^2 \tag{4}$$

and the optimal values are $\bar{u}^1 = +1$, $\bar{u}^2 = u*$.

Now

$$H^n = \gamma_1{}^n[x_1{}^{n-1} + x_1{}^{n-1}x_2{}^{n-1} - \tfrac{1}{2}(u^n)^2]$$
$$+ \gamma_2{}^n(4x_1{}^{n-1} - 2x_2{}^{n-1} + u^n) \tag{5}$$

$$\frac{\partial H^1}{\partial u^1} = -u^1\gamma_1{}^1 + \gamma_2{}^1 \tag{6}$$

$$\frac{\partial^2 H^1}{\partial (u^1)^2} = -\gamma_1{}^1 \tag{7}$$

But

$$\gamma_1{}^1 = \frac{\partial H^2}{\partial x_1{}^1} = -(1 + x_2{}^1) = -(-2 + \bar{u}^1) \tag{8a}$$

$$\gamma_2{}^1 = \frac{\partial H^2}{\partial x_2{}^1} = -x_1{}^1 = \tfrac{1}{2}\,[1 + (\bar{u}^1)^2] \tag{8b}$$

Thus, $\partial H^1/\partial u^1$ does indeed vanish at $\bar{u}^1 = 1$. But from Eqs. (7) and (8a),

$$\frac{\partial^2 H^1}{\partial (u^1)^2} = \bar{u}^1 - 2 = -1 < 0 \tag{9}$$

which corresponds to a maximum of H^1 rather than a minimum.

7.8 SECOND-ORDER VARIATIONAL EQUATIONS

We shall now examine circumstances under which a strong minimum principle, in which the stage hamiltonian is in fact minimized by the optimum values of $\mathbf{u}^n$, can be shown to exist. The significance of such a result is clear in the light of the equivalent results for continuous systems discussed in the previous chapter, and we shall find subsequently that there are other computational implications. Our procedure is again the application of Picard's iteration method.

We suppose that $\bar{\mathbf{x}}^0$ and $\bar{\mathbf{u}}^0$ have been specified and that we now consider variations $\boldsymbol{\delta}u^n$, $n = 1, 2, \ldots , N$ such that $|\delta u_i{}^n| \leq \epsilon$. $\boldsymbol{\delta}\mathbf{x}^0$ will be taken to be zero. The variational equations corresponding to Eqs.

(1) of Sec. 7.1 are then, to second order in ϵ,

$$
\delta x_i^n = \sum_{j=1}^{S} \frac{\partial f_i^n}{\partial x_j^{n-1}} \delta x_j^{n-1} + \sum_{k=1}^{R} \frac{\partial f_i^n}{\partial u_k^n} \delta u_k^n
$$

$$
+ \frac{1}{2} \sum_{j,k=1}^{S} \frac{\partial^2 f_i^n}{\partial x_j^{n-1} \partial x_k^{n-1}} \delta x_j^{n-1} \delta x_k^{n-1} + \sum_{j=1}^{S} \sum_{k=1}^{R} \frac{\partial^2 f_i^n}{\partial x_j^{n-1} \partial u_k^n} \delta u_k^n \delta x_j^{n-1}
$$

$$
+ \frac{1}{2} \sum_{j,k=1}^{R} \frac{\partial^2 f_i^n}{\partial u_j^n \partial u_k^n} \delta u_j^n \delta u_k^n + o(\epsilon^2) \quad (1)
$$

If we look upon this as a linear nonhomogeneous system for which the appropriate Green's functions Γ_{ij}^{nm} satisfy the equation

$$
\Gamma_{ij}^{n,m-1} = \sum_{k=1}^{S} \Gamma_{ik}^{nm} \frac{\partial f_k^m}{\partial x_j^{n-1}} \quad (2)
$$

then Green's identity becomes

$$
\delta x_i^n = \sum_{m=1}^{n} \sum_{j=1}^{S} \sum_{k=1}^{R} \Gamma_{ij}^{nm} \left(\frac{\partial f_j^m}{\partial u_k^m} \delta u_k^m + \frac{1}{2} \sum_{p=1}^{R} \frac{\partial^2 f_j^m}{\partial u_k^m \partial u_p^m} \delta u_k^m \delta u_p^m \right)
$$

$$
+ \frac{1}{2} \sum_{m=1}^{n} \sum_{j,k,p=1}^{S} \Gamma_{ij}^{nm} \frac{\partial^2 f_j^m}{\partial x_k^{m-1} \partial x_p^{m-1}} \delta x_k^{m-1} \delta x_p^{m-1}
$$

$$
+ \sum_{m=1}^{n} \sum_{j,k=1}^{S} \sum_{p=1}^{R} \Gamma_{ij}^{nm} \frac{\partial^2 f_j^m}{\partial x_k^{m-1} \partial u_p^m} \delta u_p^m \delta x_k^{m-1} + o(\epsilon^2) \quad (3)
$$

Evaluating Eq. (3) for $n = m - 1$ and substituting back into the right-hand side, we obtain, finally, an explicit representation for $\delta \mathbf{x}^N$

$$
\delta x_i^N = \sum_{n=1}^{N} \sum_{j=1}^{S} \sum_{k=1}^{R} \Gamma_{ij}^{Nn} \left(\frac{\partial f_j^n}{\partial u_k^n} \delta u_k^n + \frac{1}{2} \sum_{p=1}^{R} \frac{\partial^2 f_j^n}{\partial u_k^n \partial u_p^n} \delta u_k^n \delta u_p^n \right)
$$

$$
+ \frac{1}{2} \sum_{n=1}^{N} \sum_{j,k,p=1}^{S} \Gamma_{ij}^{Nn} \frac{\partial^2 f_j^n}{\partial x_k^{n-1} \partial x_p^{n-1}} \left(\sum_{m=1}^{n-1} \sum_{q=1}^{S} \sum_{v=1}^{R} \Gamma_{kq}^{n-1,m} \frac{\partial f_q^m}{\partial u_v^m} \delta u_v^m \right)
$$

$$
\times \left(\sum_{r=1}^{n-1} \sum_{w=1}^{S} \sum_{z=1}^{R} \Gamma_{pw}^{n-1,r} \frac{\partial f_w^r}{\partial u_z^r} \delta u_z^r \right) + \sum_{n=1}^{N} \sum_{j,k=1}^{S} \sum_{p=1}^{R} \Gamma_{ij}^{Nn} \frac{\partial^2 f_j^n}{\partial x_k^{n-1} \partial u_p^n} \delta u_p^n
$$

$$
\times \left(\sum_{m=1}^{n-1} \sum_{q=1}^{S} \sum_{v=1}^{R} \Gamma_{kq}^{n-1,m} \frac{\partial f_q^m}{\partial u_v^m} \delta u_v^m \right) + o(\epsilon^2) \quad (4)
$$

We shall assume, for simplicity, that the objective to be minimized, $\mathcal{E}(\mathbf{x}^N)$, is linear in the components x_i^N. By suitable definition of additional variables we can always accomplish this. If we further assume that $\mathbf{x}^N$ is completely unconstrained, we obtain

$$\delta\mathcal{E} = \sum_{i=1}^{S} \frac{\partial\mathcal{E}}{\partial x_i^N} \delta x_i^N = \sum_{i=1}^{S} \gamma_i^N \delta x_i^N \geq 0 \tag{5}$$

and multiplying Eqs. (2) and (3) by γ_i^N and summing we obtain, with the usual definition of the stage hamiltonian,

$$\begin{aligned}
\delta\mathcal{E} = {}& \sum_{n=1}^{N} \left(\sum_{k=1}^{R} \frac{\partial H^n}{\partial u_k^n} \delta u_k^n + \frac{1}{2} \sum_{j,k=1}^{R} \frac{\partial^2 H^n}{\partial u_j^n \, \partial u_k^n} \delta u_j^n \, \delta u_k^n \right) \\
&+ \frac{1}{2} \sum_{n=1}^{N} \sum_{k,p=1}^{S} \frac{\partial^2 H^n}{\partial x_k^{n-1} \, \partial x_p^{n-1}} \left(\sum_{m=1}^{n-1} \sum_{j=1}^{S} \sum_{v=1}^{R} \Gamma_{kj}^{n-1,m} \frac{\partial f_j^m}{\partial u_v^m} \delta u_v^m \right) \\
&\times \left(\sum_{r=1}^{n-1} \sum_{q=1}^{S} \sum_{z=1}^{R} \Gamma_{pq}^{n-1,r} \frac{\partial f_q^r}{\partial u_z^r} \delta u_z^r \right) + \sum_{n=1}^{N} \sum_{k=1}^{S} \sum_{p=1}^{R} \frac{\partial^2 H^n}{\partial x_k^{n-1} \, \partial u_p^n} \delta u_p^n \\
&\times \left(\sum_{m=1}^{n-1} \sum_{q=1}^{S} \sum_{j=1}^{R} \Gamma_{kq}^{n-1,m} \frac{\partial f_q^m}{\partial u_j^m} \delta u_j^m \right) + o(\epsilon^2) \geq 0 \tag{6}
\end{aligned}$$

Unlike the continuous case, in which Δt can be made arbitrarily small, there is no general way in which the last two terms can be made to vanish, and this is the source of the difference between discrete and continuous systems. Several special cases, however, can be treated. Because we are interested in the character of stationary points, we shall restrict nonvanishing δu_k^n to those corresponding to interior values of the optimal $\bar{\mathbf{u}}^n$.

Linear separable equations have the form

$$x_i^n = \sum_{j=1}^{S} A_{ij}^n x_j^{n-1} + b_i^n(\mathbf{u}^n) \tag{7}$$

where the A_{ij}^n are constants. In that case both the second partial derivatives of H^n with respect to components only of $\mathbf{x}^{n-1}$ and $\mathbf{x}^{n-1}$ and $\mathbf{u}^n$ vanish identically, and Eq. (6) reduces to

$$\delta\mathcal{E} = \sum_{n=1}^{N} \sum_{k=1}^{R} \frac{\partial H^n}{\partial u_k^n} \delta u_k^n + \frac{1}{2} \sum_{n=1}^{N} \sum_{j,k=1}^{R} \frac{\partial^2 H^n}{\partial u_j^n \, \partial u_k^n} \delta u_j^n \, \delta u_k^n + o(\epsilon^2) \geq 0 \tag{8}$$

The first term is zero by virtue of the weak minimum principle and the assumption concerning admissible $\delta\mathbf{u}^n$. It follows, then, that the hessian matrix of second derivatives of H^n must be positive definite, corresponding to a minimum. Furthermore, since the $\delta\mathbf{u}^n$ are arbitrary, Eq. (8)

establishes the sufficiency of the strong minimum principle for the linear separable case as well.

Linear nonseparable equations are of the form

$$x_i{}^n = \sum_{j=1}^{S} A_{ij}{}^n(\mathbf{u}^n) x_j{}^{n-1} + b_i{}^n(\mathbf{u}^n) \tag{9}$$

Here the mixed second partial derivatives do not vanish. By choosing the special variation, however,

$$\delta \mathbf{u}^m = \begin{cases} \delta \mathbf{u}^* & m = n^* \\ \mathbf{0} & m \neq n^* \end{cases} \tag{10}$$

for some specified n^* the last term in Eq. (6) also vanishes, and we obtain, using the weak minimum principle,

$$\delta \mathcal{E} = \frac{1}{2} \sum_{j,k=1}^{R} \frac{\partial^2 H^{n*}}{\partial u_j{}^{n*} \partial u_k{}^{n*}} \delta u_j^* \, \delta u_k^* + o(\epsilon^2) \geq 0 \tag{11}$$

which establishes the necessity of the strong minimum principle for this case also. That it is not sufficient is apparent from Eq. (6).

If there is a *single state variable*, with the scalar equation

$$x^n = f^n(x^{n-1}, \mathbf{u}^n) \tag{12}$$

then, since γ^n will be nonzero, the vanishing of $\partial H^n / \partial u_k{}^n$ implies the vanishing of $\partial f^n / \partial u_k{}^n$. It follows from Eq. (6), then, that the strong minimum principle is both necessary and sufficient. Furthermore, f^n always takes on an extreme value, and if $\partial f^n / \partial x^{n-1}$ is positive for all n, then γ^n is always of one sign. If, in addition, x^n must always be positive for physical reasons, the policy is disjoint, which means that minimizing H^n in order to minimize or maximize x^N is equivalent to minimizing or maximizing x^n at each stage by choice of $\mathbf{u}^n$. This should recall the discussion of optimal temperature profiles for single reactions in Sec. 3.5, and, indeed, an identical result holds for the choice of temperatures for a single reaction occurring in a sequence of staged reactors.

We emphasize that the results obtained here are applicable only for a linear objective.† If $\mathcal{E}$ is nonlinear, an additional term is required in Eq. (6). In particular, a linear system with a nonlinear objective is formally equivalent to a nonlinear system with a linear objective, for which the strong minimum principle does not apply. A counterexample

---

† With the exception of those for a single state variable.

which is linear and separable is

$$x_1{}^n = -3x_1{}^{n-1} - \tfrac{1}{2}(u^n)^2 \qquad x_1{}^0 = \text{given} \qquad (13a)$$
$$x_2{}^n = x_2{}^{n-1} + u^n \qquad x_2{}^0 = 0 \qquad (13b)$$
$$\mathcal{E} = -x_1{}^2 - (x_2{}^2)^2 \qquad (13c)$$

We leave the details of the demonstration to the reader.

7.9 MIXED AND STATE-VARIABLE CONSTRAINTS

For staged systems it is not necessary to distinguish between constraints on the state of the system

$$Q_k{}^n(\mathbf{x}^n) \geq 0 \qquad (1)$$

and mixed constraints

$$Q_k{}^n(\mathbf{x}^n,\mathbf{u}^n) \geq 0 \qquad (2)$$

Through the use of the state equation both types of constraint can always be put in the form

$$Q_k{}^n(\mathbf{x}^{n-1},\mathbf{u}^n) \geq 0 \qquad (3)$$

If the constraint is not violated by the optimal trajectory, the theory developed in Secs. 7.4 and 7.8 remains valid. Thus, we need only consider the case where equality must hold in Eq. (3).

As with the continuous case, we first suppose that there is a single stage decision variable u^n and only one constraint is at equality. We then have

$$Q^n(\mathbf{x}^{n-1},u^n) = 0 \qquad (4)$$

and the variational relationship

$$\delta Q^n = \sum_{i=1}^{S} \frac{\partial Q^n}{\partial x_i{}^{n-1}} \delta x_i{}^{n-1} + \frac{\partial Q^n}{\partial u^n} \delta u^n = 0 \qquad (5)$$

Coupled with the variational equation

$$\delta x_i{}^n = \sum_{j=1}^{S} \frac{\partial f_i{}^n}{\partial x_j{}^{n-1}} \delta x_j{}^{n-1} + \frac{\partial f_i{}^n}{\partial u^n} \delta u^n \qquad (6)$$

we may solve for δu^n and write

$$\delta x_i{}^n = \sum_{j=1}^{S} \left[\frac{\partial f_i{}^n}{\partial x_j{}^{n-1}} - \frac{\partial f_i{}^n}{\partial u^n} \left(\frac{\partial Q^n}{\partial u^n} \right)^{-1} \frac{\partial Q^n}{\partial x_j{}^{n-1}} \right] \delta x_j{}^{n-1} \qquad (7)$$

The decision u^n is determined by Eq. (4), and the Green's function corresponding to the difference equation (7) satisfies

$$\gamma_j^{n-1} = \sum_{i=1}^{S} \left[\frac{\partial f_i^n}{\partial x_j^{n-1}} - \frac{\partial f_i^n}{\partial u^n} \left(\frac{\partial Q^n}{\partial u^n} \right)^{-1} \frac{\partial Q^n}{\partial x_j^{n-1}} \right] \gamma_i^n \tag{8}$$

Equation (3) of Sec. 7.4 for $\delta\mathcal{E}$ remains unchanged provided we exclude values of n at which Eq. (4) holds, and the weak minimum principle must be satisfied for all stages where a constraint is not at equality.

In the general case of K_1 independent constraints and $R_1 \geq K_1$ components of $\mathbf{u}^n$ we find, as in Sec. 6.15, that the Green's vector must satisfy the equation

$$\gamma_i^{n-1} = \sum_{j=1}^{S} \left(\frac{\partial f_j^n}{\partial x_i^{n-1}} - \sum_{p,k=1}^{K_1} \frac{\partial f_j^n}{\partial u_p^n} S_{pk}^n \frac{\partial Q_k^n}{\partial x_i^{n-1}} \right) \gamma_j^n \tag{9}$$

where S_{pk}^n is the matrix such that

$$\sum_{k=1}^{K_1} S_{pk}^n \frac{\partial Q_k^n}{\partial u_i^n} = \delta_{ip} \qquad i, p = 1, 2, \ldots, K_1 \tag{10}$$

It follows then that the $R_1 - K_1$ independent components of $\mathbf{u}^n$ must be chosen to satisfy the weak minimum principle.

BIBLIOGRAPHICAL NOTES

Section 7.1: Finite difference representations might arise either because of time or space discretization of continuous processes or because of a natural staging. For the former see, for example,

G. A. Bekey: in C. T. Leondes (ed.), "Modern Control Systems Theory," McGraw-Hill Book Company, New York, 1965
J. Coste, D. Rudd, and N. R. Amundson: *Can. J. Chem. Eng.*, **39**:149 (1961)

Staged reaction and separation processes are discussed in many chemical engineering texts, such as

R. Aris: "Introduction to the Analysis of Chemical Reactors," Prentice-Hall, Inc., Englewood-Cliffs, N.J., 1965
B. D. Smith: "Design of Equilibrium Stage Processes," McGraw-Hill Book Company, New York, 1963

Section 7.2: The linear analysis for difference equations used here is discussed in

M. M. Denn and R. Aris: *Ind. Eng. Chem. Fundamentals*, **4**:7 (1965)
T. Fort: "Finite Differences and Difference Equations in the Real Domain," Oxford University Press, Fair Lawn, N.J., 1948
L. A. Zadeh and C. A. Desoer: "Linear System Theory," McGraw-Hill Book Company, New York, 1963

Sections 7.3 and 7.4: The analysis follows the paper by Denn and Aris cited above. Similar variational developments may be found in

S. S. L. Chang: *IRE Intern. Conv. Record,* **9**(4):48 (1961)
L. T. Fan and C. S. Wang: "Discrete Maximum Principle," John Wiley & Sons, Inc., New York, 1964
S. Katz: *Ind. Eng. Chem. Fundamentals,* **1**:226 (1962)

The paper by Katz contains a subtle error and erroneously proves a strong minimum principle. This incorrect strong result is cited, but not necessarily required, in several of the examples and references in the book by Fan and Wang.

Section 7.5: The lagrangian development was used by Horn prior to the work cited above:

F. Horn: *Chem. Eng. Sci.,* **15**:176 (1961)

Section 7.6: This reactor problem has been studied in the papers cited above by Horn and Denn and Aris and in the book

R. Aris: "The Optimal Design of Chemical Reactors," Academic Press, Inc., New York, 1961

A number of other applications are collected in these references and in the book by Fan and Wang.

Section 7.7: The observation that a strong minimum principle is not available in general for discrete systems is due to Rozenoer:

L. Rozenoer: *Automation Remote Contr.,* **20**:1517 (1959)

A counterexample to a strong minimum principle was first published by Horn and Jackson in

F. Horn and R. Jackson: *Ind. Eng. Chem. Fundamentals,* **4**:110 (1965)

Section 7.8: The development follows

M. M. Denn and R. Aris: *Chem. Eng. Sci.,* **20**:373 (1965)

Similar results are obtained in

F. Horn and R. Jackson: *Intern. J. Contr.,* **1**:389 (1965)

Several workers, notably Holtzman and Halkin, have examined the set-theoretic foundations of the necessary conditions for optimal discrete systems with care. Some of their work and further references are contained in

H. Halkin: *SIAM J. Contr.,* **4**:90 (1966)
J. M. Holtzman and H. Halkin: *SIAM J. Contr.,* **4**:263 (1966)

Section 7.9: These results are obtained in the paper by Denn and Aris cited for Sec. 7.2.

PROBLEMS

7.1. Consider the necessity and sufficiency of a strong minimum principle for

$$x_i^n = \sum_{j=1}^{S} a_{ij} x_j^{n-1} + b_i u^n$$

$$\min \varepsilon = \frac{1}{2} \sum_{n=1}^{N} \left[\sum_{i,j=1}^{S} x_i^n Q_{ij} x_j^n + \rho(u^n)^2 \right]$$

7.2. Denbigh has introduced the pseudo-first-order reaction sequence

$$\begin{array}{ccc} X_1 \xrightarrow{k_1} X_2 \xrightarrow{k_3} X_3 \\ \downarrow k_2 & \downarrow k_4 \\ Y_1 & Y_2 \end{array}$$

where X_3 is the desired product. Taking the reaction-rate coefficients as

$$k_i(T) = k_{i0} \exp\left(\frac{-E_i'}{T}\right)$$

where T is the temperature, the equations describing reaction in a sequence of stirred-tank reactors are

$$x_1^{n-1} = x_1^n[1 + u_1^n(1 + u_2^n)]$$

$$x_2^{n-1} = -u_1^n x_1^n + x_2^n \left[1 + 0.01 u_1^n \left(1 + \frac{1}{u_2^n} \right) \right]$$

$$x_3^{n-1} = -0.01 u_1^n x_2^n + x_3^n$$

Here, u_1 is the product of reactor residence time and k_1, and u_2 is the ratio k_2/k_1. Taken together, they uniquely define temperature and residence time. The following values of Denbigh's have been used:

$$k_2 = 10^4 k_1 \exp\left(\frac{-3,000}{T}\right) \qquad k_3 = 10^{-2} k_1 \qquad k_4 = 10^{-6} k_1 \exp\left(\frac{3,000}{T}\right)$$

Aris has introduced bounds

$$0 \le T \le 394 \qquad 0 \le u_1 \le 2,100$$

For initial conditions $x_1^0 = 1$, $x_2^0 = x_3^0 = 0$ determine the maximum conversion to X_3 in a three-stage reactor. *Hint:* Computation will be simplified by deriving bounds on initial values γ_i^0 and/or final values x_i^3. The optimal conversion is 0.549.

7.3. Obtain the equations defining an approximate nonlinear feedback control for the system described by the nonlinear difference equation

$$x_{n+2} = F(x_{n+1}, x_n, u_n)$$

$$\min \varepsilon = \frac{1}{2} \sum_{n=1}^{N} (x_n^2 + \rho u_n^2)$$

7.4. Repeat Prob. 1.13 using the formalism of the discrete minimum principle.   Extend to the case in which T_n is bounded from above and below.

7.5. Consider the reaction $X \to Y \to Z$ in a sequence of stirred tanks.   This is defined by Eqs. (1) of Sec. 7.6 with the added equation

$$x^n + y^n + z^n = \text{const}$$

Derive the procedure for maximizing z^N.   Specialize to the case

$$F(x^n) = x^n \qquad G(y^n) = y^n$$

and compare your procedure for computational complexity with the one used in the book by Fan and Wang for studying an enzymatic reaction.

8
Optimal and Feedback Control

8.1 INTRODUCTION

Many of the examples considered in the preceding chapters have been
problems in process control, where we have attempted to use the optimization theory in order to derive a feedback control system. Even in
some of the simple examples which we have studied the derivation of a
feedback law has been extremely difficult or possible only as an approximation; in more complicated systems it is generally impossible. Furthermore, we have seen that the optimal feedback control for the same system under different objectives will have a significantly different form,
although the objectives might appear physically equivalent.

In this chapter we shall briefly touch upon some practical attacks
on these difficulties. We first consider the linear servomechanism problem and show how, for a linear system, the optimization theory completely determines the feedback and feedforward gains in a linear control
system and how classical three-mode feedback control may be viewed as
a natural consequence of optimal control. The problem of ambiguity of

objective will be similarly attacked by considering another particularly easily implementable feedback system and solving the inverse problem, thus defining a class of problems for which a reasonable optimum control is known.

8.2 LINEAR SERVOMECHANISM PROBLEM

The linear servomechanism problem is the control of a linear system subject to outside disturbances in such a way that it follows a prescribed motion as closely as possible.   We shall limit our discussion here to systems with a single control variable and disturbances which are approximately constant for a time much longer than the controlled system response time, and we shall assume that the "motion" we wish to follow is the equilibrium value $\mathbf{x} = \mathbf{0}$.   Thus, the state is described by

$$\dot{x}_i = \sum_j A_{ij}x_j + b_iu + d_i \tag{1}$$

with $\mathbf{A}$, $\mathbf{b}$, and $\mathbf{d}$ consisting of constant components, and we shall use a quadratic-error criterion,

$$\mathcal{E} = \frac{1}{2} \int_0^\theta \left(\sum_{i,j} x_iC_{ij}x_j + u^2 \right) dt \tag{2}$$

$\mathbf{C}$ is symmetric, with $C_{ij} = C_{ji}$.   The u^2 term may be looked upon as a penalty function or simply a mathematical device in order to obtain the desired result.   Multiplying by a constant does not alter the result, so that there is no loss of generality in setting the coefficient of u^2 to unity. The *linear regulator problem* is the special case with $\mathbf{d} = \mathbf{0}$.

The hamiltonian for this system is

$$H = \frac{1}{2} \sum_{i,j} x_iC_{ij}x_j + \frac{1}{2}u^2 + \sum_{i,j} \gamma_iA_{ij}x_j + u \sum_i \gamma_ib_i + \sum_j \gamma_id_i \tag{3}$$

with multiplier equations

$$\dot{\gamma}_i = -\frac{\partial H}{\partial x_i} = -\sum_j C_{ij}x_j - \sum_k \gamma_kA_{ki} \tag{4}$$

The partial derivatives of H with respect to u are

$$\frac{\partial H}{\partial u} = u + \sum_i \gamma_ib_i \tag{5a}$$

$$\frac{\partial^2 H}{\partial u^2} = 1 > 0 \tag{5b}$$

Equation (5a) defines the optimum when set to zero

$$u = -\sum_i \gamma_i b_i \tag{6}$$

while Eq. (5b) ensures a minimum.   We have already established in Sec. 6.20 that the minimum principle is sufficient for an optimum here.

We seek a solution which is a linear combination of the state variables and forcing functions.   Thus,

$$\gamma_i = \sum_j M_{ij} x_j + \sum_j D_{ij} d_j \tag{7}$$

Differentiating,

$$\dot{\gamma}_i = \sum_j \dot{M}_{ij} x_j + \sum_j M_{ij} \dot{x}_j + \sum_j \dot{D}_{ij} d_j \tag{8}$$

and substituting Eqs. (1), (6), and (7),

$$\dot{\gamma}_i = \sum_j \dot{M}_{ij} x_j + \sum_{j,k} M_{ij} A_{jk} x_k$$
$$- \sum_j M_{ij} b_j \left(\sum_{k,l} M_{lk} x_k b_l + \sum_{k,l} D_{lk} d_k b_l \right) + \sum_j M_{ij} d_j + \sum_j \dot{D}_{ij} d_j \tag{9a}$$

while Eq. (4) becomes

$$\dot{\gamma}_i = -\sum_j C_{ij} x_j - \sum_k \left(\sum_l M_{kl} x_l + \sum_l D_{kl} d_l \right) A_{ki} \tag{9b}$$

Equating the coefficients of each component of **x** and of **d**, we then obtain the two equations

$$\dot{M}_{ij} + \sum_k (M_{ik} A_{kj} + M_{kj} A_{ki}) - \left(\sum_k M_{ik} b_k \right) \left(\sum_l b_l M_{lj} \right) + C_{ij} = 0 \tag{10}$$

$$\dot{D}_{ij} + \sum_k A_{ki} D_{kj} - \left(\sum_k M_{ik} b_k \right) \left(\sum_l b_l D_{lj} \right) + M_{ij} = 0 \tag{11}$$

Equation (10) is a quadratic Riccati differential equation, while Eq. (11) is a linear nonhomogeneous differential equation with variable coefficients. It should be observed that the symmetry of **C** implies a symmetric solution to Eq. (10), $M_{ij} = M_{ji}$.

We shall not consider transient solutions of Eqs. (10) and (11), for it can be established that as $\theta \to \infty$, we may obtain solutions which are constants.   Thus, the optimal control is

$$u = -\sum_{i,j} b_i M_{ij} x_j - \sum_{i,j} b_i D_{ij} d_j \tag{12}$$

where $\mathbf{M}$ and $\mathbf{D}$ are solutions of the algebraic equations

$$-\sum_k (M_{ik}A_{kj} + M_{kj}A_{ki}) + \left(\sum_k M_{ik}b_k\right)\left(\sum_l M_{lj}b_l\right) = C_{ij} \tag{13}$$

$$\sum_k \left(A_{ki} - b_k \sum_l M_{il}b_l\right) D_{kj} = -M_{ij} \tag{14}$$

We have previously solved special cases of these equations.

The obvious difficulty of implementing the feedback control defined by Eqs. (12) to (14) is that the matrix $\mathbf{C}$ is not known with certainty. It will often be possible to estimate $\mathbf{C}$ to within a scale factor, however (the coefficient of u^2!), in which case Eqs. (13) and (14) define the interactions between variables and only a single parameter needs to be chosen.

8.3 THREE-MODE CONTROL

An interesting application of the linear servomechanism problem arises when one postulates that it is not the u^2 term which should be retained in the objective but a term $\dot{u}^2$. This would follow from the fact that the rate of change of controller settings is often the limiting factor in a design, rather than the magnitude of the setting. We might have, for example, for the forced second-order system

$$\ddot{x} + b\dot{x} + ax = u + d \tag{1}$$

or

$$\dot{x}_1 = x_2 \tag{2a}$$
$$\dot{x}_2 = -ax_1 - bx_2 + u + d \tag{2b}$$

the objective

$$\mathcal{E} = \frac{1}{2}\int_0^\infty (C_1 x_1^2 + C_2 x_2^2 + \dot{u}^2)\, dt \tag{3}$$

By defining u as a new state variable x_3, we may rewrite Eqs. (2) and (3) as

$$\dot{x}_1 = x_2 \tag{4a}$$
$$\dot{x}_2 = -ax_1 - bx_2 + x_3 + d \tag{4b}$$
$$\dot{x}_3 = w \tag{4c}$$

$$\mathcal{E} = \frac{1}{2}\int_0^\theta (C_1 x_1^2 + C_2 x_2^2 + w^2)\, dt \tag{5}$$

where w is taken as the control variable. This is of the form of the linear servomechanism problem, with the degenerate solution

$$w = -M_{11}x_1 - M_{12}x_2 - M_{13}x_3 - D_{32}d \tag{6}$$

Equations (13) and (14) of Sec. 8.2 reduce to the nine equations

$$C_1 = M_{13}{}^2 + 2bM_{12} \tag{7a}$$
$$0 = M_{13}M_{23} - M_{11} + aM_{12} + bM_{22} \tag{7b}$$
$$0 = M_{13}M_{33} - M_{12} + bM_{23} \tag{7c}$$
$$C_2 = M_{23}{}^2 - 2M_{12} + 2aM_{22} \tag{7d}$$
$$0 = M_{23}M_{33} - M_{13} + aM_{23} - M_{22} \tag{7e}$$
$$0 = M_{33}{}^2 - 2M_{23} \tag{7f}$$
$$M_{12} = bD_{22} + M_{13}D_{32} \tag{7g}$$
$$M_{22} = -D_{12} + aD_{22} + M_{23}D_{32} \tag{7h}$$
$$M_{23} = -D_{22} + M_{33}D_{32} \tag{7i}$$

The solution may be shown to be

$$D_{32} = M_{33} \tag{8a}$$
$$M_{13} = -bM_{33} + \sqrt{C_1} \tag{8b}$$
$$M_{23} = \tfrac{1}{2}M_{33}{}^2 \tag{8c}$$

where M_{33} is the unique positive solution of the quartic equation

$$M_{33}{}^4 + 4aM_{33}{}^3 + 4(a^2 + b)M_{33}{}^2 + 8(ab - \sqrt{C_1})M_{33} - 4C_2$$
$$- 8a\sqrt{C_1} = 0 \quad (9)$$

Thus,

$$w = \dot{u} = -(\sqrt{C_1} - bM_{33})x - \tfrac{1}{2}M_{33}{}^2\dot{x} - M_{33}u - M_{33}d \tag{10}$$

We may substitute for $u + d$ in Eq. (10) from Eq. (1) to obtain

$$\dot{u} = -\sqrt{C_1}\,x - (\tfrac{1}{2}M_{33}{}^2 + aM_{33})\dot{x} - M_{33}\ddot{x} \tag{11}$$

or, integrating,

$$u(t) = -\sqrt{C_1}\int_0^t x(\tau)\,d\tau - (\tfrac{1}{2}M_{33}{}^2 + aM_{33})x(t) - M_{33}\dot{x}(t) \tag{12}$$

This is a three-mode controller, in which the control action is taken as a weighted sum of the offset and the derivative and the integral of the offset. Such controls are widely used in industrial applications, the integral mode having the effect of forcing the system back to zero offset in the presence of a persistent disturbance. Standard control notation would be of the form

$$u(t) = -K\left[x(t) + \frac{1}{\tau_I}\int_0^t x(\tau)\,d\tau + \tau_D\frac{dx(t)}{dt}\right] \tag{13}$$

where

$$K = \tfrac{1}{2}M_{33}(M_{33} + 2a) \tag{14a}$$

$$\tau_I = \frac{M_{33}(M_{33} + 2a)}{2\sqrt{C_1}} \tag{14b}$$

$$\tau_D = \frac{2}{M_{33} + 2a} \tag{14c}$$

These equations can be rearranged to yield

$$K = \frac{2}{\tau_D^2}(1 - a\tau_D) \tag{15}$$

and the nonnegativity of K then bounds the derivative time constant

$$a\tau_D \leq 1 \tag{16}$$

The governing equation for the controlled system subject to step disturbances is

$$\dddot{x} + (a + K\tau_D)\ddot{x} + (b + K)\dot{x} + \frac{K}{\tau_I}x = 0 \tag{17}$$

or, defining the ratio of time constants by

$$\alpha = \frac{\tau_I}{\tau_D} \tag{18}$$

and substituting Eq. (15) into (17), the governing equation becomes

$$\dddot{x} + \left[\frac{2 - a\tau_D}{\tau_D}\right]\ddot{x} + \left[b + \frac{2}{\tau_D^2}(1 - a\tau_D)\right]\dot{x}$$
$$+ \left[\frac{2}{\alpha\tau_D^3}(1 - a\tau_D)\right]x = 0 \quad (19)$$

For a stable uncontrolled system $(a, b \geq 0)$ the necessary and sufficient condition that Eq. (19) have characteristic roots with negative real parts and that the controlled system be stable is

$$(\text{coefficient of } \ddot{x})(\text{coefficient of } \dot{x}) \geq (\text{coefficient of } x) \tag{20a}$$

or

$$\frac{2 - a\tau_D}{\tau_D}\left[b + \frac{2}{\tau_D^2}(1 - a\tau_D)\right] \geq \left[\frac{2}{\alpha\tau_D^3}(1 - a\tau_D)\right] \tag{20b}$$

The inequality can be converted to equality by introducing a *stability factor* $\sigma \geq 1$ and writing

$$\frac{2 - a\tau_D}{\tau_D}\left[b + \frac{2}{\tau_D^2}(1 - a\tau_D)\right] = \sigma\left[\frac{2}{\alpha\tau_D^3}(1 - a\tau_D)\right] \tag{21}$$

Here $\sigma = 1$ corresponds to marginal stability.

Equation (21) can be rewritten as a cubic equation in τ_D

$$-\frac{ab}{2}\tau_D{}^3 + (a^2 + b)\tau_D{}^2 + \left(\frac{\sigma}{\alpha} - 3\right)a\tau_D + \left(2 - \frac{\sigma}{\alpha}\right) = 0 \qquad (22)$$

It will often happen in practice that $a^2 \gg b$. For example, in the reactor-control problem used in Chaps. 4 and 5 and again in a subsequent section of this chapter $a = 0.295$, $b = 0.005$. In that case Eq. (22) approximately factors to

$$\left(-\frac{b}{2}\tau_D{}^2 + a\tau_D - 1\right)\left(a\tau_D - 2 + \frac{\sigma}{\alpha}\right) = 0 \qquad (23)$$

and the unique root satisfying the inequality $a\tau_D \leq 1$ is

$$a\tau_D = 2 - \frac{\sigma}{\alpha} \qquad (24)$$

The nonnegativity of τ_D and the upper bound on $a\tau_D$ then limits allowable values of σ/α to

$$2 \geq \frac{\sigma}{\alpha} \geq 1 \qquad (25)$$

Equation (15) for the gain can now be rewritten in terms of the single parameter σ/α as

$$\frac{K}{a^2} = \frac{2\alpha(\sigma - \alpha)}{(2\alpha - \sigma)^2} \qquad (26)$$

Further restrictions on the allowable controller design parameters are obtained by substituting Eqs. (14), (18), (24), and (26) into the quartic equation (9) and rearranging to solve for C_2, yielding

$$C_2 = \frac{2K}{\alpha^2\tau_D{}^2}\left[-\alpha^2 - \sigma + \alpha\sigma + \frac{b}{a^2}(2\alpha - \sigma)^2\right] \qquad (27)$$

A meaningful optimum requires that the objective be positive definite, or that C_2 be nonnegative. Using the approximation $b \gg a^2$ for consistancy, α and σ are then further related by the inequality

$$\alpha^2 - \alpha\sigma + \sigma \leq 0 \qquad (28a)$$

An equivalent form obtained by completing the square is

$$(\alpha - \tfrac{1}{2}\sigma)^2 + \sigma(1 - \tfrac{1}{4}\sigma) \leq 0 \qquad (28b)$$

from which it follows that

$$\sigma \geq 4 \qquad (29)$$

That is, optimality requires at least 4 times the minimum stability con-

dition! It further follows, then, from Eq. (25) that α is bounded from
below

$$\alpha \geq \tfrac{1}{2}\sigma \geq 2 \tag{30}$$

A sharp upper bound on the allowable ratio of time constants can
be obtained by rewriting Eq. (28b) as

$$(\alpha - \tfrac{1}{2}\sigma)^2 \leq \frac{\sigma^2}{4}\left(1 - \frac{4}{\sigma}\right) \tag{28c}$$

or, taking the square root,

$$\alpha - \tfrac{1}{2}\sigma \leq \frac{\sigma}{2}\sqrt{1 - \frac{4}{\sigma}} \tag{31}$$

Thus, the ratio τ_I/τ_D is strictly bounded by the one-parameter inequality

$$2 \leq \frac{\sigma}{2} \leq \frac{\tau_I}{\tau_D} \leq \frac{\sigma}{2}\left(1 + \sqrt{1 - \frac{4}{\sigma}}\right) \tag{32}$$

For σ in the range 4 to 6 the allowable values of τ_I/τ_D are restricted
to lie between 2 and 4.7. Standard control practice for settings for three-
mode systems is a value of τ_I/τ_D of approximately 4.

8.4 INSTANTANEOUSLY OPTIMAL RELAY CONTROL

In order to circumvent the difficulty of constructing a feedback control
system from optimization theory a number of investigators have used
ad hoc methods to construct feedback systems which are, in some sense,
instantaneously optimal, rather than optimal in an overall sense. For
example, suppose that the positive definite quadratic form

$$E = \frac{1}{2}\sum_{i,j} x_i Q_{ij} x_j \tag{1}$$

is used as a measure of the deviation from the equilibrium $\mathbf{x} = \mathbf{0}$, being
positive for any offset. Then an instantaneously optimal control would
be one which drives E to zero as rapidly as possible, disregarding the
future consequences of such action.

We shall restrict attention to systems which are linear in a single
bounded control variable

$$\dot{x}_i = f_i(\mathbf{x}) + b_i(\mathbf{x})u \tag{2}$$
$$u_* \leq u \leq u^* \tag{3}$$

(Linearity is essential, but the single control is not.) This form is typical of many processes. The criterion of driving E to zero as rapidly as

possible is equivalent to minimizing the time derivative $\dot{E}$

$$\dot{E} = \sum_{i,j} x_i Q_{ij} \dot{x}_j = \sum_{i,j} x_i Q_{ij} f_j + \left(\sum_{i,j} x_i Q_{ij} b_j \right) u \qquad (4)$$

(We have made use of the symmetry of $\mathbf{Q}$.)   Since Eq. (4) is linear in u, the minimum will always occur at an extreme when the coefficient of u does not vanish identically

$$u = \begin{cases} u_* & \sum_{i,j} x_i Q_{ij} b_j > 0 \\[2ex] u^* & \sum_{i,j} x_i Q_{ij} b_j < 0 \end{cases} \qquad (5)$$

Thus we immediately obtain a feedback control law.   Note that when $\mathbf{b}$ is independent of $\mathbf{x}$, the switching criterion is linear.

This mode of control can be illustrated by returning to the example of the stirred-tank chemical reactor, defined in Sec. 4.6 as

$$\frac{d}{dt}(A - A_s) = \frac{q}{V}(A_f - A) - kA \qquad (6)$$

$$\frac{d}{dt}(T - T_s) = \frac{q}{V}(T_f - T) + \frac{(-\Delta H)}{C_p \rho} kA - \frac{T - T_c}{V C_p \rho} u \qquad (7)$$

where x_1 is taken as $A - A_s$, x_2 as $T - T_s$, control is by coolant flow rate q_c, and we have defined u as the monotonic function $U K q_c/(1 + K q_c)$. All parameters will be identical to those used in Secs. 4.6 and 5.6, in which case the bounds on u are

$$0 \leq u \leq 8 \qquad (8)$$

Comparison of Eqs. (6) and (7) with Eq. (2) indicates that the switching function has the form

$$\sum_{i,j} x_i Q_{ij} b_j = -\frac{T - T_c}{V C_p \rho} [Q_{12}(A - A_s) + Q_{22}(T - T_s)] \qquad (9)$$

As T will always exceed the coolant temperature T_c, the coefficient of the bracketed term in Eq. (9) is always negative and Eq. (5) for the control law reduces to

$$u = \begin{cases} 8 & (T - T_s) + \alpha(A - A_s) > 0 \\ 0 & (T - T_s) + \alpha(A - A_s) < 0 \end{cases} \qquad (10)$$

where $\alpha = Q_{12}/Q_{22}$.   This is a linear switching law for the nonlinear process.   An even simpler result is obtained when cross terms are excluded from the objective, in which case $\alpha = 0$.   Then the control is

$$u = \begin{cases} 8 & T > T_s \\ 0 & T < T_s \end{cases} \qquad (11)$$

That is, adiabatic operation when the temperature is below the steady-state value, full cooling when above, irrespective of the relative weighting placed on concentration and temperature deviations. It is interesting to compare this result with the time-optimal control shown in Fig. 5.9, in which the switching curve does not differ significantly from the steady-state temperature.

Paradis and Perlmutter have computed the response of this system under the control equation (11) with an initial offset of $T - T_s = -20$, $A - A_s = 2 \times 10^{-4}$. The phase plane in Fig. 5.9 indicates that away from equilibrium the temperature should approach the steady-state value immediately, first slowly and then quite rapidly, while the concentration deviation should first grow and then approach zero. Figures 8.1 and 8.2 show that this is precisely what happens, where curve a is the controlled response and curve b the uncontrolled. The first two switches occur at 23.70 and 25.65 sec, after which the controller switches between extremes rapidly. Such "chattering" near the steady state is a common characteristic of relay controllers. It should be noted that the system is asymptotically stable and returns to steady state eventually even in the absence of control.

In order to avoid chattering† some criterion must be introduced for

† Chattering may sometimes be desirable, as discussed in detail in the book by Flügge-Lotz.

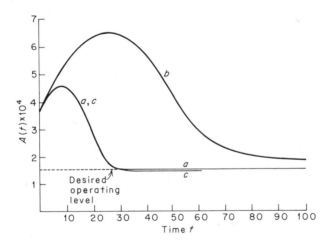

Fig. 8.1 Concentration response of the controlled and uncontrolled reactor using instantaneously optimal control. [*From W. O. Paradis and D. D. Perlmutter, AIChE J.,* **12:**876 (1966). *Copyright 1966 by the American Institute of Chemical Engineers. Reprinted by permission of the copyright owner.*]

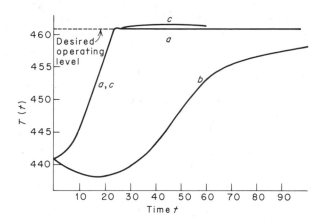

Fig. 8.2 Temperature response of the controlled and uncontrolled reactor using instantaneously optimal control. [*From W. O. Paradis and D. D. Perlmutter, AIChE J.*, **12**:876 (1966). *Copyright 1966 by the American Institute of Chemical Engineers. Reprinted by permission of the copyright owner.*]

changing from relay to another form of control. In analogy with the time-optimal case Paradis and Perlmutter simply set u to its steady-state value of 5 at the time of the second switch and allowed the natural stability of the system to complete the control. These results are shown as curve c in Figs. 8.1 and 8.2 and indicate quite satisfactory performance.

8.5 AN INVERSE PROBLEM

The results of the preceding section suggest the fruitfulness from a practical control point of view of pursuing the subject of instantaneously optimal controls somewhat further. In particular, since the one serious drawback is the possibility that rapid return toward equilibrium might cause serious future difficulties, we are led to enquire whether this ad hoc policy might also be the solution of a standard optimal-control problem, in which case we would know precisely what overall criterion is being minimized, if any. To that end we are motivated to study the inverse problem.

We shall restrict our attention for simplicity to linear systems with constant coefficients

$$\dot{x}_i = \sum_j A_{ij}x_j + b_i u \tag{1}$$

and we shall take u to be a deviation from steady-state control with symmetric bounds, in which case **b** may be normalized so that the bounds are unity

$$|u| \leq 1 \tag{2}$$

If the bounds on u are not symmetric, the subsequent algebra is slightly more cumbersome but the essential conclusions are unchanged. Under these assumptions the feedback control law of Eq. (5) of the preceding section is

$$u = -\operatorname{sgn}\left(\sum_{i,j} b_i Q_{ij} x_j\right) \tag{3a}$$

or

$$u = -\operatorname{sgn}\left(\sum_j \alpha_j x_j\right) \tag{3b}$$

a linear switching law. We shall suppose that the cost of control is negligible and consider an overall objective of the form

$$\mathcal{E} = \int_0^\theta \mathcal{F}(\mathbf{x})\, dt \tag{4}$$

where

$$\mathcal{F}(\mathbf{x}) > \mathcal{F}(0) \geq 0 \qquad \mathbf{x} \neq \mathbf{0} \tag{5}$$

The inverse problem which we wish to solve is for a function $\mathcal{F}(\mathbf{x})$ satisfying Eq. (5) such that a control of the form of Eq. (3) minimizes $\mathcal{E}$.

The hamiltonian for the linear stationary system described by Eq. (1) and the objective equation (4) is

$$H = \mathcal{F}(\mathbf{x}) + \sum_{i,j} \gamma_i A_{ij} x_j + \sum_i \gamma_i b_i u \tag{6}$$

and, since u enters linearly, whenever the coefficient of u does not vanish, the optimal control is

$$u = -\operatorname{sgn}\left(\sum_i \gamma_i b_i\right) \tag{7}$$

Equations (3) and (7) will define the same control if (but not only if!) we take

$$\gamma_i = \sum_j Q_{ij} x_j \tag{8}$$

We need, then, to determine whether such a relation is compatable with the minimum principle and, if so, what the resulting function $\mathcal{F}$ is.

The equations for the Green's functions are

$$\dot{\gamma}_i = -\frac{\partial H}{\partial x_i} = -\frac{\partial \mathcal{F}}{\partial x_i} - \sum_j \gamma_j A_{ji} \tag{9}$$

or, upon substituting Eq. (8),

$$\dot{\gamma}_i = -\frac{\partial \mathfrak{F}}{\partial x_i} - \sum_{j,k} x_k Q_{kj} A_{ji} \tag{10}$$

But differentiating Eq. (8),

$$\dot{\gamma}_i = \sum_j Q_{ij}\dot{x}_j = \sum_{j,k} Q_{ij} A_{jk} x_k - \sum_j Q_{ij} b_j \operatorname{sgn}\left(\sum_{k,l} x_k Q_{kl} b_l\right) \tag{11}$$

where we have used Eqs. (1), (7), and (8). The right-hand sides of Eqs. (10) and (11) must be identical, leading to a family of partial differential equations for $\mathfrak{F}(\mathbf{x})$:

$$\frac{\partial \mathfrak{F}}{\partial x_i} = -\sum_{j,k}(Q_{ij}A_{jk}x_k + x_k Q_{kj}A_{ji}) + \sum_j Q_{ij}b_j \operatorname{sgn}\left(\sum_{k,l} x_k Q_{kl} b_l\right) \tag{12}$$

Integration of Eq. (12) is straightforward, yielding

$$\mathfrak{F}(\mathbf{x}) = -\frac{1}{2}\sum_{i,j,k} x_i(Q_{ij}A_{jk} + Q_{kj}A_{ji})x_k + \left|\sum_{i,j} x_i Q_{ij} b_j\right| + \text{const} \tag{13}$$

If θ is fixed, the value of the constant is irrelevant. If θ is unspecified, the condition $H = 0$ establishes that the constant is zero. We obtain, then,

$$\mathfrak{F}(\mathbf{x}) = \frac{1}{2}\sum_{i,j} x_i C_{ij} x_j + \left|\sum_{i,j} x_i Q_{ij} b_j\right| \tag{14}$$

where

$$C_{ij} = -\sum_k (Q_{ik}A_{kj} + Q_{jk}A_{ki}) \tag{15}$$

The absolute-value term in Eq. (14) is a linear combination and can vanish for $\mathbf{x} \neq \mathbf{0}$, so that we ensure satisfaction of Eq. (5) by requiring that the quadratic form $\sum_{i,j} x_i C_{ij} x_j$ be positive definite. This places an interesting restriction on the uncontrolled system, for the time derivative of the positive definite quadratic form $E = \frac{1}{2}\sum_{i,j} x_j Q_{ij} x_j$ without control is simply

$$\dot{E} = -\sum_{i,j} x_i C_{ij} x_j \tag{16}$$

which is required to be negative definite. According to Liapunov stability theory, then (see Appendix 8.1), the quadratic form E must be a Liapunov function for the uncontrolled system and the uncontrolled system must be asymptotically stable. In that case $\dot{E}$ is also negative definite for the controlled system, and the controlled system is also stable.

The asymptotic stability of the uncontrolled system is sufficient to ensure the existence of a positive definite solution $\mathbf{Q}$ to Eq. (15) for arbitrary positive definite $\mathbf{C}$, although the converse is not true.

We now have a solution to the inverse problem, which establishes that for an asymptotically stable linear stationary system the relay controller with linear switching corresponds to an objective which is the integral of a positive definite quadratic form plus a second positive semidefinite term. The use of a quadratic form as an integrand is, of course, now well established, and we anticipate that the additional term will simply have the effect of bending the trajectories somewhat. Thus, the instantaneously optimal policy does correspond to a very meaningful overall objective. We are not quite finished, however, for we must still establish that the minimum of the integral of Eq. (14) does indeed occur for the control defined by Eq. (3). We must do this because of the possibility that a singular solution may exist or that the minimum principle leads to multiple solutions, another of which is in fact the optimum.

The possibility of a singular solution in which the switching criterion vanishes for a finite time interval is most easily dismissed by considering the second-order system

$$\dot{x}_1 = x_2 \tag{17a}$$
$$\dot{x}_2 = A_{21}x_1 + A_{22}x_2 + bu \tag{17b}$$

for which the switching criterion defined by Eq. (3) is

$$Q_{12}x_1 + Q_{22}x_2 = 0 \tag{18}$$

If this is differentiated with respect to time and Eqs. (17) substituted, the resulting control is

$$u = -\left(A_{21} - \frac{Q_{12}^2}{Q_{22}^2}\right)x_1 - A_{22}x_2 \tag{19}$$

On the other hand, the vanishing of the switching function means that the integrand of the objective is simply $\frac{1}{2}(C_{11}x_1^2 + 2C_{12}x_1x_2 + C_{22}x_2^2)$, and the criterion for singular control was found in Sec. 5.9 for this case to be

$$u = -\left(A_{21} - \frac{C_{11}}{C_{22}}\right)x_1 - A_{22}x_2 \tag{20}$$

and

$$\sqrt{C_{11}}\, x_1 + \sqrt{C_{22}}\, x_2 = 0 \tag{21}$$

Thus, singular control is possible and, in fact, optimal if and only if

$$\frac{C_{11}}{C_{22}} = \frac{Q_{12}^2}{Q_{22}^2} \tag{22}$$

Together with Eq. (15) this yields only discrete values of the ratio C_{11}/C_{22}, and since the **C** matrix is at the disposal of the designer, only infinitesimal changes are needed to avoid the possibility of intermediate control. The generalization to higher dimensions is straightforward and yields the same result.

Finally, it remains to be shown that there cannot be another control policy which satisfies the minimum principle. Here, for the first time, we make use of our wish to avoid chattering and presume that within some neighborhood of the origin we want to switch from the relay controller to some other form of control, perhaps linear. We shall choose that region to be an ellipsoidal surface such that the control effort is to terminate upon some manifold

$$g(\mathbf{x}) = \sum_{i,j} x_i(\theta) Q_{ij} x_j(\theta) - \text{const} = 0 \tag{23}$$

where the final time θ is unspecified. The boundary condition for the Green's functions is then

$$\gamma_i(\theta) = \nu \frac{\partial g}{\partial x_i} = 2\nu \sum_j Q_{ij} x_j \tag{24}$$

where ν is some constant and the condition $H = 0$ immediately establishes that $2\nu = 1$. Hence, for any final condition $\mathbf{x}(\theta)$ a complete set of conditions is available for the coupled differential equations (1) and (9), and the solution over any interval between switches is unique. The control given by Eq. (3) must, then, be the minimizing control for the objective defined by Eq. (14).

Any asymptotically stable linear system, then, may be controlled optimally with respect to Eq. (14) by a particularly simple feedback policy. Because of the ambiguity of defining a precise mathematical objective it will often be the case that $\mathfrak{F}$ as defined here meets all physical requirements for a control criterion, in which case the instantaneously optimal policy will provide excellent control to within some predetermined region of the desired operating conditions.

The extension to nonlinear systems of the form

$$\dot{x}_i = f_i(\mathbf{x}) + b_i(\mathbf{x})u \tag{25}$$

is trivial, and again the condition for optimality is that the quadratic form E be a Liapunov function for the uncontrolled system and, therefore, for the controlled system. Now, however, this requirement is more restrictive than simply demanding asymptotic stability, for the existence of a quadratic Liapunov function is ensured only in the region in which linearization is valid, and large regions of asymptotic stability may exist in which no quadratic Liapunov function can be found.

One final comment is required on the procedure used to solve the inverse problem. Equation (8) can be shown to be the consequence of any linear relation between γ_i and $\sum_j Q_{ij}x_j$, but any sign-preserving non-linear relation would also retain the same control law. Thus, by considering nonlinear transformations for the one-dimensional system

$$\dot{x} = Ax + bu \tag{26}$$

Thau has found that the control

$$u = -\,\text{sgn}\,Qbx \tag{27}$$

is optimal not only for the objective found here

$$\mathcal{E} = \int_0^\theta (AQx^2 + |Qbx|)\,dt \tag{28}$$

but also for

$$\mathcal{E} = \int_0^\theta (Ab^2Q^3x^4 + |b^3Q^3x^3|)\,dt \tag{29}$$

and

$$\mathcal{E} = \int_0^\theta \left(\frac{A}{b}\,x\sinh bQx + |\sinh bQx|\right)dt \tag{30}$$

One way in which this equivalence of objectives can be established is to use the relation

$$\text{sgn}\,bQx = \text{sgn}\left[\sum_{k=K_1}^{k=K_2}(bQx)^{2k+1}\right]\qquad K_2 \geq K_1 \geq 0 \tag{31}$$

and then to procede in the manner of this section. The desirability of extending this approach to multidimensional systems is obvious, as are the difficulties.

8.6 DISCRETE LINEAR REGULATOR

We have observed earlier that we may often be interested in controlling systems which evolve discretely in time and are described by difference equations. The difference-equation representation might be the natural description, or it might represent an approximation resulting from computational considerations. The procedures discussed in this chapter can all be extended to the control of discrete systems, but we shall restrict ourselves to a consideration of the analog of the problem of Sec. 8.2.

We consider a system described by the linear difference equations

$$x_i{}^n = \sum_j A_{ij}x_j{}^{n-1} + b_iu^n \tag{1}$$

The coefficients A_{ij} and b_i are taken as constants, though the extension to functions of n is direct. We have not included a disturbance term, though this, too, causes no difficulties, and we seek only to find the sequence of controls $\{u^n\}$ which regulates the system following an initial upset or change in desired operating point. The minimum-square-error criterion is again used

$$\mathcal{E} = \frac{1}{2} \sum_{n=1}^{N} \left[\sum_{i,j} x_i{}^n C_{ij} x_j{}^n + R(u^n)^2 \right] \tag{2}$$

Unlike the continuous case, we include a coefficient of the $(u^n)^2$ term in the objective, for R can be allowed to go to zero for discrete control.

A Lagrange multiplier formulation is the most convenient for solving this problem. The lagrangian is

$$\mathcal{L} = \frac{1}{2} \sum_{n} \left[\sum_{i,j} x_i{}^n C_{ij} x_j{}^n + R(u^n)^2 \right]$$
$$+ \sum_{n} \left[\sum_{i} \lambda_i{}^n \left(\sum_{j} A_{ij} x_j{}^{n-1} + b_i u^n - x_i{}^n \right) \right] \tag{3}$$

Setting partial derivatives with respect to $x_i{}^n$ and u^n to zero, respectively, we obtain

$$\sum_{i} x_i{}^n C_{ij} - \lambda_j{}^n + \sum_{i} \lambda_i{}^{n+1} A_{ij} = 0 \tag{4}$$

$$Ru^n + \sum_{i} \lambda_i{}^n b_i = 0 \tag{5}$$

Equation (4) allows us to define a variable $\lambda_i{}^{N+1}$ as zero. Unlike the analysis for the continuous problem, we shall not solve Eq. (5) for u^n at this point, for further manipulation will enable us to obtain a feedback solution which will be valid as $R \to 0$.

If Eq. (4) is multiplied by b_j and summed over j, we obtain

$$\sum_{i,j} x_i{}^n C_{ij} b_j - \sum_{j} \lambda_j{}^n b_j + \sum_{i,j} \lambda_i{}^{n+1} A_{ij} b_j = 0 \tag{6}$$

Substitution of Eq. (5) then leads to

$$\sum_{i,j} x_i{}^n C_{ij} b_j + Ru^n + \sum_{i,j} \lambda_i{}^{n+1} A_{ij} b_j = 0 \tag{7}$$

We now seek to express the multipliers in terms of the state variables as

$$\lambda_i{}^{n+1} = \sum_{k} M_{ik}{}^n x_k{}^n \tag{8}$$

Since $\lambda_i{}^{N+1}$ is zero, $M_{ik}{}^N$ must be zero, except for $N \to \infty$, in which case $x_k{}^N \to 0$. Substitution of Eq. (8) into Eq. (7) then leads to the required feedback form

$$u^n = \sum_j K_j^n x_j^{n-1} \tag{9}$$

where

$$K_j^n = -\frac{\sum_{i,k} C_{ik} b_k A_{ij} + \sum_{i,l,k} b_l A_{kl} M_{ki}^n A_{ij}}{R + \sum_{i,k} b_i C_{ik} b_k + \sum_{i'l,k} b_l A_{kl} M_{ki}^n b_i} \tag{10}$$

The feedback dependence must be on $\mathbf{x}^{n-1}$, since the state at the beginning of the control interval is the quantity that can be measured.

We still need a means of calculating M_{ij}^n in order to compute the feedback gains K_j^n. This is done by first substituting Eq. (9) into Eq. (1)

$$x_i^n = \sum_j (A_{ij} + b_i K_j^n) x_j^{n-1} \tag{11}$$

Substitution of Eq. (11) into Eq. (4) yields

$$\sum_i \left(\sum_k A_{ik} + b_i K_k^n \right) C_{ij} x_k^{n-1} - \sum_k M_{jk}^{n-1} x_k^{n-1}$$
$$+ \sum_i A_{ij} \left[\sum_l M_{il}^n \sum_k (A_{lk} + b_l K_k^n) \right] x_k^{n-1} = 0 \tag{12}$$

If Eq. (12) is to hold for all values of $\mathbf{x}^{n-1}$, the coefficient of x_k^{n-1} must be identically zero, in which case M_{ij}^n must be a solution to the difference equation

$$M_{jk}^{n-1} = \sum_i C_{ij} A_{ik} + \sum_{i,l} A_{ij} M_{il}^n A_{lk} + \left(\sum_i b_i C_{ij} + \sum_{i,l} A_{ij} M_{il}^n b_l \right) K_k^n$$
$$M_{jk}^N = 0 \tag{13}$$

Equation (13), with K_j^n defined by Eq. (10), is the generalization of the Riccati difference equation first encountered in Sec. 1.7. For the important case that $N \to \infty$, a constant solution is obtained. Clearly there is no difficulty here in allowing R to go to zero.

The instantaneously optimal approach of Sec. 8.5 may be applied to the discrete system to obtain an interesting result. We define a positive definite quadratic error over the next control interval

$$E = \frac{1}{2} \sum_{i,j} x_i^n Q_{ij} x_j^n + \frac{1}{2} \rho (u^n)^2 \tag{14}$$

The control which makes E as small as possible over the following interval is found by setting the derivative of E with respect to u^n to zero

$$\frac{\partial E}{\partial u^n} = \sum_{i,j} x_i^n Q_{ij} \frac{\partial x_j^n}{\partial u^n} + \rho u^n = 0 \tag{15}$$

From Eq. (1),

$$\frac{\partial x_j{}^n}{\partial u^n} = b_j \tag{16}$$

so that Eq. (15) becomes

$$\sum_{i,j} b_i Q_{ij} x_j{}^n + \rho u^n = 0 \tag{17}$$

The required feedback form for u^n is obtained by substituting Eq. (1) for $x_j{}^n$ into Eq. (17) and solving

$$u^n = \sum_j k_j x_j{}^{n-1} \tag{18}$$

$$k_j = -\frac{\sum_k \sum_i b_i Q_{ik} A_{kj}}{\sum_k \sum_i b_i Q_{ik} b_k + \rho} \tag{19}$$

This linear feedback control can be conveniently compared to the discrete-regulator solution for $N \to \infty$ by defining a new variable

$$\mu_{ik} = \sum_l A_{lk} M_{li} \tag{20}$$

It can be shown from Eqs. (10) and (13) that μ_{ik} is symmetric ($\mu_{ik} = \mu_{ki}$). Equation (10) for the regulator feedback gain can then be written

$$K_j = -\frac{\sum_k \sum_i b_i (C_{ik} + \mu_{ik}) A_{kj}}{\sum_k \sum_i b_i (C_{ik} + \mu_{ik}) b_j + R} \tag{21}$$

The results are identical if we make the following identifications:

$$Q_{ik} = C_{ik} + \mu_{ik} \tag{22}$$
$$\rho = R \tag{23}$$

Q_{ik} defined by Eq. (22) is symmetric, as it must be. We find, therefore, that the discrete-regulator problem is disjoint and has a solution corresponding to an instantaneous optimum over each control cycle. The parameters in the instantaneous objective will have physical significance, however, only when computed by means of Eq. (22).

APPENDIX 8.1 LIAPUNOV STABILITY

We shall review briefly here the elementary principles of Liapunov stability needed in this chapter. Consider any function $V(\mathbf{x})$ which is positive definite in a neighborhood of $\mathbf{x} = \mathbf{0}$. The values of V define a distance from the origin, although the closed contours might be quite irregular. Thus, if the system is displaced from equilibrium to a value $\mathbf{x}_0$ and there exists any positive definite function $V(\mathbf{x})$ such that $\dot{V}(\mathbf{x}) \leq 0$ in a region containing both $\mathbf{x}_0$ and the origin, the system can never escape beyond the contour $V(\mathbf{x}) = V(\mathbf{x}_0)$ and the origin is *stable*. Furthermore, if there exists a function such that $\dot{V}(\mathbf{x}) < 0$ for $\mathbf{x} \neq \mathbf{0}$, the system must continue to pass through smaller and smaller values of V and ultimately return to equilibrium. In that case the origin is *asymptotically stable*.

A function $V(\mathbf{x})$ such as that described above, positive definite with a negative semidefinite derivative, is called a *Liapunov function*. For a system satisfying the differential equations

$$\dot{x}_i = F_i(\mathbf{x}) \qquad F_i(\mathbf{0}) = 0 \tag{1}$$

the derivative of V is computed by

$$\dot{V}(\mathbf{x}) = \sum_i \frac{\partial V}{\partial x_i} \dot{x}_i = \sum_i \frac{\partial V}{\partial x_i} F_i(\mathbf{x}) \tag{2}$$

If a Liapunov function can be found in some region including the origin, the system is stable with respect to disturbances within that region. If $\dot{V}$ is negative definite, the system is asymptotically stable. Construction of a Liapunov function is generally quite difficult.

BIBLIOGRAPHICAL NOTES

Section 8.1: The conventional approach to the design of feedback control systems is treated extensively in texts such as

P. S. Buckley: "Techniques of Process Control," John Wiley & Sons, Inc., New York, 1964
D. R. Coughanowr and L. B. Koppel: "Process Systems Analysis and Control," McGraw-Hill Book Company, New York, 1965
D. D. Perlmutter: "Chemical Process Control," John Wiley & Sons, Inc., New York, 1965
J. Truxal: "Automatic Feedback Control System Synthesis," McGraw-Hill Book Company, New York, 1957

The design of optimal control systems based on a classical frequency-domain analysis is treated in, for example,

S. S. L. Chang: "Synthesis of Optimum Control Systems," McGraw-Hill Book Company, New York, 1961

A modern point of view somewhat different from that adopted here is utilized in

C. W. Merriam: "Optimization Theory and the Design of Feedback Control Systems," McGraw-Hill Book Company, New York, 1964

The two approaches are reconciled in our Chap. 12; see also

L. B. Koppel: "Introduction to Control Theory with Applications to Process Control," Prentice-Hall, Inc., Englewood Cliffs, N.J., 1968

L. Lapidus and R. Luus: "Optimal Control of Engineering Processes," Blaisdell
Publishing Company, Waltham, Mass., 1967

*and a forthcoming book by J. M. Douglas for parallel discussions pertinent to this entire
chapter.*

*Section 8.2: The properties of the linear system with quadratic-error criterion have been
investigated extensively by Kalman, with particular attention to the asymptotic
properties of the Riccati equation.   In particular see*

R. E. Kalman: *Bol. Soc. Mat. Mex.*, **5**:102 (1960)
————: in R. Bellman (ed.), "Mathematical Optimization Techniques," University
of California Press, Berkeley, 1963
————: *J. Basic Eng.*, **86**:51 (1964)

A detailed discussion is contained in

M. Athans and P. Falb: "Optimal Control," McGraw-Hill Book Company, New
York, 1966

and some useful examples are treated in the books by Koppel and Lapidus and Luus and

A. R. M. Noton: "Introduction to Variational Methods in Control Engineering,"
Pergamon Press, New York, 1965

*A computer code for the solution of the Riccati equation, as well as an excellent and
detailed discussion of much of the basic theory of linear control, is contained in*

R. E. Kalman and T. S. Englar: "A User's Manual for the Automatic Synthesis
Program," *NASA Contractor Rept. NASA CR*-475, June, 1966, available from
Clearinghouse for Federal Scientific and Technical Information, Springfield,
Va. 22151

Numerical solution of the Riccati equation by successive approximations is discussed in

N. N. Puri and W. A. Gruver: *Preprints 1967 Joint Autom. Contr. Conf., Philadelphia,*
p. 335

*Though not readily apparent, the procedure used in this paper is equivalent to that dis-
cussed in Sec. 9.6 for the numerical solution of nonlinear differential equations.*

*Section 8.3: The general relationship between the linear servomechanism problem with
a $\dot{u}^2$ cost-of-control term and classical control is part of a research program being
carried out in collaboration with G. E. O'Connor. See*

G. E. O'Connor: "Optimal Linear Control of Linear Systems: An Inverse Problem,"
M.Ch.E. Thesis, University of Delaware, Newark, Del., 1969

Section 8.4: The particular development is based upon

W. O. Paradis and D. D. Perlmutter: *AIChE J.*, **12**:876, 883 (1966)

Similar procedures, generally coupled with Liapunov stability theory (Appendix 8.1), have been applied by many authors; see, for example,

C. D. Brosilow and K. R. Handley: *AIChE J.*, **14**:467 (1968)
R. E. Kalman and J. E. Bertram: *J. Basic Eng.*, **82**:371 (1960)
D. P. Lindorff: *Preprints 1967 Joint Autom. Contr. Conf.*, *Philadelphia*, p. 394
A. K. Newman: *Preprints 1967 Joint Autom. Contr. Conf.*, *Philadelphia*, p. 91

The papers by Lindorff and Newman contain additional references. A detailed discussion of the properties of relay control systems will be found in

I. Flügge-Lotz: "Discontinuous and Optimal Control," McGraw-Hill Book Company, New York, 1968

Section 8.5: This section is based on

M. M. Denn: *Preprints 1967 Joint Autom. Contr. Conf.*, *Philadelphia*, p. 308; *AIChE J.*, **13**:926 (1967)

The generalization noted by Thau was presented in a prepared discussion of the paper at the Joint Automatic Control Conference. The inverse problem has been studied in the context of classical calculus of variations since at least 1904; see

O. Bolza: "Lectures on the Calculus of Variations," Dover Publications, Inc., New York, 1960
J. Douglas: *Trans. Am. Math. Soc.*, **50**:71 (1941)
P. Funk: "Variationsrechnung und ihr Anwendung in Physik und Technik," Springer-Verlag OHG, Berlin, 1962
F. B. Hildebrand: "Methods of Applied Mathematics," Prentice-Hall, Inc., Englewood Cliffs, N.J., 1952

The consideration of an inverse problem in control was first carried out by Kalman for the linear-quadratic case,

R. E. Kalman: *J. Basic Eng.*, **86**:51 (1964)

See also

A. G. Aleksandrov: *Eng. Cybernetics*, **4**:112 (1967)
P. Das: *Automation Remote Contr.*, **27**:1506 (1966)
R. W. Obermayer and F. A. Muckler: *IEEE Conv. Rec.*, pt. 6, 153 (1965)
Z. V. Rekazius and T. C. Hsia: *IEEE Trans. Autom. Contr.*, **AC9**:370 (1964)
F. E. Thau: *IEEE Trans. Autom. Contr.*, **AC12**:674 (1967)

Several authors have recently studied the related problem of comparing performance of simple feedback controllers to the optimal control for specified performance indices. See

A. T. Fuller: *Intern. J. Contr.*, **5**:197 (1967)
M. G. Millman and S. Katz: *Ind. Eng. Chem. Proc. Des. Develop.*, **6**:477 (1967)

Section 8.6: The Riccati equation for the feedback gains is obtained in a different manner in the monograph by Kalman and Englar cited for Sec. 8.2, together with a computer code for solution. See also the books by Koppel and Lapidus and Luus and

S. M. Roberts: "Dynamic Programming in Chemical Engineering and Process Control," Academic Press, Inc., New York, 1964

W. G. Tuel, Jr.: *Preprints 1967 Joint Autom. Contr. Conf., Philadelphia*, p. 549

J. Tou: "Optimum Design of Digital Control Systems," Academic Press, Inc., New York, 1963

————: "Modern Control Theory," McGraw-Hill Book Company, New York, 1964

The book by Roberts contains further references. Instantaneously optimal methods have been applied to discrete systems by

R. Koepcke and L. Lapidus: *Chem. Eng. Sci.*, **16**:252 (1961)

W. F. Stevens and L. A. Wanniger: *Can. J. Chem. Eng.*, **44**:158 (1966)

Appendix 8.1: A good introductory treatment of Liapunov stability theory can be found in most of the texts on control noted above and in

J. P. LaSalle and S. Lefschetz: "Stability by Liapunov's Direct Method with Applications," Academic Press, Inc., New York, 1961

For an alternative approach see

M. M. Denn: "A Macroscopic Condition for Stability," *AIChE J*, in press

PROBLEMS

8.1. For systems described by the equations

$$\dot{x}_i = f_i(\mathbf{x}) + b_i(\mathbf{x})u$$

use the methods of Secs. 8.2 and 4.8 to obtain the linear and quadratic terms in the nonlinear feedback control which minimizes

$$\varepsilon = \frac{1}{2} \int_0^\theta \left(\sum_{i,j} x_i C_{ij} x_j + u^2 \right) dt$$

Extend to the feedback-feedforward control for a step disturbance d which enters as

$$\dot{x}_i = f_i(\mathbf{x}) + b_i(\mathbf{x})u + g_i(\mathbf{x})d$$

8.2. Extend the results of Sec. 8.3 to a second-order system with *numerator dynamics,*

$$\ddot{x} + a\dot{x} + bx = u + e\dot{u} + d$$

8.3. Extend the control approach of Sec. 8.4 and the optimization analysis to the case when $E(\mathbf{x})$ is an arbitrary convex positive definite function.

8.4. The unconstrained control problem

$$\dot{\mathbf{x}} = \mathbf{f}(\mathbf{x},u,\mathbf{p})$$
$$\mathbf{x}(0) = \mathbf{x}_0$$
$$\min \varepsilon = \int_0^\theta \mathcal{F}(\mathbf{x},u,\mathbf{c}) \, dt$$

where $\mathbf{p}$ and $\mathbf{c}$ are parameters, has been solved for a given set of values of $\mathbf{x}_0$, $\mathbf{p}$, and $\mathbf{c}$. Obtain equations for the change $\delta u(t)$ in the optimal control when $\mathbf{x}_0$, $\mathbf{p}$, and $\mathbf{c}$ are changed by small amounts $\delta\mathbf{x}_0$, $\delta\mathbf{p}$, and $\delta\mathbf{c}$, respectively. In particular, show that δu

may be expressed as

$$\delta u = \sum_j K_j(t)\, \delta x_j + \sum_k g_{1k}(t)\, \delta p_k + \sum_m g_{2m}(t)\, \delta c_m$$

where $\delta \mathbf{x}$ is the change in $\mathbf{x}$ and $K_j(t)$ and $g_{ij}(t)$ are solutions of initial-value problems. (*Hint:* The method of Sec. 8.2 can be used to solve the coupled equations for changes in state and Green's functions.) Comment on the application of this result to the following control problems:

(*a*) Feedback control when the system state is to be maintained near an optimal trajectory.

(*b*) Feedback-feedforward control when small, relatively constant disturbances can enter the system and the optimal feedback control for the case of no disturbance can be obtained.

8.5. The system

$$\ddot{x}(t) + a\dot{x}(t) + bx(t) = u(t)$$

is to be regulated by piecewise constant controls with changes in u every τ time units to minimize

$$\varepsilon = \frac{1}{2} \int_0^\theta (x^2 + c\dot{x}^2)\, dt$$

Obtain the equivalent form

$$x_1{}^n = x_2{}^{n-1}$$
$$x_2{}^n = -\beta x_1{}^{n-1} - \alpha x_2{}^{n-1} + w^n$$

$$\min \varepsilon = \frac{1}{2} \sum_{n=1}^N [(x_1{}^n)^2 + C(x_2{}^n)^2]$$

Obtain explicit values for the parameters in the optimal control for $N \to \infty$

$$w^n = -K_1 x_1{}^{n-1} - K_2 x_2{}^{n-1}$$

Extend these results to the system with pure delay T

$$\ddot{x}(t) + a\dot{x}(t) + bx(t) = u(t - T)$$

(The analytical solution to this problem has been obtained by Koppel.)

9
Numerical Computation

9.1 INTRODUCTION

The optimization problems studied in the preceding six chapters are prototypes which, because they are amenable to analytical solution or simple computation, help to elucidate the structure to be anticipated in certain classes of variational problems. As in Chaps. 1 and 2, however, where we dealt with optimization problems involving only differential calculus, we must recognize that the necessary conditions for optimality will lead to serious computational difficulties if rational procedures for numerical solution are not developed. In this chapter we shall consider several methods of computation which are analogs of the techniques introduced in Chap. 2. In most cases we shall rely heavily upon the Green's function treatment of linear differential and difference equations developed in Secs. 6.2 and 7.2.

9.2 NEWTON–RAPHSON BOUNDARY ITERATION

The computational difficulties to be anticipated in the solution of a variational problem are best illustrated by recalling the necessary conditions for minimizing the function $\mathcal{E}[\mathbf{x}(\theta)]$ in a continuous system when $\mathbf{x}(\theta)$ is unconstrained, $\mathbf{x}(0) = \mathbf{x}_0$ is given, and θ is specified. We must simultaneously solve the S state equations with *initial* conditions

$$\dot{x}_i = f_i(\mathbf{x},\mathbf{u}) \qquad \begin{array}{l} 0 < t \le \theta \\ i = 1, 2, \ldots, S \end{array}$$

$$\mathbf{x}(0) = \mathbf{x}_0 \tag{1}$$

the S multiplier equations with *final* conditions

$$\dot{\gamma}_i = -\sum_{j=1}^{S} \gamma_j \frac{\partial f_j}{\partial x_i} \qquad \begin{array}{l} 0 \le t < \theta \\ i = 1, 2, \ldots, S \end{array}$$

$$\gamma_i(\theta) = \frac{\partial \mathcal{E}}{\partial x_i} \tag{2}$$

together with the minimization of the hamiltonian at each value of t

$$\min_{\mathbf{u}(t)} H = \sum_{i=1}^{S} \gamma_i f_i \tag{3}$$

An obvious procedure for obtaining the solution is to assume either $\mathbf{x}(\theta)$ or $\boldsymbol{\gamma}(0)$ and integrate Eqs. (1) and (2), choosing $\mathbf{u}$ at each value of t to satisfy the minimum principle. By choosing $\mathbf{x}(\theta)$ we compute an optimal solution for whatever the resulting value of $\mathbf{x}(0)$, but we must continue the process until we find the $\mathbf{x}(\theta)$ which leads to the required $\mathbf{x}_0$. Let us suppose that the possible range over which each component $x_i(\theta)$ may vary can be adequately covered by choosing M values, $x_{i(1)}(\theta)$, $x_{i(2)}(\theta)$, \ldots, $x_{i(M)}(\theta)$. There are, then, M^S possible combinations of final conditions to be evaluated, and for each the $2S$ differential equations (1) and (2) must be integrated, or a total of $2SM^S$ differential equations are to be solved. A very modest number for M might be 10, in which case a system described by only three differential equations would require the numerical integration of 6,000 differential equations in order to reduce the interval of uncertainty by a factor of 10. For $S = 4$ the number is 80,000. This exponential dependence of the number of computational steps on the dimension of the system is generally referred to as *the curse of dimensionality*.

For certain types of problems the curse may be exorcised by a linearization procedure for improving upon estimates of $\mathbf{x}(\theta)$. We make a first estimate $\bar{\mathbf{x}}(\theta)$ and integrate Eqs. (1) and (2) with respect to t from $t = \theta$ to $t = 0$, determining $\bar{\mathbf{u}}$ at each step in the numerical integration from the minimum condition. The value of $\mathbf{x}$ so computed at $t = 0$, $\bar{\mathbf{x}}(0)$, will generally not correspond to the required value $\mathbf{x}_0$. A new value of $\mathbf{x}(\theta)$ will produce new functions $\mathbf{x}(t)$, $\mathbf{u}(t)$ in the same way, and the first-order variational equation must be

$$\delta \dot{x}_i = \sum_{j=1}^{S} \frac{\partial f_i}{\partial x_j} \delta x_j + \sum_{k=1}^{R} \frac{\partial f_i}{\partial u_k} \delta u_k \tag{4}$$

where $\delta \mathbf{x} = \mathbf{x} - \bar{\mathbf{x}}$, $\delta \mathbf{u} = \mathbf{u} - \bar{\mathbf{u}}$, and partial derivatives are evaluated along the trajectory determined by $\bar{\mathbf{u}}$. Defining the Green's functions $\Gamma_{ij}(\theta, t)$ by

$$\dot{\Gamma}_{ij}(\theta, t) = - \sum_{k=1}^{S} \Gamma_{ik}(\theta, t) \frac{\partial f_k}{\partial x_j} \tag{5}$$

$$\Gamma_{ij}(\theta, \theta) = \delta_{ij} = \begin{cases} 1 & i = j \\ 0 & i \neq j \end{cases} \tag{6}$$

Green's identity, Eq. (15) of Sec. 6.2, may be written

$$\delta x_i(\theta) = \sum_{j=1}^{S} \Gamma_{ij}(\theta, 0)\, \delta x_j(0) + \int_0^\theta \sum_{j=1}^{S} \sum_{k=1}^{R} \Gamma_{ij}(\theta, t) \frac{\partial f_j}{\partial u_k} \delta u_k\, dt \tag{7}$$

At this point we make the critical assumption that the optimal policy does not differ significantly for neighboring values of $\mathbf{x}(\theta)$, in which case $\delta \mathbf{u}$ will be approximately zero. Equation (7) then provides an algorithm for determining the new estimate of $\mathbf{x}(\theta)$

$$x_i(\theta) = \bar{x}_i(\theta) + \sum_{j=1}^{S} \Gamma_{ij}(\theta, 0)[x_{j0} - \bar{x}_j(0)] \tag{8}$$

Equation (8) may be looked upon as an approximation to the first two terms of a Taylor series expansion, in which case we might interpret $\Gamma(\theta, 0)$ as an array of partial derivatives

$$\Gamma_{ij}(\theta, 0) = \frac{\partial x_i(\theta)}{\partial x_j(0)} \tag{9}$$

which is quite consistent with the notion of an influence function introduced earlier. Equation (8) is then analogous to the Newton-Raphson procedure of Sec. 2.2, where the function to be driven to zero by choice of $\mathbf{x}(\theta)$ is $\mathbf{x}(0) - \mathbf{x}_0$.

This procedure is equally applicable to systems described by difference equations

$$x_i{}^n = f_i{}^n(\mathbf{x}^{n-1}, \mathbf{u}^n) \qquad \begin{array}{l} n = 1, 2, \ldots, N \\ i = 1, 2, \ldots, S \end{array} \tag{10}$$

The variational equations are

$$\delta x_i{}^n = \sum_{j=1}^{S} \frac{\partial f_i{}^n}{\partial x_j{}^{n-1}} \delta x_i{}^{n-1} + \sum_{k=1}^{R} \frac{\partial f_i{}^n}{\partial u_k{}^n} \delta u_k{}^n \tag{11}$$

with Green's functions defined by

$$\Gamma_{ij}^{N,n-1} = \sum_{k=1}^{S} \Gamma_{ik}^{Nn} \frac{\partial f_k^n}{\partial x_j^{n-1}} \tag{12}$$

$$\Gamma_{ij}^{NN} = \delta_{ij} = \begin{cases} 1 & i = j \\ 0 & i \neq j \end{cases} \tag{13}$$

The first-order correction to an estimate $\bar{x}^N$ is then

$$x_i^N = \bar{x}_i^N + \sum_{j=1}^{S} \Gamma_{ij}^{N0}[x_j^0 - \bar{x}_j^0] \tag{14}$$

As outlined, this Newton-Raphson procedure requires the calculation of $S^2 + S$ Green's functions, the S^2 functions $\Gamma_{ij}(\theta,t)(\Gamma_{ij}^{Nn})$ needed for the iteration and the S functions $\gamma_i(t)(\gamma_i^n)$ for the optimization problem. By use of Eq. (18) of Sec. 6.2 or Eq. (8) of Sec. 7.2 it follows from the linearity and homogeneity of the Green's function equations that γ can be calculated from Γ

$$\gamma_i(t) = \sum_j \frac{\partial \mathcal{E}}{\partial x_j} \Gamma_{ji}(\theta,t) \tag{15}$$

with an identical relation for staged systems. Thus, only S^2 equations need be solved on each iteration.

It is likely that the linearization used in the derivation of the algorithm will be poor during the early stages of computation, so that using the full correction called for in Eqs. (8) and (14) might not result in convergence. To overcome this possibility the algorithm must be written as

$$x_i(\theta) = \bar{x}_i(\theta) + \frac{1}{r} \sum_{j=1}^{S} \Gamma_{ij}(\theta,0)[x_{jo} - \bar{x}_j(0)] \tag{16}$$

$$x_i^N = \bar{x}_i^N + \frac{1}{r} \sum_{j=1}^{S} \Gamma_{ij}^{N0}(x_j^0 - \bar{x}_j^0) \tag{17}$$

where $r \geq 1$ is a parameter controlling the step size. r must be taken large initially and allowed to approach unity during the final stages of convergence.

9.3 OPTIMAL TEMPERATURE PROFILE BY NEWTON–RAPHSON BOUNDARY ITERATION

As an example of the Newton-Raphson boundary-iteration algorithm we shall consider the problem of computing the optimal temperature profile in a plug-flow tubular reactor or batch reactor for the consecutive-reaction sequence

$$X_1 \rightarrow X_2 \rightarrow \text{products}$$

We have examined this system previously in Secs. 4.12 and 6.10 and have some appreciation of the type of behavior to be anticipated. Taking $\nu = 1$, $F(x_1) = x_1{}^2$ (second-order reaction), and $G(x_2) = x_2$ (first-order reaction), the state is described by the two equations

$$\dot{x}_1 = -k_{10}e^{-E_1'/u}x_1{}^2 \qquad x_1(0) = x_{10} \tag{1a}$$
$$\dot{x}_2 = k_{10}e^{-E_1'/u}x_1{}^2 - k_{20}e^{-E_2'/u}x_2 \qquad x_2(0) = x_{20} \tag{1b}$$

with $u(t)$ bounded from above and below

$$u_* \leq u \leq u^* \tag{2}$$

The objective is the maximization of value of the product stream, or minimization of

$$\mathcal{E} = -c[x_1(\theta) - x_{10}] - [x_2(\theta) - x_{20}] \tag{3}$$

where c reflects the value of feed x_1 relative to desired product x_2.

The Green's functions $\Gamma_{ij}(\theta,t)$ for the iterative algorithm are defined by Eqs. (5) and (6) of the previous section, which become

$$\dot{\Gamma}_{11} = -\Gamma_{11}\frac{\partial f_1}{\partial x_1} - \Gamma_{12}\frac{\partial f_2}{\partial x_1} = 2x_1k_{10}e^{-E_1'/u}(\Gamma_{11} - \Gamma_{12})$$
$$\Gamma_{11}(\theta,\theta) = 1 \quad (4a)$$

$$\dot{\Gamma}_{12} = -\Gamma_{11}\frac{\partial f_1}{\partial x_2} - \Gamma_{12}\frac{\partial f_2}{\partial x_2} = k_{20}e^{-E_2'/u}\Gamma_{12} \qquad \Gamma_{12}(\theta,\theta) = 0 \quad (4b)$$

$$\dot{\Gamma}_{21} = -\Gamma_{21}\frac{\partial f_1}{\partial x_1} - \Gamma_{22}\frac{\partial f_2}{\partial x_1} = 2x_1k_{10}e^{-E_1'/u}(\Gamma_{21} - \Gamma_{22})$$
$$\Gamma_{21}(\theta,\theta) = 0 \quad (4c)$$

$$\dot{\Gamma}_{22} = -\Gamma_{21}\frac{\partial f_1}{\partial x_2} - \Gamma_{22}\frac{\partial f_2}{\partial x_2} = k_{20}e^{-E_2'/u}\Gamma_{22} \qquad \Gamma_{22}(\theta,\theta) = 1 \quad (4d)$$

It easily follows that $\Gamma_{12}(\theta,t) \equiv 0$ and will not enter the computation, although we shall carry it along in the discussion for the sake of generality.

The hamiltonian for the optimization problem is

$$H = -\gamma_1 k_{10}e^{-E_1'/u}x_1{}^2 + \gamma_2(k_{10}e^{-E_1'/u}x_1{}^2 - k_{20}e^{-E_2'/u}x_2) \tag{5}$$

where the γ_i are computed from the Γ_{ij} by

$$\gamma_1 = \frac{\partial \mathcal{E}}{\partial x_1}\Gamma_{11} + \frac{\partial \mathcal{E}}{\partial x_2}\Gamma_{21} = -c\Gamma_{11} - \Gamma_{21} \tag{6a}$$

$$\gamma_2 = \frac{\partial \mathcal{E}}{\partial x_1}\Gamma_{12} + \frac{\partial \mathcal{E}}{\partial x_2}\Gamma_{22} = -c\Gamma_{12} - \Gamma_{22} \tag{6b}$$

Minimization of the hamiltonian leads to the equation for the optimal $u(t)$

$$u(t) = \begin{cases} u^* & v(t) \geq u^* \\ v(t) & u_* < v(t) < u^* \\ u_* & v(t) \leq u_* \end{cases} \tag{7}$$

where the unconstrained optimum is

$$v(t) = \frac{E_2' - E_1'}{\ln \dfrac{\gamma_2 x_2 k_{20}}{(\gamma_2 - \gamma_1) x_1^2 k_{10}}} \tag{8}$$

The computational procedure is now to choose trial values $\bar{x}_1(\theta)$, $\bar{x}_2(\theta)$ and integrate the six equations (1) and (4) numerically from $t = \theta$ to $t = 0$, evaluating u at each step of the integration by Eq. (7) in conjunction with Eqs. (8) and (6). At $t = 0$ the computed values $\bar{x}_1(0)$, $\bar{x}_2(0)$ are compared with the desired values x_{10}, x_{20} and a new trial carried out with values

$$x_1(\theta) = \bar{x}_1(\theta) + \frac{1}{r}\{\Gamma_{11}(\theta,0)[x_{10} - \bar{x}_1(0)] + \Gamma_{12}(\theta,0)[x_{20} - \bar{x}_2(0)]\} \tag{9a}$$

$$x_2(\theta) = \bar{x}_2(\theta) + \frac{1}{r}\{\Gamma_{21}(\theta,0)[x_{10} - \bar{x}_1(0)] + \Gamma_{22}(\theta,0)[x_{20} - \bar{x}_2(0)]\} \tag{9b}$$

In carrying out this process it is found that convergence cannot be obtained when x_1 and x_2 are allowed to penetrate too far into physically impossible regions (negative concentrations), so that a further practical modification is to apply Eqs. (9) at any value of t for which a preset bound on one of the state variables has been exceeded and begin a new iteration, rather than integrate all the way to $t = 0$.

The values of the parameters used in the calculation are as follows:

$$k_{10} = 5 \times 10^{10} \qquad k_{20} = 3.33 \times 10^{17}$$
$$E_1' = 9 \times 10^3 \qquad E_2' = 15 \times 10^3$$
$$u_* = 335 \qquad u^* = 355$$
$$\theta = 6 \qquad c = 0.3$$
$$x_{10} = 1 \qquad x_{20} = 0$$

The iteration parameter r was taken initially as 2 and increased by 1 each time the error, defined as $|x_1(0) - x_{10}| + |x_2(0) - x_{20}|$, did not decrease. In the linear region, where the ratio of errors on successive iterations is approximately $1 - 1/r$, the value of r was increased by 0.5 for each iteration for which an improvement was obtained. Starting values of $x_1(\theta)$ and $x_2(\theta)$ were taken as the four combinations of

$x_1(\theta) = 0.254$, 0.647 and $x_2(\theta) = 0$, 0.746, calculated from the limiting isothermal policies as defining the extreme values.

The approach to the values of $x_1(\theta)$ and $x_2(\theta)$ which satisfy the two-point boundary-value problem ($x_1 = 0.421$, $x_2 = 0.497$) are shown in Fig. 9.1, where the necessity of maintaining $r > 1$ during the early stages to prevent serious overshoot or divergence is evident. Successive values of the optimal temperature profile for the iterations starting from the point $x_1 = 0.647$, $x_2 = 0$ are shown in Fig. 9.2, where the ultimate profile, shown as a broken line, is approached with some oscillation. In all four cases convergence to within 0.1 in $u(t)$ at all t was obtained with between 12 and 20 iterations.

As described in this section and the preceding one, the algorithm is restricted to problems with unconstrained final values. This restriction can be removed by the use of penalty functions, although it is found that the sensitivity is too great to obtain convergence for large values of the penalty constant, so that only approximate solutions can be realized. The usefulness of the Newton-Raphson method rests in part upon the ability to obtain an explicit representation of the optimal decision, such as Eq. (7), for if the hamiltonian had to be minimized by use of the search methods of Chap. 2 at each integration step of every iteration to find the proper $\mathbf{u}(t)$, the computing time would be excessive.

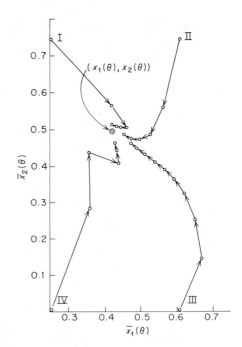

Fig. 9.1 Successive approximations to final conditions using Newton-Raphson boundary iteration. [*From M. M. Denn and R. Aris, Ind. Eng. Chem. Fundamentals,* **4**:7 (1965). *Copyright 1965 by the American Chemical Society. Reprinted by permission of the copyright owner.*]

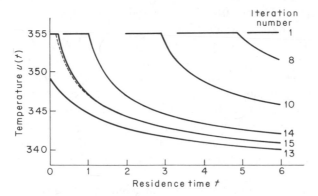

Fig. 9.2 Successive approximations to the optimal temperature profile using Newton-Raphson boundary iteration. [*From M. M. Denn and R. Aris, Ind. Eng. Chem. Fundamentals*, **4:**7 (1965). *Copyright* 1965 *by the American Chemical Society. Reprinted by permission of the copyright owner.*]

It is this latter consideration, rather than convergence difficulty, which has proved to be the primary drawback in our application of this method to several optimization problems.

9.4 STEEP-DESCENT BOUNDARY ITERATION

The Newton-Raphson boundary-iteration algorithm converges well near the solution to the two-point boundary-value problem resulting from the minimum principle, but it requires the solution of $S(S-1)$ additional differential or difference equations for each iteration. An approach which requires the solution of fewer equations per iteration, but which might be expected to have poorer convergence properties, is solution of the boundary-value problem by steep descent. The error in satisfying the specified final (initial) conditions is taken as a function of the assumed initial (final) conditions, and the minimum of this error is then found.

To demonstrate this procedure we shall consider the optimal-pressure-profile problem for the gas-phase reaction

$$X_1 \rightarrow 2X_2 \rightarrow \text{decomposition products}$$

where the conversion to intermediate X_2 is to be maximized. This problem was studied in Sec. 4.13, where the system equations were written as

$$\dot{x}_1 = -2k_1 u \frac{x_1}{A + x_2} \qquad x_1(0) = x_{10} \tag{1a}$$

$$\dot{x}_2 = 4k_1 u \frac{x_1}{A + x_2} - 4k_2 u^2 \frac{x_2{}^2}{(A + x_2)^2} \qquad x_2(0) = x_{20} \tag{1b}$$

Here $A = 2x_{10} + x_{20}$, and the objective is to minimize

$$\varepsilon = -x_2(\theta) \tag{2}$$

For simplicity $u(t)$ is taken to be unconstrained.

The hamiltonian for optimization is

$$H = \gamma_1\left(-2k_1u\,\frac{x_1}{A + x_2}\right) + \gamma_2\left[4k_1u\,\frac{x_1}{A + x_2} - 4k_2u^2\,\frac{x_2{}^2}{(A + x_2)^2}\right] \tag{3}$$

with multiplier equations

$$\dot{\gamma}_1 = -\frac{\partial H}{\partial x_1} = \frac{2k_1u}{A + x_2}\,(\gamma_1 - 2\gamma_2) \qquad \gamma_1(\theta) = \frac{\partial\varepsilon}{\partial x_1} = 0 \tag{4a}$$

$$\dot{\gamma}_2 = -\frac{\partial H}{\partial x_2} = -\frac{2k_1ux_1}{(A + x_2)^2}\,(\gamma_1 - 2\gamma_2) + \frac{8k_2u^2\gamma_2Ax_2}{(A + x_2)^3}$$

$$\gamma_2(\theta) = \frac{\partial\varepsilon}{\partial x_2} = -1 \tag{4b}$$

The optimal pressure $u(t)$ is obtained by setting $\partial H/\partial u$ to zero to obtain

$$u = -\frac{A + x_2}{4}\,\frac{k_1x_1(\gamma_1 - 2\gamma_2)}{k_2\gamma_2x_2{}^2} \tag{5}$$

It is readily verified that $\gamma_2 \leq 0$, in which case $\partial^2H/\partial u^2 \geq 0$ and the condition for a minimum is met. We shall attempt to choose initial values $\gamma_1(0)$, $\gamma_2(0)$ in order to match the boundary conditions on γ at $t = \theta$, and we shall do this by using Box's approximate (complex) steep-descent method, discussed in Sec. 2.9, to find $\gamma_1(0)$, $\gamma_2(0)$ which minimize

$$E = [\gamma_1(\theta)]^2 + [\gamma_2(\theta) + 1]^2 \tag{6}$$

We begin with at least three pairs $\gamma_1(0)$, $\gamma_2(0)$ and integrate Eqs. (1) and (3), calculating u at each step of the numerical integration from Eq. (5), and determine the value of the error E from Eq. (6) in each case. By reflecting the worst point through the centroid new values are found until E is minimized.

The parameters used for computation are

$k_1 = 1.035 \times 10^{-2}$ $\qquad$ $k_2 = 4.530 \times 10^{-2}$

$x_{10} = 0.010$ $\qquad\qquad$ $x_{20} = 0.002$

$\theta = 8.0$ $\qquad\qquad\quad$ $A = 0.022$

For these parameters the solution of the boundary-value problem is at $\gamma_1(0) = -0.8201$, $\gamma_2(0) = -0.4563$, with a value of $x_2(\theta)$ of 0.0132. Table 9.1 shows a sequence of iterations using a three-point simplex with one starting point close to the minimizing point, and convergence to 1 part in 10^4 in $\gamma(\theta)$ is obtained in 17 iterations. Computations from other initial triangles are shown in Tables 9.2 and 9.3, and it is evident that poor results might be obtained without a good first estimate.

Table 9.1 Successive approximations to the initial values of multipliers using the complex method for steep-descent boundary iteration to minimize final error in boundary conditions

$Iteration$	$-\gamma_1(0)$	$-\gamma_2(0)$	$E \times 10^6$	$x_2(\theta) \times 10^2$	$-\gamma_1(0)$	$-\gamma_2(0)$	$E \times 10^6$	$x_2(\theta) \times 10^2$	$-\gamma_1(0)$	$-\gamma_2(0)$	$E \times 10^6$	$x_2(\theta) \times 10^2$
	0.8000	**0.4500**	5.2×10^4	1.130	**0.5000**	**1.0000**	3.9×10^{10}	0.759	**0.2000**	**0.9000**	$\mathbf{6.4 \times 10^{10}}$	**0.732**
1	0.8000	0.4500	5.2×10^4	1.130	**0.5000**	**1.0000**	$\mathbf{3.9 \times 10^{10}}$	**0.759**	0.8900	0.6317	2.3×10^8	0.955
2	0.8000	0.4500	5.2×10^4	1.130	**0.8879**	**0.4837**	$\mathbf{1.4 \times 10^5}$	**1.123**	**0.8900**	**0.6317**	$\mathbf{2.3 \times 10^8}$	**0.955**
3	0.8000	0.4500	5.2×10^4	1.130	0.8385	0.4685	1.2×10^4	1.123	0.8470	0.4778	1.1×10^5	1.129
4	0.8000	0.4500	5.2×10^4	1.130	0.8385	0.4685	1.2×10^4	1.132	**0.8470**	**0.4778**	$\mathbf{1.1 \times 10^5}$	**1.129**
5	**0.8000**	**0.4500**	$\mathbf{5.2 \times 10^4}$	**1.130**	**0.8385**	**0.4685**	$\mathbf{1.2 \times 10^4}$	1.132	0.8045	0.4494	6.3×10^3	1.132
6	0.8330	0.4637	7.6×10^2	1.132	0.8082	0.4502	6.0×10^2	1.132	0.8045	0.4494	6.3×10^3	1.132
7	0.8330	0.4637	7.6×10^2	1.132	0.8082	0.4502	6.0×10^2	1.132	**0.8045**	**0.4494**	$\mathbf{6.3 \times 10^3}$	**1.132**
8	**0.8330**	**0.4637**	$\mathbf{7.6 \times 10^2}$	**1.132**	**0.8082**	**0.4502**	$\mathbf{6.0 \times 10^2}$	**1.132**	0.8291	0.4610	133	1.132
9	0.8110	0.4512	95.6	1.132	0.8167	0.4544	22.3	1.132	0.8291	0.4610	133	1.132
10	0.8110	0.4512	95.6	1.132	0.8167	0.4544	22.3	1.132	**0.8291**	**0.4610**	**133**	**1.132**
11	**0.8110**	**0.4512**	**95.6**	**1.132**	**0.8167**	**0.4544**	**22.3**	**1.132**	0.8174	0.4547	8.6	1.132
12	0.8203	0.4563	6.5	1.132	0.8200	0.4561	0.14	1.132	0.8174	0.4547	8.6	1.132
13	0.8203	0.4563	6.5	1.132	0.8200	0.4561	0.14	1.132	**0.8174**	**0.4547**	**8.6**	**1.132**
14	**0.8203**	**0.4563**	**6.5**	**1.132**	0.8200	0.4561	0.14	1.132	0.8194	0.4558	0.78	1.132
15	0.8199	0.4561	0.63	1.132	0.8200	0.4561	0.14	1.132	**0.8194**	**0.4558**	**0.78**	**1.132**
16	**0.8199**	**0.4561**	**0.63**	**1.132**	0.8200	0.4561	0.14	1.132	0.8202	0.4563	0.04	1.132
17	0.8200	0.4562	0.01	1.132	0.8200	0.4561	**0.14**	1.132	0.8202	0.4563	0.04	1.132

Table 9.2 Successive approximations to the initial values of multipliers using the complex method for steep-descent boundary iteration to minimize final error in boundary conditions

Iteration	$-\gamma_1(0)$	$-\gamma_2(0)$	$E \times 10^6$	$x_2(\theta) \times 10^2$	$-\gamma_1(0)$	$-\gamma_2(0)$	$E \times 10^6$	$x_2(\theta) \times 10^2$	$-\gamma_1(0)$	$-\gamma_2(0)$	$E \times 10^6$	$x_2(\theta) \times 10^2$
	0.7000	0.4000	2.5×10^5	1.123	0.6000	0.5000	6.1×10^{10}	0.882	**1.0000**	**1.0000**	$\mathbf{6.8 \times 10^{11}}$	**0.834**
1	0.7000	0.4000	2.5×10^5	1.123	**0.6000**	**0.5000**	$\mathbf{6.1 \times 10^{10}}$	**0.882**	0.4633	0.1567	7.9×10^5	−1.916
2	0.7000	0.4000	2.5×10^5	1.123	0.5869	0.3414	5.8×10^5	1.111	**0.4633**	**0.1567**	$\mathbf{7.9 \times 10^5}$	**−1.916**
3	0.7000	0.4000	2.5×10^5	1.123	**0.5869**	**0.3414**	$\mathbf{5.8 \times 10^5}$	**1.111**	0.6014	0.3208	4.4×10^5	1.087
4	0.7000	0.4000	2.5×10^5	1.123	0.6847	0.3705	2.5×10^5	1.116	**0.6014**	**0.3208**	$\mathbf{4.4 \times 10^5}$	**1.087**
5	0.7000	0.4000	2.5×10^5	1.123	**0.6847**	**0.3705**	$\mathbf{2.5 \times 10^5}$	**1.116**	0.7409	0.4197	1.2×10^5	1.127
6	**0.7000**	**0.4000**	$\mathbf{2.5 \times 10^5}$	**1.123**	0.7121	0.4007	3.7×10^4	1.130	0.7409	0.4197	1.2×10^5	1.127
7	0.7406	0.4156	2.2×10^4	1.131	**0.7121**	**0.4007**	$\mathbf{3.7 \times 10^4}$	**1.130**	**0.7409**	**0.4197**	$\mathbf{1.2 \times 10^5}$	**1.127**
8	0.7406	0.4156	2.2×10^4	1.131	0.7390	0.4131	9.7×10^3	1.131	0.7186	0.4020	1.4×10^4	1.131
9	**0.7406**	**0.4156**	$\mathbf{2.2 \times 10^4}$	**1.131**	**0.7390**	**0.4131**	$\mathbf{9.7 \times 10^3}$	**1.131**	**0.7186**	**0.4020**	$\mathbf{1.4 \times 10^4}$	**1.131**
10	0.7225	0.4032	1.1×10^4	1.132	0.7390	0.4131	9.7×10^3	1.131	0.7186	0.4020	1.4×10^4	1.131
11	**0.7225**	**0.4032**	$\mathbf{1.1 \times 10^4}$	**1.132**	**0.7390**	**0.4131**	$\mathbf{9.7 \times 10^3}$	**1.132**	0.7186	0.4020	1.4×10^4	1.131
12	0.7464	0.4171	8.3×10^3	1.132	0.7390	0.4131	9.7×10^3	1.132	0.7372	0.4115	8.1×10^3	1.132
13	**0.7464**	**0.4171**	$\mathbf{8.3 \times 10^3}$	**1.132**	0.7433	0.4149	7.2×10^3	1.132	**0.7372**	**0.4115**	$\mathbf{8.1 \times 10^3}$	**1.132**
14	0.7370	0.4111	7.9×10^3	1.132	0.7433	0.4149	7.2×10^3	1.132	0.7372	0.4115	8.1×10^3	1.132
15	**0.7370**	**0.4111**	$\mathbf{7.9 \times 10^3}$	**1.132**	**0.7433**	**0.4149**	$\mathbf{7.2 \times 10^3}$	**1.132**	0.7372	0.4115	8.1×10^3	1.132
16	0.7454	0.4161	6.9×10^3	1.132	0.7433	0.4149	7.2×10^3	1.132	0.7417	0.4138	7.2×10^3	1.132
17	0.7454	0.4161	6.9×10^3	1.132	0.7437	0.4150	6.9×10^3	1.132	**0.7417**	**0.4138**	$\mathbf{7.2 \times 10^3}$	**1.132**

Table 9.3 Successive approximations to the initial values of multipliers using the complex method for steep-descent boundary iteration to minimize final error in boundary conditions

Iteration	$-\gamma_1(0)$	$-\gamma_2(0)$	$E \times 10^6$	$x_2(\theta) \times 10^2$	$-\gamma_1(0)$	$-\gamma_2(0)$	$E \times 10^6$	$x_2(\theta) \times 10^2$	$-\gamma_1(0)$	$-\gamma_2(0)$	$E \times 10^6$	$x_2(\theta) \times 10^2$
	0.4000	0.6000	8.4×10^9	0.780	0.5000	0.8000	1.7×10^{10}	0.774	0.7000	0.7000	3.3×10^9	0.834
1	0.4000	0.6000	8.4×10^9	0.780	0.5767	0.5700	2.1×10^9	0.836	0.7000	0.7000	3.3×10^9	0.834
2	0.7654	0.6537	1.2×10^9	0.874	0.5767	0.5700	2.1×10^9	0.836	0.7000	0.7000	3.3×10^9	0.834
3	0.7654	0.6537	1.2×10^9	0.874	0.5767	0.5700	2.1×10^9	0.836	0.6566	0.5648	9.8×10^8	0.871
4	0.7654	0.6537	1.2×10^9	0.874	0.7819	0.6302	7.7×10^8	0.894	0.6566	0.5648	9.8×10^8	0.871
5	0.6939	0.5675	6.9×10^8	0.889	0.7819	0.6302	7.7×10^8	0.894	0.6566	0.5648	9.8×10^8	0.871
6	0.6939	0.5675	6.9×10^8	0.889	0.7819	0.6302	7.7×10^8	0.894	0.7818	0.6170	6.2×10^8	0.902
7	0.6939	0.5675	6.9×10^8	0.889	0.7143	0.5720	6.0×10^8	0.897	0.7818	0.6170	6.2×10^8	0.902
8	0.7769	0.6089	5.7×10^8	0.905	0.7143	0.5720	6.0×10^8	0.897	0.7818	0.6170	6.2×10^8	0.902
9	0.7769	0.6089	5.7×10^8	0.905	0.7143	0.5720	6.0×10^8	0.897	0.7263	0.5763	5.7×10^8	0.900
10	0.7769	0.6089	5.7×10^8	0.905	0.7716	0.6035	5.5×10^8	0.906	0.7263	0.5763	5.7×10^8	0.900
11	0.7340	0.5799	5.5×10^8	0.902	0.7716	0.6035	5.5×10^8	0.906	0.7263	0.5763	5.7×10^8	0.900
12	0.7340	0.5799	5.5×10^8	0.902	0.7716	0.6035	5.5×10^8	0.906	0.7669	0.6000	5.5×10^8	0.906
13	0.7340	0.5799	5.5×10^8	0.902	0.7392	0.5826	5.5×10^8	0.903	0.7669	0.6000	5.5×10^8	0.906
14	0.7632	0.5974	5.4×10^8	0.906	0.7392	0.5826	5.5×10^8	0.903	0.7669	0.6000	5.5×10^8	0.906
15	0.7632	0.5974	5.4×10^8	0.906	0.7788	0.6072	5.4×10^8	0.908	0.7669	0.6000	5.5×10^8	0.906
16	0.7632	0.5974	5.4×10^8	0.906	0.7788	0.6072	5.4×10^8	0.908	0.7732	0.6035	5.4×10^8	0.907
17	0.7632	0.5974	5.4×10^8	0.906	0.7625	0.5968	5.4×10^8	0.906	0.7732	0.6035	5.4×10^8	0.907

For problems of the specific type considered here, where all initial values of the state variables are known and final values unspecified, we can use a more direct approach to steep-descent boundary iteration. The value of the objective $\mathcal{E}$ depends only upon the choice of $\gamma(0)$, for everything else is determined from the minimum-principle equations if all initial conditions are specified. Instead of minimizing E, the error in final conditions, it is reasonable simply to seek the minimum of $\mathcal{E}$ directly by steep-descent iteration on the initial conditions. Tables 9.4 and 9.5 show the results of such a calculation using the simplex-complex procedure. In neither case are the values of $\gamma_1(0) = -0.8201$, $\gamma_2(0) = -0.4563$ approached, although the same ratio 1.80 of these values is obtained. It is evident from Eq. (5) that only the ratio γ_1/γ_2 is required for defining the optimum, and Eqs. (4) can be combined to give a single equation for this ratio. Hence the optimum is obtained for any initial pair in the ratio 1.80 and, had we so desired, we might have reduced this particular problem to a one-dimensional search.

9.5 NEWTON–RAPHSON FUNCTION ITERATION: A SPECIAL CASE

We have already seen how a computational scheme of the Newton-Raphson type can be applied to boundary-value problems by linearization of the boundary conditions. The theory of linear differential and difference equations is highly developed, and we know that the principle of superposition can be used to solve linear boundary-value problems. Thus, we might anticipate that a Newton-Raphson linearization approach to the solution of differential equations would be practical. Before developing the procedure in general it is helpful in this case to study a specific simple example.

A convenient demonstration problem of some importance in physics is the minimization of the integral

$$\mathcal{E} = \int_1^2 \frac{[1 + (\dot{x})^2]^{\frac{1}{2}}}{x}\, dt \tag{1}$$

where the function $x(t)$ is to be chosen in the interval $1 \leq t \leq 2$ subject to boundary conditions $x(1) = 1$, $x(2) = 2$. This is a special case of Fermat's minimum-time principle for the path of a light ray through an optically inhomogeneous medium. In the notation of Sec. 3.2 we write

$$\mathcal{F}(x,\dot{x},t) = x^{-1}[1 + (\dot{x})^2]^{\frac{1}{2}} \tag{2}$$

and the Euler equation

$$\frac{d}{dt}\frac{\partial \mathcal{F}}{\partial \dot{x}} = \frac{\partial \mathcal{F}}{\partial x} \tag{3}$$

Table 9.4 Successive approximations to the initial values of multipliers using the complex method for steep-descent boundary iteration to minimize the objective

Iteration	$-\gamma_1(0)$	$-\gamma_2(0)$	$\dfrac{\gamma_1(0)}{\gamma_2(0)}$	$-\varepsilon \times 10^2$	$-\gamma_1(0)$	$-\gamma_2(0)$	$\dfrac{\gamma_1(0)}{\gamma_2(0)}$	$-\varepsilon \times 10^2$	$-\gamma_1(0)$	$-\gamma_2(0)$	$\dfrac{\gamma_1(0)}{\gamma_2(0)}$	$-\varepsilon \times 10^2$
	0.5000	0.5000	1.00	0.834	0.5000	1.0000	0.50	0.759	**0.2000**	**0.9000**	**0.22**	**0.732**
1	0.5000	0.5000	1.00	0.834	**0.5000**	**1.0000**	**0.50**	**0.759**	0.6600	0.6700	0.99	0.831
2	0.5000	0.5000	1.00	0.834	0.6227	0.3637	1.71	1.107	**0.6600**	**0.6700**	**0.99**	**0.831**
3	**0.5000**	**0.5000**	**1.00**	**0.834**	0.6227	0.3637	1.71	1.107	0.5087	0.3048	1.67	1.086
4	0.5739	0.3136	1.83	1.126	**0.6227**	**0.3637**	**1.71**	**1.107**	**0.5087**	**0.3048**	**1.67**	**1.086**
5	0.5739	0.3136	1.83	1.126	0.6227	0.3637	1.71	1.107	0.6460	0.3567	1.81	1.131
6	**0.5739**	**0.3136**	**1.83**	**1.126**	0.6129	0.3418	1.79	1.132	**0.6460**	**0.3567**	**1.81**	**1.131**
7	0.6591	0.3682	1.79	1.132	0.6129	0.3418	1.79	1.132	0.6460	0.3567	1.81	1.131
8	**0.6591**	**0.3682**	**1.79**	**1.132**	**0.6129**	**0.3418**	**1.79**	**1.132**	**0.6384**	**0.3544**	**1.80**	**1.132**
9	0.6078	0.3381	1.80	1.132	0.6129	0.3418	1.79	1.132	0.6384	0.3544	1.80	1.132
10	0.6078	0.3381	1.80	1.132	0.6243	0.3474	1.80	1.132	0.6384	0.3544	1.80	1.132

Table 9.5 Successive approximations to the initial values of multipliers using the complex method for steep-descent boundary iteration to minimize the objective

Iteration	$-\gamma_1(0)$	$-\gamma_2(0)$	$\dfrac{\gamma_1(0)}{\gamma_2(0)}$	$-\varepsilon \times 10^2$	$-\gamma_1(0)$	$-\gamma_2(0)$	$\dfrac{\gamma_1(0)}{\gamma_2(0)}$	$-\varepsilon \times 10^2$	$-\gamma_1(0)$	$-\gamma_2(0)$	$\dfrac{\gamma_1(0)}{\gamma_2(0)}$	$-\varepsilon \times 10^2$
	0.9000	0.5500	1.64	1.068	0.5000	1.0000	0.50	0.759	0.2000	0.9000	0.22	0.732
1	0.9000	0.5500	1.64	1.068	0.5000	1.0000	0.50	0.759	0.9667	0.7083	1.36	0.937
2	0.9000	0.5500	1.64	1.068	0.9873	0.5830	1.69	1.098	0.9667	0.7083	1.36	0.937
3	0.9000	0.5500	1.64	1.068	0.9873	0.5830	1.69	1.098	0.9408	0.5489	1.71	1.108
4	0.9982	0.5744	1.74	1.118	0.9873	0.5830	1.69	1.098	0.9408	0.5489	1.71	1.108
5	0.9982	0.5744	1.74	1.118	0.9600	0.5503	1.75	1.121	0.9408	0.5489	1.71	1.108
6	0.9982	0.5744	1.74	1.118	0.9600	0.5503	1.75	1.121	0.9995	0.5695	1.75	1.124
7	0.9699	0.5521	1.76	1.125	0.9600	0.5503	1.75	1.121	0.9995	0.5695	1.75	1.124
8	0.9699	0.5521	1.76	1.125	0.9979	0.5665	1.76	1.126	0.9995	0.5695	1.75	1.124
9	0.9699	0.5521	1.76	1.125	0.9979	0.5665	1.76	1.126	0.9756	0.5539	1.76	1.126
10	0.9957	0.5645	1.76	1.127	0.9979	0.5665	1.76	1.126	0.9756	0.5539	1.76	1.126

reduces to the nonlinear second-order equation

$$x\ddot{x} + (\dot{x})^2 + 1 = 0 \qquad \begin{aligned} x(1) &= 1 \\ x(2) &= 2 \end{aligned} \tag{4}$$

By noting that

$$x\ddot{x} + (\dot{x})^2 = \frac{d}{dt}\, x\dot{x} = \frac{1}{2}\frac{d}{dt}\left(\frac{d}{dt}\,x^2\right) \tag{5}$$

Eq. (4) can be integrated directly to obtain the solution satisfying the boundary conditions

$$x(t) = (6t - t^2 - 4)^{\frac{1}{2}} \tag{6}$$

We wish now to solve Eq. (4) iteratively by a Newton-Raphson expansion analogous to that developed for nonlinear algebraic equations in Sec. 2.2. We suppose that we have an estimate of the solution, $x^{(k)}(t)$, and that the solution is the result of the $(k+1)$st iteration, $x^{(k+1)}(t)$. The nonlinear terms in Eq. (4) may be written

$$x^{(k+1)}\ddot{x}^{(k+1)} = x^{(k)}\ddot{x}^{(k)} + \ddot{x}^{(k)}(x^{(k+1)} - x^{(k)}) + x^{(k)}(\ddot{x}^{(k+1)} - \ddot{x}^{(k)})$$
$$+ \text{ higher-order terms} \tag{7a}$$

$$(\dot{x}^{(k+1)})^2 = (\dot{x}^{(k)})^2 + 2\dot{x}^{(k)}(\dot{x}^{(k+1)} - \dot{x}^{(k)})$$
$$+ \text{ higher-order terms} \tag{7b}$$

With some rearranging and the dropping of higher-order terms Eq. (5) then becomes a linear ordinary differential equation in $x^{(k+1)}$

$$\ddot{x}^{(k+1)} + \frac{2\dot{x}^{(k)}}{x^{(k)}}\,\dot{x}^{(k+1)} + \frac{\ddot{x}^{(k)}}{x^{(k)}}\,x^{(k+1)} = \ddot{x}^{(k)} + \frac{(\dot{x}^{(k)})^2 - 1}{x^{(k)}}$$
$$\begin{aligned} x^{(k+1)}(1) &= 1 \\ x^{(k+1)}(2) &= 2 \end{aligned} \tag{8}$$

A particularly convenient starting estimate which satisfies both boundary conditions is

$$x^{(0)}(t) = t \tag{9}$$

in which case Eq. (8) for $x^{(1)}$ simplifies to

$$\ddot{x}^{(1)} + \frac{2}{t}\,\dot{x}^{(1)} = 0 \tag{10}$$

This linear homogeneous equation has two solutions, $x^{(1)} = 1/t$ and $x^{(1)} = 1$. Using the principle of superposition, the general solution is a linear combination of the two

$$x^{(1)} = c_1 + c_2 t^{-1} \tag{11}$$

and the constants are evaluated from the boundary conditions at $t = 1$

and $t = 2$

$$x^{(1)}(1) = 1 = c_1 + c_2 \tag{12a}$$
$$x^{(1)}(2) = 2 = c_1 + \tfrac{1}{2}c_2 \tag{12b}$$

The solution is then

$$x^{(1)}(t) = 3 - 2t^{-1} \tag{13}$$

Table 9.6 shows the agreement between the exact solution and these first two Newton-Raphson approximations.

The starting approximation need not satisfy all or any of the boundary conditions, though by use of superposition all subsequent approximations will. For example, with the constant starting value $x^{(0)} = 1.5$ Eq. (8) for $x^{(1)}$ becomes

$$\ddot{x}^{(1)} = -\tfrac{2}{3} \tag{14}$$

The homogeneous solutions are $x^{(1)} = 1$, $x^{(1)} = t$, while the particular solution obtained from the method of undetermined coefficients is $-\tfrac{1}{3}t^2$. By superposition, then, the general solution is

$$x^{(1)} = c_1 + c_2 t - \tfrac{1}{3}t^2 \tag{15}$$

Solving for the constants from the boundary conditions,

$$x^{(1)}(1) = 1 = c_1 + c_2 - \tfrac{1}{3} \tag{16a}$$
$$x^{(1)}(2) = 2 = c_1 + 2c_2 - \tfrac{4}{3} \tag{16b}$$

so that the solution is

$$x^{(1)} = -\tfrac{2}{3} + 2t - \tfrac{1}{3}t^2 \tag{17}$$

**Table 9.6 Comparison of exact
solution and approximation using
Newton-Raphson function iteration**

t	$x^{(0)}(t)$	$x^{(1)}(t)$	$x(t)$
1.0	1.000	1.000	1.000
1.1	1.100	1.182	1.179
1.2	1.200	1.334	1.327
1.3	1.300	1.462	1.453
1.4	1.400	1.572	1.562
1.5	1.500	1.667	1.658
1.6	1.600	1.750	1.744
1.7	1.700	1.824	1.819
1.8	1.800	1.889	1.887
1.9	1.900	1.947	1.947
2.0	2.000	2.000	2.000

**Table 9.7 Comparison of exact
solution and approximation using
Newton-Raphson function iteration**

t	$x^{(0)}(t)$	$x^{(1)}(t)$	$x(t)$
1.0	1.500	1.000	1.000
1.1	1.500	1.130	1.179
1.2	1.500	1.253	1.327
1.3	1.500	1.370	1.453
1.4	1.500	1.480	1.562
1.5	1.500	1.583	1.658
1.6	1.500	1.680	1.744
1.7	1.500	1.770	1.819
1.8	1.500	1.853	1.887
1.9	1.500	1.930	1.947
2.0	1.500	2.000	2.000

Table 9.7 shows the start of convergence for this sequence of Newton-Raphson approximations, and considering the crude starting value, the agreement on the first iteration is excellent.

It is helpful to observe here that the principle of superposition can be used in such a way as to reduce the subsequent algebra. We can construct a particular solution satisfying the boundary condition at $t = 1$,

$$x^{(1p)} = \tfrac{4}{3} - \tfrac{1}{3}t^2 \tag{18}$$

To this we add a multiple of a nontrivial homogeneous solution which vanishes at $t = 1$

$$x^{(1h)} = t - 1 \tag{19}$$

so that the solution is written

$$x^{(1)} = x^{(1p)} + c_1 x^{(1h)} = \tfrac{4}{3} - \tfrac{1}{3}t^2 + c_1(1 - t) \tag{20}$$

The boundary condition at $t = 1$ is automatically satisfied, and the *single* coefficient is now evaluated from the one remaining boundary condition at $t = 2$, leading again to the result in Eq. (17).

9.6 NEWTON-RAPHSON FUNCTION ITERATION: GENERAL ALGORITHM

The example of the preceding section is a graphic demonstration of the use of a Newton-Raphson linearization of the differential equation to obtain an approximate solution to a nonlinear boundary-value problem, but it is peculiar in that explicit analytical solutions are available for both the exact and approximate problems. The general variational problem in which the decision function $\mathbf{u}(t)$ is unconstrained leads to $2S$ nonlinear first-order equations with S conditions specified at $t = 0$ $(n = 0)$ and S conditions at $t = \theta$ $(n = N)$. We now consider the extension of the Newton-Raphson approach to this problem. This is frequently referred to as *quasilinearization*.

We shall suppose that we have a system of $2S$ differential equations

$$\dot{y}_i = F_i(\mathbf{y}) \qquad i = 1, 2, \ldots, 2S \tag{1}$$

In a variational problem S of the variables would correspond to the state variables $x_i(t)$ and the remaining S to the Green's functions $\gamma_i(t)$. It is assumed that the condition $\partial H / \partial u = 0$ has been used to obtain $\mathbf{u}$ explicitly in terms of $\mathbf{x}$ and $\mathbf{\gamma}$. If we number the components of $\mathbf{y}$ such that the first S components are specified at $t = 0$, we have the condition

$$y_i(0) = \begin{cases} y_{i0} & i = 1, 2, \ldots, S \\ \text{unspecified} & i = S + 1, S + 2, \ldots, 2S \end{cases} \tag{2}$$

Any S variables or combination might be specified at $t = \theta$, depending upon the nature of the problem.

We now assume that we have an nth approximation to the solution $\mathbf{y}^{(n)}$. If the $(n + 1)$st result is assumed to be the exact solution, we write Eq. (1) as

$$\dot{y}_i^{(n+1)} = F_i(\mathbf{y}^{(n+1)}) \tag{3}$$

and the right-hand side may be expanded in a Taylor series about the function $\mathbf{y}^{(n)}$ as

$$\dot{y}_i^{(n+1)} = F_i(\mathbf{y}^{(n)}) + \sum_{j=1}^{2S} \frac{\partial F_i(\mathbf{y}^{(n)})}{\partial y_j} (y_j^{(n+1)} - y_j^{(n)}) + \cdots \tag{4}$$

or, regrouping and neglecting higher-order terms,

$$\dot{y}_i^{(n+1)} = \sum_{j=1}^{2S} \frac{\partial F_i(\mathbf{y}^{(n)})}{\partial y_j} y_j^{(n+1)} + \left[F_i(\mathbf{y}^{(n)}) - \sum_{j=1}^{2S} \frac{\partial F_i(\mathbf{y}^{(n)})}{\partial y_j} y_j^{(n)} \right] \tag{5}$$

Equation (5) is a set of linear nonhomogeneous ordinary differential equations to which the principle of superposition applies. Thus, we need obtain only a single particular solution which satisfies the S boundary conditions at $t = 0$ and add to this, with undetermined coefficients, S solutions of the homogeneous equation

$$\dot{y}_i^{(n+1),h} = \sum_{j=1}^{2S} \frac{F_i(\mathbf{y}^{(n)})}{y_j} y_j^{(n+1),h} \tag{6}$$

The S coefficients are then found from the S boundary conditions at $t = \theta$.

To be more precise, let us denote by $y_i^{(n+1),p}$ the solution of Eq. (5) which satisfies the initial condition

$$y_i^{(n+1),p}(0) = \begin{cases} y_{i0} & i = 1, 2, \ldots, S \\ \text{arbitrary} & i = S + 1, S + 2, \ldots, 2S \end{cases} \qquad (7)$$

and by $y_i^{(n+1),h_k}$, $k = 1, 2, \ldots, S$, the linearly independent solutions of Eq. (6) with zero initial conditions for $i = 1, 2, \ldots, S$. The general solution of Eq. (5) may then be written

$$y_i^{(n+1)}(t) = y_i^{(n+1),p}(t) + \sum_{k=1}^{S} c_k y_i^{(n+1),h_k}(t) \qquad i = 1, 2, \ldots, 2S \quad (8)$$

and the S constants c_k are determined from the S specified conditions at $t = \theta$. For example, if $y_1(\theta)$, $y_2(\theta)$, and the last $S - 2$ values are specified, the c_k would be obtained from the S linear algebraic equations

$$y_i(\theta) = y_i^{(n+1),p}(\theta) + \sum_{k=1}^{S} c_k y_i^{(n+1),h_k}(\theta)$$
$$i = 1, 3, S + 2, S + 3, \ldots, 2S \quad (9)$$

As outlined, the method is confined to variational problems in which there are no trajectory constraints on $\mathbf{u}$ or $\mathbf{x}$ but only specifications at $t = 0$ and θ. Since $\mathbf{u}$ must be completely eliminated by means of the stationary condition for the hamiltonian, constraints on $\mathbf{u}$ can be imposed only if they are convertable to constraints on $\mathbf{x}$ and $\boldsymbol{\gamma}$, which can be incorporated by means of penalty functions. Finally, we note that a completely analogous treatment can be applied to difference equations.

9.7 OPTIMAL PRESSURE PROFILE BY NEWTON–RAPHSON FUNCTION ITERATION

In order to demonstrate the use of Newton-Raphson function iteration we shall again consider the optimal-pressure problem. The equations are given in Sec. 9.4 as

$$\dot{x}_1 = -2k_1 u \frac{x_1}{A + x_2} \qquad x_1(0) = x_{10} \qquad (1a)$$

$$\dot{x}_2 = 4k_1 u \frac{x_1}{A + x_2} - 4k_2 u^2 \frac{x_2^2}{(A + x_2)^2} \qquad x_2(0) = x_{20} \qquad (1b)$$

$$\mathcal{E} = -x_2(\theta) \qquad (2)$$

The Green's functions satisfy

$$\dot{\gamma}_1 = \frac{2k_1 u}{A + x_2} (\gamma_1 - 2\gamma_2) \qquad \gamma_1(\theta) = 0 \tag{3a}$$

$$\dot{\gamma}_2 = -\frac{2k_1 u x_1}{(A + x_2)^2} (\gamma_1 - 2\gamma_2) + \frac{8k_2 u^2 \gamma_2 A x_2}{(A + x_2)^3} \qquad \gamma_2(\theta) = -1 \tag{3b}$$

and the optimal pressure satisfies

$$u(t) = -\frac{A + x_2}{4} \frac{k_1 x_1 (\gamma_1 - 2\gamma_2)}{k_2 \gamma_2 x_2^2} \tag{4}$$

Substitution of Eq. (4) into Eqs. (1) and (3) leads to the four equations in four variables

$$\dot{x}_1 = \frac{k_1^2 x_1^2}{k_2 x_2^2} \left(-1 + \frac{\gamma_1}{2\gamma_2} \right) \tag{5a}$$

$$\dot{x}_2 = \frac{k_1^2 x_1^2}{k_2 x_2^2} \left(1 - \frac{\gamma_1^2}{4\gamma_2^2} \right) \tag{5b}$$

$$\dot{\gamma}_1 = -\frac{k_1^2 x_1}{2k_2 \gamma_2 x_2^2} (\gamma_1 - 2\gamma_2)^2 \tag{5c}$$

$$\dot{\gamma}_2 = \frac{k_1^2 x_1^2}{2k_2 \gamma_2 x_2^3} (\gamma_1 - 2\gamma_2)^2 \tag{5d}$$

In the notation of the previous section x_1, x_2, γ_1, and γ_2 correspond, respectively, to y_1, y_2, y_3, y_4. Values at $t = 0$ are given for x_1 and x_2 (y_1 and y_2) and at $t = \theta$ for γ_1 and γ_2 (y_3 and y_4). The linearized equations for the iteration become, after some simplification,

$$\dot{x}_1^{(n+1)} = \frac{k_1^2 (x_1^{(n)})^2}{k_2 (x_2^{(n)})^2} \left[\frac{2}{x_1^{(n)}} \left(-1 + \frac{\gamma_1^{(n)}}{2\gamma_2^{(n)}} \right) x_1^{(n+1)} \right.$$

$$- \frac{2}{x_2^{(n)}} \left(-1 + \frac{\gamma_1^{(n)}}{2\gamma_2^{(n)}} \right) x_2^{(n+1)} + \frac{1}{2\gamma_2^{(n)}} \gamma_1^{(n+1)}$$

$$\left. - \frac{\gamma_1^{(n)}}{2(\gamma_1^{(n)})^2} \gamma_2^{(n+1)} \right] + \frac{k_1^2 (x_1^{(n)})^2}{k_2 (x_2^{(n)})^2} \left(-1 + \frac{\gamma_1^{(n)}}{\gamma_2^{(n)}} \right) \tag{6a}$$

$$\dot{x}_2^{(n+1)} = \frac{k_1^2 (x_1^{(n)})^2}{k_2 (x_2^{(n)})^2} \left\{ \frac{2}{x_1^{(n)}} \left[1 - \frac{(\gamma_1^{(n)})^2}{4(\gamma_2^{(n)})^2} \right] x_1^{(n+1)} \right.$$

$$- \frac{2}{x_2^{(n)}} \left[1 - \frac{(\gamma_1^{(n)})^2}{4(\gamma_2^{(n)})^2} \right] x_2^{(n+1)} - \frac{\gamma_1^{(n)}}{2(\gamma_2^{(n)})^2} \gamma_1^{(n+1)}$$

$$\left. + \frac{(\gamma_1^{(n)})^2}{2(\gamma_2^{(n)})^3} \gamma_2^{(n+1)} \right\} + \frac{k_1^2 (x_1^{(n)})^2}{k_2 (x_2^{(n)})^2} \left[1 - \frac{(\gamma_1^{(n)})^2}{4(\gamma_2^{(n)})^2} \right] \tag{6b}$$

$$
\dot{\gamma}_1^{(n+1)} = \frac{k_1{}^2 x_1^{(n)}}{k_2 (x_2^{(n)})^2 \gamma_2^{(n)}} \left[- \frac{(\gamma_1^{(n)} - 2\gamma_2^{(n)})^2}{2 x_1^{(n)}} x_1^{(n+1)} \right.
$$

$$
+ \frac{(\gamma_1^{(n)} - 2\gamma_2^{(n)})^2}{2 x_2^{(n)}} x_2^{(n+1)} - (\gamma_1^{(n)} - 2\gamma_2^{(n)}) \gamma_1^{(n+1)}
$$

$$
\left. + \frac{(\gamma_1^{(n)})^2 - 4(\gamma_2^{(n)})^2}{2 \gamma_2^{(n)}} \gamma_2^{(n+1)} \right]
$$

$$
- \frac{k_1{}^2 x_1^{(n)}}{2 k_2 (x_2^{(n)})^2 \gamma_2^{(n)}} (\gamma_1^{(n)} - 2\gamma_2^{(n)})^2 \qquad (6c)
$$

$$
\dot{\gamma}_2^{(n+1)} = \frac{k_1{}^2 (x_1^{(n)})^2}{k_2 (x_2^{(n)})^3 \gamma_2^{(n)}} \left[\frac{(\gamma_1^{(n)} - 2\gamma_2^{(n)})^2}{x_1^{(n)}} x_1^{(n+1)} \right.
$$

$$
- \frac{3}{2 x_2^{(n)}} (\gamma_1^{(n)} - 2\gamma_2^{(n)})^2 x_2^{(n+1)} + (\gamma_1^{(n)} - 2\gamma_2^{(n)}) \gamma_1^{(n+1)}
$$

$$
\left. + \frac{4(\gamma_2^{(n)})^2 - (\gamma_1^{(n)})^2}{2 \gamma_2^{(n)}} \gamma_2^{(n+1)} \right]
$$

$$
+ \frac{k_1{}^2 (x_1^{(n)})^2}{2 k_2 (x_2^{(n)})^3 \gamma_2^{(n)}} (\gamma_1^{(n)} - 2\gamma_2^{(n)})^2 \qquad (6d)
$$

The *particular solution*, with components $x_1^{(n+1),p}$, $x_2^{(n+1),p}$, $\gamma_1^{(n+1),p}$, $\gamma_2^{(n+1),p}$, is the solution of Eqs. (6) with initial conditions x_{10}, x_{20}, 0, 1. (The latter two are arbitrary.)   The homogeneous equations are obtained by deleting the last term in each of Eqs. (6), and the homogeneous solutions, with components $x_1^{(n+1),h_k}$, $x_2^{(n+1),h_k}$, $\gamma_1^{(n+1),h_k}$, $\gamma_2^{(n+1),h_k}$, $k = 1, 2$, are solutions of the homogeneous equations with initial conditions 0, 0, 1, 0 and 0, 0, 1, 1.   (The latter two are arbitrary but must be independent for the two solutions and not identically zero.)   The general solution can then be written

$$
x_1^{(n+1)}(t) = x_1^{(n+1),p}(t) + c_1 x_1^{(n+1),h_1}(t) + c_2 x_1^{(n+1),h_2}(t) \qquad (7a)
$$

$$
x_2^{(n+1)}(t) = x_2^{(n+1),p}(t) + c_1 x_2^{(n+1),h_1}(t) + c_2 x_2^{(n+1),h_2}(t) \qquad (7b)
$$

$$
\gamma_1^{(n+1)}(t) = \gamma_1^{(n+1),p}(t) + c_1 \gamma_1^{(n+1),h_1}(t) + c_2 \gamma_1^{(n+1),h_2}(t) \qquad (7c)
$$

$$
\gamma_2^{(n+1)}(t) = \gamma_2^{(n+1),p}(t) + c_1 \gamma_2^{(n+1),h_1}(t) + c_2 \gamma_2^{(n+1),h_2}(t) \qquad (7d)
$$

The initial conditions on x_1 and x_2 are automatically satisfied, and the coefficients c_1 and c_2 are evaluated from the boundary conditions on γ_1 and γ_2 at $t = \theta$

$$
\gamma_1(\theta) = \quad 0 = \gamma_1^{(n+1),p}(\theta) + c_1 \gamma_1^{(n+1),h_1}(\theta) + c_2 \gamma_1^{(n+1),h_2}(\theta) \qquad (8a)
$$

$$
\gamma_2(\theta) = -1 = \gamma_2^{(n+1),p}(\theta) + c_1 \gamma_2^{(n+1),h_1}(\theta) + c_2 \gamma_2^{(n+1),h_2}(\theta) \qquad (8b)
$$

We show here some calculations of Lee using the parameters in Sec. 9.4 with the following constant initial choices:

$$
x_1^{(0)} = 0.01 \qquad x_2^{(0)} = 0.01
$$

$$
\gamma_1^{(0)} = 0 \qquad \gamma_2^{(0)} = -1.0
$$

Convergence of x_1 and x_2 is shown in Fig. 9.3, and the results of the first five iterations are listed in Table 9.8. The rather poor starting values result in rapid convergence, though with some oscillation, and it can, in fact, be established that when convergence occurs for this Newton-Raphson procedure, it is quadratic.

9.8 GENERAL COMMENTS ON INDIRECT METHODS

The computational techniques discussed thus far make use of the necessary conditions for the optimal decision function at each iteration and are often called *indirect methods*. Reasonable starting estimates can often be obtained by first assuming a function $\mathbf{u}(t)$ and then solving state and multiplier equations to obtain functions or boundary conditions, as required, or by the use of the direct methods to be developed in the following sections.

A common feature of indirect methods is the necessity of solving state and Green's function equations simultaneously, integrating both

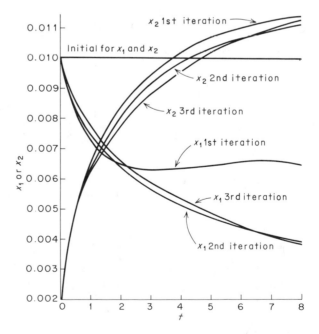

Fig. 9.3 Successive approximations to optimal concentration profiles using Newton-Raphson function iteration. [*From E. S. Lee, Chem. Eng. Sci.,* **21**:183 (1966). *Copyright 1966 by Pergamon Press. Reprinted by permission of the copyright owner.*]

Table 9.8 Successive approximations to optimal concentration and multiplier profiles using Newton-Raphson function iteration†

t	$x_1^{(0)} \times 10^2$	$x_1^{(1)} \times 10^2$	$x_1^{(2)} \times 10^2$	$x_1^{(3)} \times 10^2$	$x_1^{(4)} \times 10^2$	$x_1^{(5)} \times 10^2$
0	1.0000	1.0000	1.0000	1.0000	1.0000	1.0000
2	1.0000	0.6434	0.6372	0.6620	0.6634	0.6634
4	1.0000	0.6370	0.5119	0.5332	0.5342	0.5343
6	1.0000	0.6607	0.4393	0.4486	0.4493	0.4493
8	1.0000	0.6602	0.3927	0.3860	0.3864	0.3864

t	$x_2^{(0)} \times 10^2$	$x_2^{(1)} \times 10^2$	$x_2^{(2)} \times 10^2$	$x_2^{(3)} \times 10^2$	$x_2^{(4)} \times 10^2$	$x_2^{(5)} \times 10^2$
0	1.0000	0.2000	0.2000	0.2000	0.2000	0.2000
2	1.0000	0.8471	0.8060	0.7795	0.7771	0.7770
4	1.0000	1.0215	0.9757	0.9610	0.9597	0.9597
6	1.0000	1.1011	1.0648	1.0651	1.0643	1.0643
8	1.0000	1.1392	1.1147	1.1324	1.1320	1.1320

t	$-\gamma_1^{(0)}$	$-\gamma_1^{(1)}$	$-\gamma_1^{(2)}$	$-\gamma_1^{(3)}$	$-\gamma_1^{(4)}$	$-\gamma_1^{(5)}$
0	0	1.2736	0.8555	0.8188	0.8202	0.8202
2	0	0.8612	0.6945	0.6785	0.6780	0.6780
4	0	0.5696	0.4868	0.4911	0.4913	0.4914
6	0	0.3032	0.2687	0.2648	0.2654	0.2654
8	0	0	0	0	0	0

t	$-\gamma_2^{(0)}$	$-\gamma_2^{(1)}$	$-\gamma_2^{(2)}$	$-\gamma_2^{(3)}$	$-\gamma_2^{(4)}$	$-\gamma_2^{(5)}$
0	1.0000	1.0908	0.5492	0.4625	0.4563	0.4562
2	1.0000	1.2665	0.7399	0.6647	0.6650	0.6650
4	1.0000	1.2592	0.8636	0.7915	0.7911	0.7911
6	1.0000	1.1320	0.9345	0.9002	0.8997	0.8997
8	1.0000	1.0000	1.0000	1.0000	1.0000	1.0000

† From E. S. Lee, *Chem. Eng. Sci.*, **21**:183 (1966). Copyright 1966 by Pergamon Press. Reprinted by permission of the copyright owner.

from either $t = 0$ to θ or θ to 0. A potentially serious computational problem not evident in the examples can be observed by considering the simplest of systems

$$\dot{x} = ax + u \tag{1}$$

for which the Green's function equation is

$$\dot{\gamma} = -a\gamma \tag{2}$$

Solving both equations,

$$x(t) = x(0)e^{at} + \int_0^t e^{a(t-\tau)} u(\tau)\, d\tau \tag{3a}$$

$$\gamma(t) = \gamma(0)e^{-at} \tag{3b}$$

Thus, when either equation has decreasing exponential behavior, the other must be an increasing exponential. It is for this reason that the statement is often made that the optimization equations are unstable. In any indirect method extreme sensitivity and poor convergence will result whenever θ is significantly greater than the smallest time constant of the system.

9.9 STEEP DESCENT

The difficulties associated with the solution of the two-point boundary-value problem may be circumvented in many instances by adopting a *direct* approach which generalizes the steep-descent technique developed in Secs. 2.5 and 2.6. The necessary conditions for optimality are not used, and instead we obtain equations for adjusting estimates of the decision function in order to improve the value of the objective. For simplicity we shall carry out the development with only a single decision variable.

The system is described by the ordinary differential equations with known initial values

$$\dot{x}_i = f_i(\mathbf{x}, u) \qquad x_i(0) = x_{i0} \tag{1}$$

where $u(t)$ may be bounded from above and below

$$u_* \leq u(t) \leq u^* \tag{2}$$

The goal is to minimize $\mathcal{E}[\mathbf{x}(\theta)]$, and for the present we shall assume that the values of $\mathbf{x}(\theta)$ are entirely unconstrained.

If we choose any function $\bar{u}(t)$ which satisfies the upper- and lower-bound constraint, we may integrate Eq. (1) and obtain a value of $\mathcal{E}$. The effect of a small change $\delta u(t)$ in the entire decision function is then described to first order by the linear variational equations

$$\delta\dot{x}_i = \sum_j \frac{\partial f_i}{\partial x_j}\, \delta x_j + \frac{\partial f_i}{\partial u}\, \delta u \qquad \delta x_i(0) = 0 \tag{3}$$

and the corresponding first-order change in $\mathcal{E}$ is

$$\delta\mathcal{E} = \sum_i \frac{\partial \mathcal{E}}{\partial x_i}\, \delta x_i(\theta) \tag{4}$$

All partial derivatives are evaluated along the solution determined by $\bar{u}$. The Green's function for Eq. (3), as discussed in Sec. 6.2, must satisfy the adjoint equation

$$\dot{\gamma}_i = -\sum_j \gamma_j \frac{\partial f_j}{\partial x_i} \tag{5}$$

Green's identity is then, noting that $\delta x_i(0) = 0$,

$$\sum_i \gamma_i(\theta)\, \delta x_i(\theta) = \int_0^\theta \sum_i \gamma_i \frac{\partial f_i}{\partial u}\, \delta u\, dt \tag{6}$$

By defining boundary conditions for Eq. (5) as

$$\gamma_i(\theta) = \frac{\partial \mathcal{E}}{\partial x_i} \tag{7}$$

we can combine Eqs. (4) and (6) to write

$$\delta \mathcal{E} = \int_0^\theta \sum_i \gamma_i \frac{\partial f_i}{\partial u}\, \delta u\, dt \tag{8}$$

Until this point the approach does not differ from the development of the weak minimum principle in Sec. 6.5. Now, as in Sec. 2.5, we make use of the fact that $\bar{u}(t)$ is *not* the optimum, and we seek $\delta u(t)$ so that $\mathcal{E}$ is made smaller, or $\delta \mathcal{E} \le 0$. An obvious choice is

$$\delta u(t) = -w(t) \sum_i \gamma_i \frac{\partial f_i}{\partial u} \qquad w(t) \ge 0 \tag{9}$$

where $w(t)$ is sufficiently small to avoid violation of the linearity assumption. Then

$$\delta \mathcal{E} = -\int_0^\theta w(t) \left(\sum_i \gamma_i \frac{\partial f_i}{\partial u} \right)^2 dt \le 0 \tag{10}$$

At a bound we must take w equal to zero to avoid leaving the allowable region.

It is helpful to examine an alternative approach which follows the geometrical development in Sec. 2.6. We define distance in the decision space as

$$\Delta^2 = \int_0^\theta g(t)[\delta u(t)]^2\, dt \qquad g(t) \ge 0 \tag{11}$$

and we seek δu which minimizes (makes as negative as possible)

$$\delta \mathcal{E} = \int_0^\theta \sum_i \gamma_i \frac{\partial f_i}{\partial u} \, \delta u(t) \, dt \tag{12}$$

This is an isoperimetric problem in the calculus of variations (Sec. 3.6) for which the Euler equation is

$$\frac{\partial}{\partial(\delta u)} \left[\sum_i \gamma_i \frac{\partial f_i}{\partial u} \, \delta u + \lambda g(\delta u)^2 \right] = 0 \tag{13}$$

Here λ is a constant Lagrange multiplier.

Differentiating and solving for δu, we obtain

$$\delta u = -\frac{1}{2\lambda g(t)} \sum_i \gamma_i \frac{\partial f_i}{\partial u} \tag{14}$$

From Eq. (11), then,

$$\Delta^2 = \frac{1}{4\lambda^2} \int_0^\theta \frac{1}{g} \left(\sum_i \gamma_i \frac{\partial f_i}{\partial u} \right)^2 dt \tag{15}$$

or, substituting back into Eq. (10),

$$\delta u = -\Delta \frac{G(t) \sum_i \gamma_i \, \partial f_i/\partial u}{\left[\int_0^\theta G(\tau) \left(\sum_j \gamma_j \, \partial f_j/\partial u \right)^2 d\tau \right]^{1/2}} \tag{16}$$

where $G(t)$ is the inverse of $g(t)$ and the positive square root has been used. For later purposes it is helpful to introduce the notation

$$I_{\varepsilon\varepsilon} = \int_0^\theta G(\tau) \left(\sum_j \gamma_j \frac{\partial f_j}{\partial u} \right)^2 d\tau \tag{17}$$

in which case

$$\delta u = -\Delta \frac{G(t) \sum_i \gamma_i \, \partial f_i/\partial u}{I_{\varepsilon\varepsilon}^{1/2}} \tag{18}$$

Thus,

$$w(t) = \Delta \frac{G(t)}{I_{\varepsilon\varepsilon}^{1/2}} \tag{19}$$

where G is the inverse of the metric (weighting function) of the space. If we agree to define the hamiltonian

$$H = \sum_i \gamma_i f_i \tag{20}$$

for any decision function, not just the optimum, the improvements in u may be conveniently written

$$\delta u = -w(t)\frac{\partial H}{\partial u} \tag{21}$$

In the case of more than one decision it easily follows that

$$\delta u_i = -\sum_j w_{ij}\frac{\partial H}{\partial u_j} \tag{22}$$

For systems described by difference equations

$$x_i{}^n = f_i{}^n(\mathbf{x}^{n-1},\mathbf{u}^n) \tag{23}$$

where $\mathcal{E}(\mathbf{x}^N)$ is to be minimized, an analogous development leads to

$$\delta u_i{}^n = -\sum_j w_{ij}{}^n\frac{\partial H^n}{\partial u_j{}^n} \tag{24}$$

with

$$H^n = \sum_i \gamma_i{}^n f_i{}^n \tag{25}$$

$$\gamma_i{}^{n-1} = \sum_j \gamma_j{}^n\frac{\partial f_j{}^n}{\partial x_i{}^{n-1}} \qquad \gamma_i{}^N = \frac{\partial \mathcal{E}}{\partial x^N} \tag{26}$$

In that case the equivalent to Eq. (19) is

$$w^n = -\Delta\frac{G^n}{I_{\mathcal{E}\mathcal{E}}{}^{\frac{1}{2}}} \tag{27}$$

$$I_{\mathcal{E}\mathcal{E}} = \sum_{n=1}^N G^n\left(\frac{\partial H^n}{\partial u^n}\right)^2 \tag{28}$$

The significant feature of steep descent should be noted here. The boundary-value problem is completely uncoupled, for the state equations are first solved from $t = 0$ to θ ($n = 0$ to N) and then the multiplier equations solved with known final condition from $t = \theta$ to $t = 0$. If the state equations are stable when integrated from $t = 0$ in the forward direction, the multiplier equations are stable in the direction in which they are integrated, from θ to zero. One practical note of precaution is in order. When stored functions are used for numerical integration, a means must be provided for interpolation, for most numerical-integration algorithms require function evaluation at several locations between grid points.

9.10 STEEP DESCENT: OPTIMAL PRESSURE PROFILE

The method of steep descent is the most important of the computational techniques we shall discuss, and we demonstrate its use with several examples. The first of these is the optimal-pressure problem for consecutive reactions studied previously in this chapter. The pertinent equations are

$$\dot{x}_1 = -2k_1 u \frac{x_1}{A + x_1} \tag{1a}$$

$$\dot{x}_2 = 4k_1 u \frac{x_1}{A + x_2} - 4k_2 u^2 \frac{x_2^2}{(A + x_2)^2} \tag{1b}$$

$$\mathcal{E} = -x_2(\theta) \tag{2}$$

$$\dot{\gamma}_1 = \frac{2k_1 u}{A + x_2} (\gamma_1 - 2\gamma_2) \qquad \gamma_1(\theta) = 0 \tag{3a}$$

$$\dot{\gamma}_2 = -\frac{2k_1 u x_1}{(A + x_2)^2} (\gamma_1 - 2\gamma_2) + \frac{8k_2 u^2 \gamma_2 A x_2}{(A + x_2)^3} \qquad \gamma_2(\theta) = -1 \tag{3b}$$

$$\frac{\partial H}{\partial u} = -2k_1 \frac{x_1}{A + x_2} (\gamma_1 - 2\gamma_2) - 8k_2 \frac{x_2^2}{(A + x_2)^2} u \tag{4}$$

The values of the parameters are those used for previous calculations. Following a choice of u, $\bar{u}(t)$, the new decision function is calculated from the equation

$$u_{\text{new}}(t) = \bar{u} - w \frac{\partial H}{\partial u} \tag{5}$$

where Eq. (4) for $\partial H/\partial u$ is evaluated for $\bar{u}$ and the solutions of Eqs. (1) and (3) are calculated using $\bar{u}$. For these calculations w was initially set equal to the constant value of 25. The program was written so as to reduce w by a factor of 2 whenever improvement was not obtained, but effective convergence was obtained before this feature was needed. Figures 9.4 and 9.5 show successive pressure profiles computed from con-

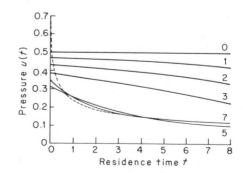

Fig. 9.4 Successive approximations to the optimal pressure profile using steep descent, starting from the constant policy $u = 0.5$.

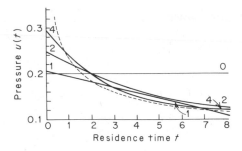

Fig. 9.5 Successive approximations to the optimal pressure profile using steep descent, starting from the constant policy $u = 0.2$.

stant starting profiles of $u = 0.5$ and $u = 0.2$, respectively, with the vertical scale on the latter expanded to avoid cluttering. The dashed line is the solution obtained from the necessary conditions by the indirect techniques discussed previously. The rapid convergence is shown in Fig. 9.6, where the lower points are those for the starting value $u = 0.5$. The near-optimal values of the objective, 0.011305 after seven iterations and 0.011292 after four, respectively, compared with an optimum of 0.011318, are obtained with profiles which differ markedly from the optimum over the first portion of the reactor. This insensitivity would be an aid in ultimate reactor design.

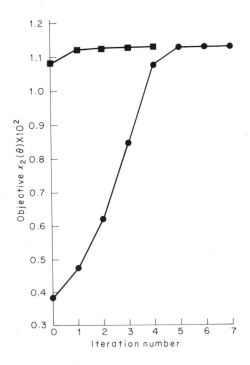

Fig. 9.6 Improvement of the objective function on successive iterations using steep descent.

9.11 STEEP DESCENT: OPTIMAL TEMPERATURE PROFILE

As a second example of steep descent we consider the tubular-reactor optimal-temperature-profile problem solved by Newton-Raphson boundary iteration in Sec. 9.3.   The equations for the system are

$$\dot{x}_1 = -k_{10}e^{-E_1'/u}x_1^2 \tag{1a}$$
$$\dot{x}_2 = k_{10}e^{-E_1'/u}x_1 - k_{20}e^{-E_2'/u}x_2 \tag{1b}$$
$$u_* \leq u \leq u^* \tag{2}$$
$$\mathcal{E} = -c[x_1(\theta) - x_{10}] - [x_2(\theta) - x_{20}] \tag{3}$$

The Green's function equations are then

$$\dot{\gamma}_1 = 2x_1k_{10}e^{-E_1'/u}(\gamma_1 - \gamma_2) \qquad \gamma_1(\theta) = -c \tag{4a}$$
$$\dot{\gamma}_2 = k_{20}e^{-E_2'/u}\gamma_2 \qquad \gamma_2(\theta) = -1 \tag{4b}$$

and the decision derivative of the hamiltonian is readily computed as

$$\frac{\partial H}{\partial u} = -\frac{k_{10}E_1'x_1^2e^{-E_1'/u}}{u^2}(\gamma_1 - \gamma_2) - \frac{k_{20}E_2'x_2e^{-E_2'/u}}{u^2}\gamma_2 \tag{5}$$

The parameters for computation are the same as those in Sec. 9.3.   The new decision is calculated from the equation

$$u_{\text{new}}(t) = \bar{u}(t) - w(t)\frac{\partial H}{\partial u} \tag{6}$$

A reasonable estimate of $w(t)$ may be obtained in a systematic way. Following the geometric interpretation, we may write, for uniform weighting,

$$w(t) = \frac{\Delta}{\left[\int_0^\theta \left(\frac{\partial H}{\partial u}\right)^2 dt\right]^{1/2}} \tag{7}$$

If data are stored at N uniform grid points spaced Δt apart, then, approximately,

$$w(t) = \frac{\Delta}{\left\{\sum_{n=1}^{N}\left[\frac{\partial H}{\partial u}(t_n)\right]^2 \Delta t\right\}^{1/2}} \tag{8}$$

or, combining $(\Delta t)^{1/2}$ into the step size,

$$w(t) = \frac{\delta}{\left\{\sum_{n=1}^{N}\left[\frac{\partial H}{\partial u}(t_n)\right]^2\right\}^{1/2}} \tag{9}$$

If δu is to be of the order of one degree in magnitude then

$$\frac{\delta(\partial H/\partial u)}{\left\{ \sum_{n=1}^{N} \left[\frac{\partial H}{\partial u}(t_n) \right]^2 \right\}^{1/2}} \approx 1 \tag{10}$$

or

$$\delta \approx N^{1/2} \tag{11}$$

For simplicity δ is taken as an integer. Equation (6) is written for computation as

$$u_{\text{new}} = \bar{u} - \delta(t) \frac{\partial H/\partial u}{\left\{ \sum_{n=1}^{N} \left[\frac{\partial H}{\partial u}(t_n) \right]^2 \right\}^{1/2}} \tag{12}$$

where $\delta(t)$ is a constant δ, unless Eq. (12) would cause violation of one of the constraints, in which case $\delta(t)$ is taken as the largest value which does not violate the constraint. The constant δ is taken initially as the smallest integer greater than $N^{1/2}$ and halved each time a step does not lead to improvement in the value of $\mathcal{E}$. For the calculations shown here $N = 60$, and the initial value of δ is 8.

Figure 9.7 shows successive iterations starting from an initial constant policy of $u = u_*$ except over the first integration step, where u is linear between u^* and u_*. This initial segment is motivated by the

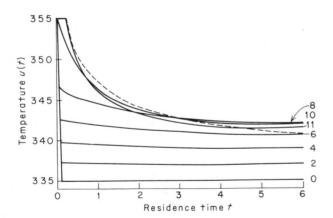

Fig. 9.7 Successive approximations to the optimal temperature profile using steep descent, starting from the constant policy $u = 335$. [*From J. M. Douglas and M. M. Denn, Ind. Eng. Chem.*, **57**(11):18 (1965). *Copyright* 1965 *by the American Chemical Society. Reprinted by permission of the copyright owner.*]

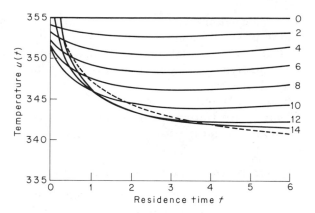

Fig. 9.8 Successive approximations to the optimal temperature profile using steep descent, starting from the constant policy $u = 355$. [*From J. M. Douglas and M. M. Denn, Ind. Eng. Chem.,* **57**(11):18 (1965). *Copyright* 1965 *by the American Chemical Society. Reprinted by permission of the copyright owner.*]

solution to the generalized Euler equation in Sec. 4.12, which requires an infinite initial slope for profiles not at an upper bound. The method of choosing the weighting $w(t)$ in Eq. (12) is seen to produce changes in $u(t)$ of order unity, as desired. The dashed line is the solution obtained using Newton-Raphson boundary iteration. Successive values of the profit are shown as the upper line in Fig. 9.9, and it can be seen that convergence is essentially obtained on the eleventh iteration, though the temperature profile differs from that obtained using the necessary conditions.

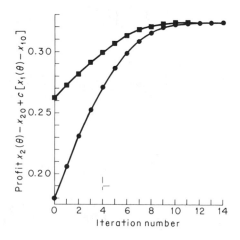

Fig. 9.9 Improvement of the objective function on successive iterations using steep descent. [*From J. M. Douglas and M. M. Denn, Ind. Eng. Chem.,* **57**(11):18 (1965). *Copyright* 1965 *by the American Chemical Society. Reprinted by permission of the copyright owner.*]

Calculations starting from an initial constant policy $u = u^*$ are shown in Fig. 9.8 and the lower curve in Fig. 9.9, and the conclusions are similar. The two calculations lead to values of the objective of 0.3236 and 0.3237, respectively, compared to a true optimum of 0.3238. It is typical of steep-descent procedures that convergence is rapid far from the optimum and very slow near an insensitive minimum or maximum. Though a frustration in calculation, this insensitivity of the optimum is, of course, an aid in actual engineering design considerations.

9.12 STEEP DESCENT: OPTIMAL STAGED TEMPERATURES

We can examine the use of steep descent for a staged system by considering the problem of specifying optimal temperatures in a sequence of continuous-flow reactors of equal residence time. The reaction sequence is the one considered for the tabular reactor

$$X_1 \rightarrow X_2 \rightarrow \text{products}$$

with all rate parameters identical to those in Sec. 9.3. We have considered this problem in Secs. 1.12 and 7.6, where, with F quadratic and G linear, the state equations and objective are written

$$0 = x_1^{n-1} - x_1^n - \theta k_{10} e^{-E_1'/u^n}(x_1^n)^2 \tag{1a}$$

$$0 = x_2^{n-1} - x_2^n + \theta k_{10} e^{-E_1'/u^n}(x_1^n)^2 - \theta k_{20} e^{-E_2'/u^n} x_2^n \tag{1b}$$

$$\mathcal{E} = -c(x_1^N - x_{10}) - (x_2^N - x_{20}) \tag{2}$$

$$u_* \leq u^n \leq u^* \tag{3}$$

By implicit differentiation the Green's function equations are shown in Sec. 7.6 to be

$$\gamma_1^{n-1} = \frac{\gamma_1^n}{1 + 2\theta k_{10} e^{-E_1'/u^n} x_1^n}$$

$$+ \frac{2\theta \gamma_2^n k_{10} e^{-E_1'/u^n} x_1^n}{(1 + 2\theta k_{10} e^{-E_1'/u^n} x_1^n)(1 + \theta k_{20} e^{-E_2'/u^n})} \qquad \gamma_1^N = -c \tag{4a}$$

$$\gamma_2^{n-1} = \frac{\gamma_2^n}{1 + \theta k_{20} e^{-E_2'/u^n}} \qquad \gamma_2^N = -1 \tag{4b}$$

The decision derivative of the stage hamiltonian is

$$\frac{\partial H^n}{\partial u^n} =$$

$$\frac{\theta}{(u^n)^2} \frac{\gamma_2^n E_1' k_{10} e^{-E_1'/u^n}(x_1^n)^2 - \gamma_1^n E_1' k_{10} e^{-E_1'/u^n}(x_1^n)^2(1 + \theta k_{20} e^{-E_2'/u^n})}{- \gamma_2^n E_2' k_{20} e^{-E_2'/u^n} x_2^n(1 + 2\theta k_{10} e^{-E_1'/u^n} x_1^n)}{(1 + \theta k_{20} e^{-E_2'/u^n})(1 + 2\theta k_{10} e^{-E_1'/u^n} x_1^n)} \tag{5}$$

Following the specification of a temperature sequence $\{\bar{u}^n\}$ and the successive solution of Eqs. (1) and Eqs. (4), the new decision is obtained from the algorithm of Eqs. (20) and (23) of Sec. 9.9

$$u^n_{\text{new}} = \bar{u}^n - \Delta \frac{\partial H^n / \partial u^n}{\left[\sum_{n=1}^{N} \left(\frac{\partial H^n}{\partial u^n} \right)^2 \right]^{1/2}} \tag{6}$$

where we have taken the weighting as uniform. If a bound is exceeded by this calculation, that value of u^n is simply moved to the bound, as in the previous section. The starting value of Δ is taken as the smallest integer greater than $N^{1/2}$ in order to obtain starting-temperature changes of the order of unity. Δ is halved each time an improvement does not occur on an iteration.

The residence time θ was chosen so that $N\theta$ equals the tubular-reactor residence time of 6 used in the preceding numerical study. Figure 9.10 shows the convergence for a three-stage reactor system ($N = 3$, $\theta = 2$) for starting values entirely at the upper and lower bounds.

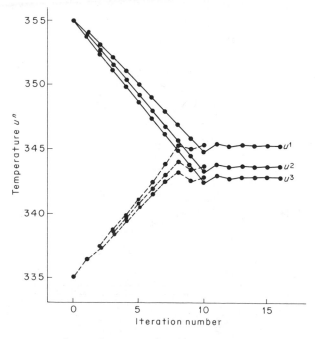

Fig. 9.10 Successive approximations to the optimal temperature sequence in three reactors using steep descent. [*From M. M. Denn and R. Aris, Ind. Eng. Chem. Fundamentals, 4:213 (1965). Copyright 1965 by the American Chemical Society. Reprinted by permission of the copyright owner.*]

The calculations in Fig. 9.11 show convergence for a 60-stage reactor ($N = 60$, $\theta = 0.1$) from the lower bounds, with the discrete profile represented as a continuous curve not only for convenience but also to emphasize the usefulness of a large but finite number of stages to approximate a tubular reactor with diffusive transport. For the latter case 22 iterations were required for convergence, defined as a step size of less than 10^{-3}.

This process is a convenient one with which to demonstrate the use of penalty functions for end-point constraints. If we take the objective as simply one of maximizing the amount of intermediate

$$\mathcal{E} = -x_2{}^N \tag{7}$$

and impose the constraint of 60 percent conversion of raw material

$$x_1{}^N = 0.4 \tag{8}$$

then the penalty-function approach described in Sec. 1.16 would substitute the modified objective

$$\tilde{\mathcal{E}} = -x_2{}^N + \tfrac{1}{2}K(x_1{}^N - 0.4)^2 \tag{9}$$

In the minimization of $\tilde{\mathcal{E}}$ the multiplier equations (4) and direction of steep descent described by Eqs. (5) and (6) do not change, but the

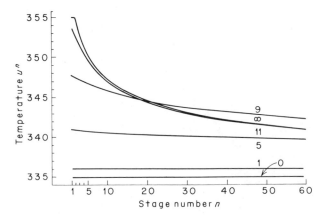

Fig. 9.11 Successive approximations to the optimal temperature sequence in a 60-stage reactor using steep descent, starting from the constant policy $u = 335$. [*From M. M. Denn and R. Aris, Ind. Eng. Chem. Fundamentals*, **4**:213 (1965). *Copyright 1965 by the American Chemical Society. Reprinted by permission of the copyright owner.*]

boundary conditions on the multipliers do.   We now have

$$\gamma_1{}^N = \frac{\partial \tilde{\mathcal{E}}}{\partial x_1{}^N} = K(x_1{}^N - 0.4) \tag{10a}$$

$$\gamma_2{}^N = \frac{\partial \tilde{\mathcal{E}}}{\partial x_2{}^N} = -1 \tag{10b}$$

Following convergence of the iterations for a given value of the penalty constant K and a corresponding error in the constraint a new value of K can be estimated by the relation

$$K_{\text{new}} = K_{\text{old}} \frac{x_1{}^N - 0.4}{\text{tolerance}} \tag{11}$$

For these calculations the tolerance in the constraint was set at 10^{-4}.

Figure 9.12 shows the result of a sequence of penalty-function calculations for $N = 60$ starting with $K = 1$.   Convergence was obtained in 17 iterations from the linearly decreasing starting-temperature sequence to the curve shown as $K = 1$, with an error in the constraint $x_1{}^N = 0.4$ of 1.53×10^{-2}.   Following Eq. (11), the new value of K was set at 1.53×10^2, and 107 iterations were required for convergence, with a constraint error of 8.3×10^{-4}.   A value of K of 1.27×10^3 was then used for 24 further iterations to the dashed line in Fig. 9.12 and an error of 10^{-4}.   This extremely slow convergence appears to be typical of the penalty-function approach and is due in part to the sensitivity of the boundary conditions for γ to small changes in the constraint error, as may be seen from Eq. (10a).

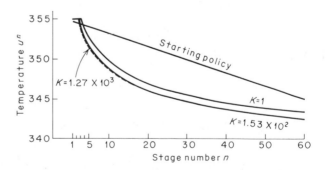

Fig. 9.12 Approximations to the optimal temperature sequence with constrained output using steep descent and penalty functions. [*From M. M. Denn and R. Aris, Ind. Eng. Chem. Fundamentals,* **4**:213 (1965). *Copyright* 1965 *by the American Chemical Society. Reprinted by permission of the copyright owner.*]

9.13 GRADIENT PROJECTION FOR CONSTRAINED END POINTS

By a direct extension of the analysis in Sec. 9.9 we can generalize the steep-descent procedure to account for constrained end points. We again consider a system described by the differential equations

$$\dot{x}_i = f_i(\mathbf{x},u) \qquad x_i(0) = x_{i0} \tag{1}$$

with bounds on u,

$$u_* \leq u \leq u^* \tag{2}$$

The objective is to minimize $\mathcal{E}[\mathbf{x}(\theta)]$, but we presume that there is also a single (for simplicity) constraint on the final state

$$\phi[\mathbf{x}(\theta)] = 0 \tag{3}$$

For any function $\bar{u}(t)$ which lies between the upper and lower bound we can integrate Eqs. (1) to obtain a value of $\mathcal{E}$ and of ϕ which will generally not be zero. The first-order effect of a small change is then described by the equations

$$\delta\dot{x}_i = \sum_j \frac{\partial f_i}{\partial x_j}\,\delta x_j + \frac{\partial f_i}{\partial u}\,\delta u \qquad \delta x_i(0) = 0 \tag{4}$$

with corresponding first-order changes in $\mathcal{E}$ and ϕ

$$\delta\mathcal{E} = \sum_i \frac{\partial \mathcal{E}}{\partial x_i}\,\delta x_i(\theta) \tag{5}$$

$$\delta\phi = \sum_i \frac{\partial \phi}{\partial x_i}\,\partial x_i(\theta) \tag{6}$$

We now define two sets of Green's functions for Eq. (4), $\boldsymbol{\gamma}$ and $\boldsymbol{\psi}$, satisfying the adjoint differential equation but different boundary conditions

$$\dot{\gamma}_i = -\sum_j \gamma_j \frac{\partial f_j}{\partial x_i} \qquad \gamma_i(0) = \frac{\partial \mathcal{E}}{\partial x_i} \tag{7}$$

$$\dot{\psi}_i = -\sum_j \psi_j \frac{\partial f_j}{\partial x_i} \qquad \psi_i(0) = \frac{\partial \phi}{\partial x_i} \tag{8}$$

Using Green's identity, Eqs. (5) and (6) will then become, respectively,

$$\delta\mathcal{E} = \int_0^\theta \sum_i \gamma_i \frac{\partial f_i}{\partial u}\,\delta u\,dt \tag{9}$$

$$\delta\phi = \int_0^\theta \sum_i \psi_i \frac{\partial f_i}{\partial u}\,\delta u\,dt \tag{10}$$

We now choose a distance in the decision space

$$\Delta^2 = \int_0^\theta g(t)[\delta u(t)]^2 \, dt \tag{11}$$

and ask for the function δu which minimizes $\delta \mathcal{E}$ in Eq. (9) for a fixed distance Δ^2 *and a specified correction to the constraint $\delta \phi$*. This is an iso-perimetric problem with two integral constraints, and hence two constant Lagrange multipliers, denoted by λ and ν. The Euler equation is

$$\frac{\partial}{\partial(\delta u)} \left[\sum_i \gamma_i \frac{\partial f_i}{\partial u} \delta u + \nu \sum_i \psi_i \frac{\partial f_i}{\partial u} \delta u + \lambda g (\delta u)^2 \right] = 0 \tag{12}$$

or

$$\delta u = - \frac{G}{2\lambda} \sum_i (\gamma_i + \nu \psi_i) \frac{\partial f_i}{\partial u} \tag{13}$$

where $G(t)$ is the inverse of $g(t)$.

We need to use the constraint equations (10) and (11) to evaluate λ and ν. By substituting Eq. (13) for δu into Eq. (10) for the fixed value $\delta \phi$ we obtain

$$\delta \phi = - \frac{1}{\lambda} \int_0^\theta G(\tau) \left(\sum_i \psi_i \frac{\partial f_i}{\partial u} \right) \left(\sum_j \gamma_j \frac{\partial f_j}{\partial u} \right) d\tau$$

$$- \frac{\nu}{\lambda} \int_0^\theta G(\tau) \left(\sum_i \psi_i \frac{\partial f_i}{\partial u} \right)^2 d\tau \tag{14}$$

It is convenient to define three integrals

$$I_{\mathcal{E}\mathcal{E}} = \int_0^\theta G(\tau) \left(\sum_i \gamma_i \frac{\partial f_i}{\partial u} \right)^2 d\tau \tag{15a}$$

$$I_{\mathcal{E}\phi} = \int_0^\theta G(\tau) \left(\sum_j \gamma_j \frac{\partial f_j}{\partial u} \right) \left(\sum_i \psi_i \frac{\partial f_i}{\partial u} \right) d\tau \tag{15b}$$

$$I_{\phi\phi} = \int_0^\theta G(\tau) \left(\sum_i \psi_i \frac{\partial f_i}{\partial u} \right)^2 d\tau \tag{15c}$$

The subscripts denote the appropriate boundary conditions for the Green's functions used in the integral. Equation (14) then becomes, slightly rearranged,

$$\nu = - \frac{I_{\mathcal{E}\phi} + \lambda \, \delta \phi}{I_{\phi\phi}} \tag{16}$$

If Eqs. (13) and (16) are now substituted into Eq. (11) for Δ^2, we obtain, after some algebraic manipulation,

$$\frac{1}{2\lambda} = \pm \left[\frac{I_{\phi\phi}\Delta^2 - (\delta\phi)^2}{I_{\phi\phi}I_{\mathcal{E}\mathcal{E}} - I_{\mathcal{E}\phi}^2} \right]^{1/2} \tag{17}$$

Finally, then, Eq. (13) for δu becomes

$$\delta u = -G(t) \left[\frac{I_{\phi\phi}\Delta^2 - (\delta\phi)^2}{I_{\phi\phi}I_{\varepsilon\varepsilon} - I_{\varepsilon\phi}{}^2} \right]^{\frac{1}{2}} \sum_i \left(\gamma_i - \frac{I_{\varepsilon\phi}}{I_{\phi\phi}} \psi_i \right) \frac{\partial f_i}{\partial u}$$

$$+ G(t) \frac{\delta\phi}{I_{\phi\phi}} \sum_i \psi_i \frac{\partial f_i}{\partial u} \quad (18)$$

The presence of the square root imposes an upper limit on the correction $\delta\phi$ which can be sought, and a full correction cannot generally be obtained in a single iteration. The procedure outlined here is geometrically equivalent to projecting the gradient onto the subspace defined by the constraint. Identical equations are obtained for systems with difference equations if integrals are replaced by sums and continuous variables by their discrete analogs.

If the gradient-projection procedure is applied to the reactor-temperature problem of the preceding section with $\phi(\mathbf{x}^N) = x_1{}^N - 0.4$, the only additional computations required are the recalculation of Eqs. (4) of that section, substituting $\psi_1{}^n$, $\psi_2{}^n$ for $\gamma_1{}^n$, $\gamma_2{}^n$, with boundary conditions 1, 0 on $\mathbf{\psi}^N$ and 0, -1 on $\mathbf{\gamma}^N$, followed by calculation of the sums $I_{\varepsilon\varepsilon}$, $I_{\varepsilon\phi}$, and $I_{\phi\phi}$. Computations were carried out using a maximum correction $\delta\phi$ of 0.01 in magnitude for each iteration. Lines I and II in Fig. 9.13

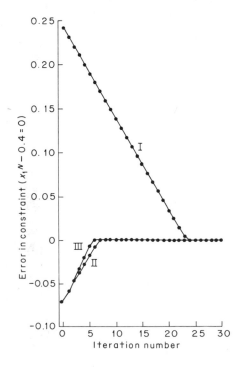

Fig. 9.13 Approach to final value constraint using steep descent and gradient projection. [*From M. M. Denn and R. Aris, Ind. Eng. Chem. Fundamentals,* **4**:213 (1965). *Copyright* 1965 *by the American Chemical Society. Reprinted by permission of the copyright owner.*]

show the convergence of the constraint for $N = 3$, $\theta = 2$ from starting policies at the lower and upper bounds, respectively. The correction is seen to be linear. Line III shows the convergence for the 60-stage system, $N = 60$, $\theta = 0.1$, from the starting policy in Fig. 9.12. Twenty-two iterations were required here for convergence to the same optimal policy as that shown in Fig. 9.12 for $K = 1.27 \times 10^3$. In all cases the final error in the constraint was less than 10^{-5}.

9.14 MIN H

Steep descent provides a rapidly convergent method of obtaining approximate solutions to variational problems, but it is evident from the examples that ultimate convergence to the solution of the minimum-principle necessary conditions is slow and computationally unfeasible. The *min-H* procedure is one which can be used only in the latter stages of solution but which will, with only minor modifications of the computer code, lead to a solution satisfying the necessary conditions.

The system and multiplier equations are

$$\dot{x}_i = f_i(\mathbf{x}, u) \qquad x_i(0) = x_{io} \tag{1}$$

$$\dot{\gamma}_i = - \sum_j \gamma_j \frac{\partial f_j}{\partial x_i} \qquad \gamma_i(\theta) = \frac{\partial \mathcal{E}}{\partial x_i} \tag{2}$$

For a given decision function $\bar{u}(t)$ the solutions to Eqs. (1) and (2) are denoted as $\bar{x}(t)$, $\bar{\gamma}(t)$. Writing the hamiltonian in terms of these latter variables, we have

$$H = \sum_i \bar{\gamma}_i f_i(\bar{x}, u) \tag{3}$$

The new value of u is chosen so that H is minimized. In that way convergence is forced to the solution of the necessary conditions. In practice severe oscillations and divergence may occur if the full correction is used, so that if we denote the function obtained by minimizing H in Eq. (3) as $\bar{u}$, the new value of u is obtained from the relation

$$u_{\text{new}} = \bar{u} + \frac{1}{r} (\bar{\bar{u}} - \bar{u}) \tag{4}$$

A value of $r = 2$ has generally been found to be satisfactory. Modifications similar to those in Sec. 9.13 can be made to accommodate end-point constraints.

The relation to steep descent may be observed by considering the special case for which the minimum of H occurs at an interior value.

Then the minimum, $\bar{u}$, satisfies the equation

$$\sum_i \bar{\gamma}_i \frac{\partial f_i(\bar{\mathbf{x}}, \bar{u})}{\partial u} = 0 \tag{5}$$

This can be expanded about $\bar{u}$ for a Newton-Raphson solution as

$$\sum_i \bar{\gamma}_i \frac{\partial f_i(\bar{\mathbf{x}}, \bar{u})}{\partial u} + \sum_i \bar{\gamma}_i \frac{\partial^2 f_i(\bar{\mathbf{x}}, \bar{u})}{\partial u^2} (\bar{\bar{u}} - \bar{u}) + \cdots = 0 \tag{6}$$

and if the higher-order terms can be neglected, the min-H procedure is equivalent to

$$\bar{\bar{u}} \equiv u_{\text{new}} = \bar{u} - \left(\frac{\partial^2 H}{\partial u^2}\right)^{-1} \frac{\partial H}{\partial u} \tag{7}$$

That is, we get nearly the same result as a steep-descent procedure with the inverse of $\partial^2 H/\partial u^2$ as the weighting factor w. At the minimum the hessian of H is positive, and so Eq. (7) will allow convergence, and this must be true by continuity arguments in a neighborhood of the minimum. For staged systems we have no strong minimum principle in general, and $\partial^2 H/\partial u^2$ need not be positive near the optimum, so that this procedure might lead to divergence arbitrarily close to the minimum in a staged system.

As an example of the convergence properties of the min-H algorithm we consider the problem of the optimal tubular-reactor temperature profile for the consecutive reactions. The state and multiplier equations (1) and (4) of Sec. 9.11 are solved sequentially to obtain $\bar{x}_1$, $\bar{x}_2$, $\bar{\gamma}_1$, $\bar{\gamma}_2$ for the specified $\bar{u}$. Using the necessary conditions in Sec. 9.3, we then find $\bar{u}$ from the relation

$$\bar{u} = \begin{cases} u^* & v(t) \geq u^* \\ v(t) & u_* < v(t) < u^* \\ u_* & v(t) \leq u_* \end{cases} \tag{8}$$

with

$$v(t) = \frac{E_2' - E_1'}{\ln \dfrac{\bar{\gamma}_2 \bar{x}_2 k_{20}}{(\bar{\gamma}_2 - \bar{\gamma}_1)\bar{x}_1{}^2 k_{10}}} \tag{9}$$

For these computations the initial profile was taken as the constant $u = 345$ and r was set at 2. The results are shown in Table 9.9, where ϵ is the integral of the absolute value of the difference between old and new temperature. The stabilizing factor $r = 2$ clearly slows convergence for the small values of t. Convergence is not uniform and, indeed, ε is not uniformly decreasing, but very rapid convergence to a solution satisfying the necessary conditions is obtained in this way. Numerical

Table 9.9 Successive approximations to optimal temperature profile using the min-H method

Iteration Number

t	0	1	2	4	6	8	10	12	14
0	345.00	350.00	352.50	354.38	354.84	354.96	354.99	355.00	355.00
0.5	345.00	350.00	352.50	354.38	354.84	354.96	354.99	355.00	355.00
1	345.00	350.00	349.88	349.20	348.86	348.76	348.70	348.74	348.74
2	345.00	345.76	344.61	344.19	344.15	344.16	344.16	344.16	344.16
3	345.00	343.89	342.50	342.10	342.11	342.12	342.13	342.13	342.13
4	345.00	342.94	341.36	340.89	340.89	340.90	340.91	340.91	340.91
5	345.00	342.38	340.67	340.09	340.06	340.07	340.07	340.07	340.07
6	345.00	342.05	340.21	339.52	339.46	339.46	339.46	339.46	339.46
$x_1(6)$	0.4147	0.4088	0.4186	0.4196	0.4196	0.4196	0.4196	0.4196	0.4196
$x_2(6)$	0.4952	0.4999	0.4977	0.4973	0.4974	0.4974	0.4974	0.4974	0.4974
$-\varepsilon$	0.3196	0.3225	0.3233	0.3232	0.3233	0.3233	0.3233	0.3233	0.3233
ϵ	29.7	18.1	4.3	0.70	0.29	0.086	0.022	0.0054	0.0013

values differ somewhat from those indicated in earlier calculations for this system because of the use of a different numerical-integration procedure and a constant value of u over an integration step instead of using interpolation.

9.15 SECOND-ORDER EFFECTS

In looking at the convergence properties of steep descent for problems in differential calculus in Sec. 2.8 we found that by retaining second-order terms in the steep-descent analysis we were led to a relation with quadratic convergence which was, in fact, equivalent to the Newton-Raphson solution. We can adopt the same approach for the variational problem in order to find what the pertinent second-order effects are.

It helps to examine the case of a single state variable first. We have

$$\dot{x} = f(x,u) \tag{1}$$

and we seek to minimize $\mathcal{E}[x(\theta)]$. If we choose $\bar{u}(t)$ and expand Eq. (1) to second order, we have

$$\delta\dot{x} = \frac{\partial f}{\partial x}\,\delta x + \frac{\partial f}{\partial u}\,\delta u + \frac{1}{2}\frac{\partial^2 f}{\partial x^2}\,\delta x^2 + \frac{\partial^2 f}{\partial x\,\partial u}\,\delta x\,\delta u + \frac{1}{2}\frac{\partial^2 f}{\partial u^2}\,\delta u^2$$
$$\delta x(0) = 0 \tag{2}$$

The corresponding second-order change in $\mathcal{E}$ is

$$\delta\mathcal{E} = \mathcal{E}'\,\delta x(\theta) + \tfrac{1}{2}\mathcal{E}''\,\delta x(\theta)^2 \tag{3}$$

where the prime denotes differentiation. Our goal is to choose δu so that we minimize $\delta\mathcal{E}$. The quadratic nature of the problem prevents an unbounded solution, and it is not necessary to specify a step size in the decision space.

Equations (2) and (3) define a variational problem of the type we have studied extensively. Here the state variable is δx and the decision δu, and we wish to choose δu to minimize a function [Eq. (3)] of $\delta x(\theta)$. The hamiltonian, which we denote by h, is

$$h = \psi\left(\frac{\partial f}{\partial x}\,\delta x + \frac{\partial f}{\partial u}\,\delta u + \frac{1}{2}\frac{\partial^2 f}{\partial x^2}\,\delta x^2 + \frac{\partial^2 f}{\partial x\,\partial u}\,\delta x\,\delta u + \frac{1}{2}\frac{\partial^2 f}{\partial u^2}\,\delta u^2\right) \tag{4}$$

where the Green's function is denoted by ψ. For unconstrained δu, then, we can set $\partial h/\partial(\delta u)$ to zero to obtain

$$\delta u = -\left(\frac{\partial^2 f}{\partial u^2}\right)^{-1}\left(\frac{\partial f}{\partial u} + \frac{\partial^2 f}{\partial x\,\partial u}\,\delta x\right) \tag{5}$$

The Newton-Raphson approximation to min H, Eq. (7) of Sec. 9.14, would correspond to Eq. (5) without the term $\partial^2 f/(\partial x\,\partial u)\,\delta x$, so that min H is clearly not the complete second-order correction for steep descent.

We can carry the analysis somewhat further. The first-order solution to Eq. (2) is

$$\delta x = \int_0^t \exp\left[\int_t^\tau \frac{\partial f}{\partial x}(\sigma)\,d\sigma\right] \frac{\partial f}{\partial u}(\tau)\,\delta u(\tau)\,d\tau \tag{6}$$

Equation (5) can then be written as a Volterra-type integral equation for the correction $\delta u(t)$

$$\delta u(t) + \left(\frac{\partial^2 f}{\partial u^2}\right)^{-1} \frac{\partial^2 f}{\partial x\,\partial u} \int_0^t \exp\left[\int_t^\tau \frac{\partial f}{\partial x}(\sigma)\right] \frac{\partial f}{\partial u}(\tau)\,\delta u(\tau)\,d\tau$$
$$= -\left(\frac{\partial^2 f}{\partial u^2}\right)^{-1} \frac{\partial f}{\partial u} \tag{7}$$

There is no computational advantage in doing so, however, for the solution of an integral equation is not an easy task and Eq. (5), although it contains the unknown δx, can be used directly. Noting that δx is simply $x - \bar{x}$, it follows that the best second-order correction δu is a known explicit function of x, which we denote as $\delta u(x)$. Equation (1) can then be written

$$\dot{x} = f[x,\,\bar{u} + \delta u(x)] \tag{8}$$

which can be integrated numerically in the usual fashion. The new value of u is then constructed from Eq. (5).

9.16 SECOND VARIATION

A complete second-order approach to the numerical solution of variational problems requires that we generalize the result of the preceding section to a system of equations. We consider

$$\dot{x}_i = f_i(\mathbf{x},u) \tag{1}$$

with $u(t)$ to be chosen to minimize $\mathcal{E}[\mathbf{x}(\theta)]$. We shall assume $\mathbf{x}(\theta)$ to be completely unspecified, though the extension to constrained final values can be carried out. Unlike the quasi-second-order min-H procedure, we must assume here that u is unconstrained. For notational convenience we shall use the hamiltonian formulation in what follows, so that we define

$$H = \sum_i \gamma_i f_i \tag{2}$$

$$\dot{\gamma}_i = -\frac{\partial H}{\partial x_i} = -\sum_j \gamma_j \frac{\partial f_j}{\partial x_i} \qquad \gamma_i(\theta) = \frac{\partial \mathcal{E}}{\partial x_i} \tag{3}$$

Now, if we expand Eq. (1) about some choice $\bar{u}(t)$ and retain terms to second order, we obtain

$$\delta \dot{x}_i = \sum_j \frac{\partial f_i}{\partial x_j} \delta x_j + \frac{\partial f_i}{\partial u} \delta u + \frac{1}{2} \sum_{j,k} \frac{\partial^2 f_i}{\partial x_j \, \partial x_k} \delta x_j \, \delta x_k + \sum_j \frac{\partial^2 f_i}{\partial x_j \, \partial u} \delta x_j \, \delta u$$

$$+ \frac{1}{2} \frac{\partial^2 f_i}{\partial u^2} \delta u^2 \qquad \delta x_i(0) = 0 \quad (4)$$

The second-order expansion of the objective is

$$\delta \mathcal{E} = \sum_i \frac{\partial \mathcal{E}}{\partial x_i} \delta x_i(\theta) + \frac{1}{2} \sum_{i,j} \frac{\partial^2 \mathcal{E}}{\partial x_i \, \partial x_j} \delta x_i(\theta) \, \delta x_j(\theta) \tag{5}$$

This defines a variational problem, for we must choose $\delta u(t)$, subject to the differential-equation restriction, Eq. (4), so that we minimize $\delta \mathcal{E}$ in Eq. (5). Denoting the hamiltonian for this subsidiary problem by h and the multiplier by ψ, we have

$$h = \sum_{i,j} \psi_i \frac{\partial f_i}{\partial x_j} \delta x_j + \sum_i \psi_i \frac{\partial f_i}{\partial u} \delta u + \frac{1}{2} \sum_{i,j,k} \psi_i \frac{\partial^2 f_i}{\partial x_j \, \partial x_k} \delta x_j \, \delta x_k$$

$$+ \sum_{i,j} \psi_i \frac{\partial^2 f_i}{\partial x_j \, \partial u} \delta x_j \, \delta u + \frac{1}{2} \sum_i \psi_i \frac{\partial^2 f_i}{\partial u^2} \delta u^2 \tag{6}$$

$$\dot{\psi}_j = - \frac{\partial h}{\partial(\delta x_j)} = - \sum_i \psi_i \frac{\partial f_i}{\partial x_j} - \sum_{i,k} \psi_i \frac{\partial^2 f_i}{\partial x_j \, \partial x_k} \delta x_k - \sum_i \psi_i \frac{\partial^2 f_i}{\partial x_j \, \partial u} \delta u \tag{7}$$

$$\psi_j(\theta) = \frac{\partial(\delta \mathcal{E})}{\partial(\delta x_j)} = \frac{\partial \mathcal{E}}{\partial x_j} + \sum_k \frac{\partial^2 \mathcal{E}}{\partial x_j \, \partial x_k} \delta x_k \tag{8}$$

By assuming that u is unconstrained we can allow δu to take on any value, so that the minimum of h is found by setting the derivative with respect to δu to zero

$$\frac{\partial h}{\partial(\delta u)} = \sum_i \psi_i \frac{\partial f_i}{\partial u} + \sum_{i,j} \psi_i \frac{\partial^2 f_i}{\partial x_j \, \partial u} \delta x_j + \sum_i \psi_i \frac{\partial^2 f_i}{\partial u^2} \delta u = 0 \tag{9}$$

or

$$\delta u = - \left(\sum_i \psi_i \frac{\partial^2 f_i}{\partial u^2} \right)^{-1} \left(\sum_i \psi_i \frac{\partial f_i}{\partial u} + \sum_{i,j} \psi_i \frac{\partial^2 f_i}{\partial x_j \, \partial u} \delta x_j \right) \tag{10}$$

There are notational advantages to using the hamiltonian and multiplier for the original problem, defined by Eqs. (2) and (3). We define $\delta \gamma$ by

$$\delta \gamma_j = \psi_j - \gamma_j \tag{11}$$

From Eqs. (3) and (7),

$$\delta\dot{\gamma}_j = -\sum_i \delta\gamma_i \frac{\partial f_i}{\partial x_j} - \sum_k \frac{\partial^2 H}{\partial x_j\,\partial x_k}\,\delta x_k - \sum_{i,k} \delta\gamma_i \frac{\partial^2 f_i}{\partial x_j\,\partial x_k}\,\delta x_k$$

$$- \frac{\partial^2 H}{\partial x_j\,\partial u}\,\delta u - \sum_i \delta\gamma_i \frac{\partial^2 f_i}{\partial x_j\,\partial u}\,\delta u \quad (12)$$

while the boundary condition is obtained from Eqs. (3) and (8)

$$\delta\gamma_j(\theta) = \frac{\partial\mathcal{E}}{\partial x_j} + \sum_k \frac{\partial^2\mathcal{E}}{\partial x_j\,\partial x_k}\,\delta x_k - \frac{\partial\mathcal{E}}{\partial x_j} = \sum_k \frac{\partial^2\mathcal{E}}{\partial x_j\,\partial x_k}\,\delta x_k(\theta) \quad (13)$$

Equation (10) for δu becomes

$$\delta u = -\left(\frac{\partial^2 H}{\partial u^2} + \sum_i \delta\gamma_i \frac{\partial f_i}{\partial u}\right)^{-1}\left(\frac{\partial H}{\partial u} + \sum_i \delta\gamma_i \frac{\partial f_i}{\partial u} + \sum_j \frac{\partial^2 H}{\partial x_j\,\partial u}\,\delta x_j\right.$$

$$\left. + \sum_{i,j} \delta\gamma_i \frac{\partial^2 f_i}{\partial x_j\,\partial u}\,\delta x_j\right) \quad (14)$$

We require a solution to the system of Eqs. (1) and (12) to (14) which can be implemented for computation without the necessity of solving a boundary-value problem.

The fact that we have only quadratic nonlinearities suggests the use of the approximation technique developed in Sec. 4.8. We shall seek a solution of the form

$$\delta\gamma_j(t) = g_j(t) + \sum_k M_{jk}(t)\,\delta x_k(t) + \text{higher-order terms} \quad (15)$$

The functions g_j and M_{jk} must be obtained, but it is evident from Eq. (13) that they must satisfy the final conditions

$$g_j(\theta) = 0 \quad (16a)$$

$$M_{jk}(\theta) = \frac{\partial^2\mathcal{E}}{\partial x_j\,\partial x_k} \quad (16b)$$

We can also express Eqs. (4), (12), and (14) to comparable order as

$$\delta\dot{x}_i = \sum_j \frac{\partial f_i}{\partial x_j}\,\delta x_j + \frac{\partial f_i}{\partial u}\,\delta u + \text{higher-order terms} \quad (17)$$

$$\delta\dot{\gamma}_j = -\sum_i \delta\gamma_i \frac{\partial f_i}{\partial x_j} - \sum_k \frac{\partial^2 H}{\partial x_j\,\partial x_k}\,\delta x_k - \frac{\partial^2 H}{\partial x_j\,\partial u}\,\delta u$$

$$+ \text{higher-order terms} \quad (18)$$

$$\delta u = -\left(\frac{\partial^2 H}{\partial u^2}\right)^{-1}\left(\frac{\partial H}{\partial u} + \sum_i \delta\gamma_i \frac{\partial f_i}{\partial u} + \sum_j \frac{\partial^2 H}{\partial x_j\,\partial u}\,\delta x_j\right)$$

$$+ \text{higher-order terms} \quad (19)$$

Substituting Eq. (15) into Eqs. (18) and (19), we obtain

$$
\delta\dot{\gamma}_i = -\sum_i g_i \frac{\partial f_i}{\partial x_j} - \sum_{i,k} M_{ik}\, \delta x_k \frac{\partial f_i}{\partial x_j} - \sum_k \frac{\partial^2 H}{\partial x_j\, \partial x_k}\, \delta x_k
$$

$$
- \frac{\partial^2 H}{\partial x_j\, \partial u} \left(-\frac{\partial^2 H}{\partial u^2}\right)^{-1}\!\left(\frac{\partial H}{\partial u} + \sum_i g_i \frac{\partial f_i}{\partial u} + \sum_{i,k} M_{ik}\, \delta x_k \frac{\partial f_i}{\partial u}\right.
$$

$$
\left. + \sum_k \frac{\partial^2 H}{\partial x_k\, \partial u}\, \delta x_k\right) \quad (20)
$$

On the other hand it must be possible to differentiate Eq. (15) with respect to t and write

$$
\delta\dot{x}_j = \dot{g}_j + \sum_k \dot{M}_{jk}\, \delta x_k + \sum_k M_{jk}\, \delta\dot{x}_k \tag{21}
$$

or, substituting from Eqs. (15) and (17) to (19),

$$
\delta\dot{x}_j = \dot{g}_j + \sum_k \dot{M}_{jk}\, \delta x_k + \sum_{k,i} M_{jk} \frac{\partial f_k}{\delta x_j}\, \delta x_i
$$

$$
+ \sum_k M_{jk} \frac{\partial f_k}{\partial u}\left(-\frac{\partial^2 H}{\partial u^2}\right)^{-1}\!\left(\frac{\partial H}{\partial u} + \sum_i g_i \frac{\partial f_i}{\partial u} + \sum_{i,p} M_{ip}\, \delta x_p \frac{\partial f_i}{\partial u}\right.
$$

$$
\left. + \sum_i \frac{\partial^2 H}{\partial x_i\, \partial u}\, \delta x_i\right) \quad (22)
$$

The coefficients of $(\delta x)^0$ and $(\delta x)^1$ in Eqs. (20) and (22) must be the same, leading to the equations

$$
\dot{M}_{ij} + \sum_k \left[\frac{\partial f_k}{\partial x_i} - \left(\frac{\partial^2 H}{\partial u^2}\right)^{-1} \frac{\partial f_k}{\partial u} \frac{\partial^2 H}{\partial u\, \partial x_i}\right] M_{kj} + \sum_k M_{ik}\left[\frac{\partial f_k}{\partial x_j}\right.
$$

$$
\left. - \left(\frac{\partial^2 H}{\partial u^2}\right)^{-1} \frac{\partial f_k}{\partial u} \frac{\partial^2 H}{\partial u\, \partial x_j}\right] - \left(\sum_k M_{ik} \frac{\partial f_k}{\partial u}\right)\left(\frac{\partial^2 H}{\partial u^2}\right)^{-1}\!\left(\sum_p \frac{\partial f_p}{\partial u} M_{pj}\right)
$$

$$
+ \frac{\partial^2 H}{\partial x_i\, \partial x_j} - \frac{\partial^2 H}{\partial x_i\, \partial u}\left(\frac{\partial^2 H}{\partial u^2}\right)^{-1} \frac{\partial^2 H}{\partial x_j\, \partial u} = 0 \qquad M_{ij}(\theta) = \frac{\partial^2 \mathcal{E}}{\partial x_i\, \partial x_j} \quad (23)
$$

$$
\dot{g}_i - \sum_j \left[\sum_k M_{ik} \frac{\partial f_k}{\partial u}\left(\frac{\partial^2 H}{\partial u^2}\right)^{-1} \frac{\partial f_j}{\partial u} - \frac{\partial f_j}{\partial x_i} + \left(\frac{\partial^2 H}{\partial u^2}\right)^{-1} \frac{\partial f_j}{\partial u} \frac{\partial^2 H}{\partial u\, \partial x_i}\right] g_j
$$

$$
- \sum_j M_{ij} \frac{\partial f_j}{\partial u}\left(\frac{\partial^2 H}{\partial u^2}\right)^{-1} \frac{\partial H}{\partial u} - \frac{\partial^2 H}{\partial x_i\, \partial u}\left(\frac{\partial^2 H}{\partial u^2}\right)^{-1} \frac{\partial H}{\partial u} = 0 \qquad g_i(\theta) = 0
$$

$$
\tag{24}
$$

Equation (23) for the "feedback gain" is simply the Riccati equation obtained for the optimal control problem in Sec. 8.2, and M_{ij} is symmetric

$(M_{ij} = M_{ij})$.   As convergence occurs and $\partial H/\partial u$ goes to zero, the forcing term in Eq. (24) also goes to zero and the "feedforward gain" vanishes. The correction to the decision, $\delta u(t)$, is then obtained from Eqs. (19) and (15) as

$$
\delta u(t) = -\left(\frac{\partial^2 H}{\partial u^2}\right)^{-1}\left[\frac{\partial H}{\partial u} + \sum_i g_i \frac{\partial f_i}{\partial u}\right.
$$

$$
\left. + \sum_i\left(\frac{\partial^2 H}{\partial x_i\,\partial u} + \sum_j \frac{\partial f_j}{\partial u} M_{ji}\right)(x_i - \bar{x}_i)\right] \quad (25)
$$

where x_i is the (as yet unknown) value of the state corresponding to the new decision function.

The computational algorithm which arises from these equations is then as follows:

1. Choose $\bar{u}(t)$, solve Eqs. (1) for $\bar{\mathbf{x}}$ and then (3) for $\bar{\boldsymbol{\gamma}}$ in succession, and evaluate all partial derivatives of H.
2. Solve Eqs. (23) and (24), where all coefficients depend upon $\bar{u}(t)$, $\bar{\mathbf{x}}(t)$, and $\bar{\boldsymbol{\gamma}}(t)$.
3. Compute $u(t)$ from the expression

$$
u_{\text{new}}(\mathbf{x},t) = \bar{u}(t) + \frac{1}{r}\,\delta u(t) \tag{26}
$$

where δu is defined by Eq. (25) and r is a relaxation parameter, $r \geq 1$.   Note that u_{new} is an explicit function of $\mathbf{x}$.
4. Solve Eq. (1) in the form

$$
\dot{x}_i = f_i[\mathbf{x}, u_{\text{new}}(\mathbf{x},t)] \tag{27}
$$

simultaneously calculating $u(t)$, and repeat until convergence is obtained.

It is interesting to note that after convergence is obtained, Eq. (25) provides the feedback gains for optimal linear control about the optimal path.

The second-variation method was applied to the optimal-pressure-profile problem studied several times earlier in this chapter.   There are two state variables, hence five auxiliary functions to be computed, M_{11}, M_{12}, M_{22}, g_1, g_2.   We shall not write down the cumbersome set of specific equations but simply present the results for an initial choice of decision function $\bar{u}(t) = \text{const} = 0.2$, one of the starting values used for steep descent, with $r = 1$.   Solutions to the Riccati equation are shown in Fig. 9.14, where differences beyond the second iteration were too small to be seen on that scale.   From the first iteration on, the

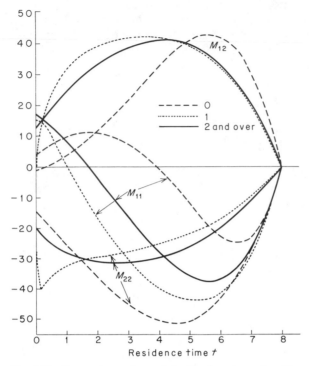

Fig. 9.14 Successive values of feedback gains using the second-variation method.

Table 9.10 Successive approximations to optimal pressure profile and outlet concentrations using the second-variation method

	Iteration number					
t	0	1	2	3	4	5
0	0.2000	1.1253	0.5713	0.7248	0.7068	0.7174
0.2	0.2000	0.6880	0.4221	0.3944	0.3902	0.3902
0.4	0.2000	0.4914	0.3527	0.3208	0.3194	0.3195
0.6	0.2000	0.3846	0.3014	0.2798	0.2786	0.2786
0.8	0.2000	0.3189	0.2666	0.2529	0.2522	0.2521
1.0	0.2000	0.2749	0.2421	0.2337	0.2332	0.2332
2.0	0.2000	0.1743	0.1819	0.1829	0.1829	0.1829
3.0	0.2000	0.1358	0.1566	0.1590	0.1590	0.1590
4.0	0.2000	0.1169	0.1421	0.1442	0.1442	0.1442
5.0	0.2000	0.1082	0.1324	0.1339	0.1339	0.1339
6.0	0.2000	0.1056	0.1251	0.1260	0.1260	0.1260
7.0	0.2000	0.1066	0.1191	0.1198	0.1198	0.1198
8.0	0.2000	0.1092	0.1140	0.1148	0.1148	0.1147
$x_1(8)$	3.3338×10^{-3}	3.6683×10^{-3}	3.8309×10^{-3}	3.8625×10^{-3}	3.8624×10^{-3}	3.8622×10^{-3}
$x_2(8)$	1.0823×10^{-2}	1.1115×10^{-2}	1.1315×10^{-2}	1.1319×10^{-2}	1.1319×10^{-2}	1.1319×10^{-2}

functions g_1 and g_2 were at least one order of magnitude smaller than γ_1 and γ_2, to which they act as correction terms, and could have been neglected without affecting the convergence. Successive pressure profiles are given in Table 9.10, where, following some initial overshoot resulting from using $r = 1$, the convergence is extremely rapid. With the exception of the value right at $t = 0$ the agreement is exact with solutions obtained earlier using the necessary conditions.

9.17 GENERAL REMARKS

Though we have not enumerated every technique available for the numerical solution of variational problems, this sketch of computational methods has touched upon the most important and frequently used classes of techniques. By far the most reliable is the method of steep descent for improving functional values of u. From extremely poor initial estimates convergence to near minimal values of the objective are obtained by a stable sequence of calculations. As a first-order method, however, it generally demonstrates poor ultimate convergence to the exact solution of the necessary conditions.

All other methods require a "good" initial estimate of either boundary values or one or more functions. Such estimates are the end product of a steep-descent solution, and a rational computational procedure would normally include this easily coded phase I regardless of the ultimate scheme to be used. The final calculational procedure, if any, must depend upon the particular problem and local circumstances. Min H, for example, is more easily coded than second variation, but it lacks the complete second-order convergence properties of the latter. Furthermore, if second derivatives are easily obtained and if a general code for integrating the Riccati equation is available, the coding differences are not significant. On the other hand, if $u(t)$ is constrained, second variation simply cannot be used. Similarly, steep-descent boundary iteration is attractive if, but only if, a highly reliable automatic routine of function minimization without calculation of derivatives is available, while all the indirect methods require relative stability of the equations.

BIBLIOGRAPHICAL NOTES

Section 9.1: We shall include pertinent references for individual techniques as they are discussed. We list here, however, several studies which parallel all or large parts of this chapter in that they develop and compare several computational procedures:

R. E. Kopp and H. G. Moyer: in C. T. Leondes (ed.), "Advances in Control Systems," vol. 4, Academic Press, Inc., New York, 1966

L. Lapidus: *Chem. Eng. Progr.*, **63**(12):64 (1967)

L. Lapidus and R. Luus: "Optimal Control of Engineering Processes," Blaisdell Publishing Company, Waltham, Mass., 1967

A. R. M. Noton: "Introduction to Variational Methods in Control Engineering,"
    Pergamon Press, New York, 1965
B. D. Tapley and J. M. Lewallen: *J. Optimization Theory Appl.*, **1**:1 (1967)

*Useful comparisons are also made in papers by Storey and Rosenbrock and Kopp,
    McGill, Moyer, and Pinkham in*

A. V. Balakrishnan and L. W. Neustadt: "Computing Methods in Optimization
    Problems," Academic Press, Inc., New York, 1964

*Most procedures have their foundation in perturbation analysis for the computation of
    ballistic trajectories by Bliss in 1919 and 1920, summarized in*

G. A. Bliss: "Mathematics for Exterior Ballistics," John Wiley & Sons, Inc., New
    York, 1944

Sections 9.2 and 9.3: The development and example follow

M. M. Denn and R. Aris: *Ind. Eng. Chem. Fundamentals*, **4**:7 (1965)

*The scheme is based on a procedure for the solution of boundary-value problems without
    decision variables by*

T. R. Goodman and G. N. Lance: *Math. Tables Other Aids Comp.*, **10**:82 (1956)

See also

J. V. Breakwell, J. L. Speyer, and A. E. Bryson: *SIAM J. Contr.*, **1**:193 (1963)
R. R. Greenly, *AIAA J.*, **1**:1463 (1963)
A. H. Jazwinski: *AIAA J.*, **1**:2674 (1963)
———: *AIAA J.*, **2**:1371 (1964)
S. A. Jurovics and J. E. McIntyre: *ARS J.*, **32**:1354 (1962)
D. K. Scharmack: in C. T. Leondes (ed.), "Advances in Control Systems," vol. 4,
    Academic Press, Inc., New York, 1966

*Section 9.4: Little computational experience is available for this rather obvious approach.
    Some pertinent remarks are contained in the reviews of Noton and Storey and
    Rosenbrock cited above; see also*

J. W. Sutherland and E. V. Bohn: *Preprints 1966 Joint Autom. Contr. Conf.*, Seattle,
    p. 177

*Neustadt has developed a different steep-descent boundary-interation procedure for
    solving linear time-optimal and similar problems. See papers by Fadden and
    Gilbert, Gilbert, and Paiewonsky and coworkers in the collection edited by Bala-
    krishnan and Neustadt and a review by Paiewonsky,*

B. Paiewonsky: in G. Leitmann (ed.), "Topics in Optimization," Academic Press,
    Inc., New York, 1967

*Sections 9.5 to 9.7: The Newton-Raphson (quasilinearization) procedure for the solution
    of boundary-value problems is developed in detail in*

R. E. Bellman and R. E. Kalaba: "Quasilinearization and Nonlinear Boundary-
    value Problems," American Elsevier Publishing Company, New York, 1965

Convergence proofs may be found in

R. E. Kalaba: *J. Math. Mech.*, **8**:519 (1959)
C. T. Leondes and G. Paine: *J. Optimization Theory Appl.*, **2**:316 (1968)
R. McGill and P. Kenneth: *Proc. 14th Intern. Astron. Federation Congr.*, *Paris*, 1963

The reactor example is from

E. S. Lee: *Chem. Eng. Sci.*, **21**:183 (1966)

and the calculations shown were done by Lee. There is now an extensive literature of applications, some recorded in the book by Bellman and Kalaba and in the reviews cited above. Still further references may be found in

P. Kenneth and R. McGill: in C. T. Leondes (ed.), "Advances in Control Systems," vol. 3, Academic Press, Inc., New York, 1966
E. S. Lee: *AIChE J.*, **14**:467 (1968)
————: *Ind. Eng. Chem. Fundamentals*, **7**:152, 164 (1968)
R. McGill: *SIAM J. Contr.*, **3**:291 (1965)
———— and P. Kenneth: *AIAA J.*, **2**:1761 (1964)
C. H. Schley and I. Lee: *IEEE Trans. Autom. Contr.*, **AC12**:139 (1967)
R. J. Sylvester and F. Meyer: *J. SIAM*, **13**:586 (1965)

The papers by Kenneth and McGill and McGill use penalty functions to incorporate state and decision constraints.

Section 9.8: Pertinent comments on computational effort are in

R. E. Kalman: *Proc. IBM Sci. Comp. Symp. Contr. Theory Appl.*, *White Plains, N.Y.*, *1966*, p. 25.

Sections 9.9 to 9.13: The idea of using steep descent to solve variational problems originated with Hadamard; see

R. Courant: *Bull. Am. Math. Soc.*, **49**:1 (1943)

The first practical application to a variational problem appears to be in

J. H. Laning, Jr., and R. H. Battin: "Random Processes in Automatic Control," McGraw-Hill Book Company, New York, 1956

Subsequently, practical implementation was accomplished independently about 1960 by Bryson, Horn, and Kelley; see

A. E. Bryson, W. F. Denham, F. J. Carroll, and K. Mikami: *J. Aerospace Sci.*, **29**:420 (1962)
F. Horn and U. Troltenier: *Chem. Ing. Tech.*, **32**:382 (1960)
H. J. Kelley: in G. Leitmann (ed.), "Optimization Techniques with Applications to Aerospace Systems," Academic Press, Inc., New York, 1962

The most comprehensive treatment of various types of constraints is contained in

W. F. Denham: Steepest-ascent Solution of Optimal Programming Problems, *Raytheon Co. Rept.* BR-2393, Bedford, Mass., 1963; also Ph.D. thesis, Harvard University, Cambridge, Mass., 1963

Some of this work has appeared as

W. F. Denham and A. E. Bryson: *AIAA J.*, **2**:25 (1964)

An interesting discussion of convergence is contained in

D. E. Johansen: in C. T. Leondes (ed.), "Advances in Control Systems," vol. 4, Academic Press, Inc., New York, 1966

Johansen's comments on convergence of singular controls are not totally in agreement with our own experiences. The formalism for staged systems was developed by Bryson, Horn, Lee, and by Denn and Aris for all classes of constraints; see

A. E. Bryson: in A. G. Oettinger (ed.), "Proceedings of the Harvard Symposium on Digital Computers and Their Applications," Harvard University Press, Cambridge, Mass., 1962

M. M. Denn and R. Aris: *Ind. Eng. Chem. Fundamentals*, **4**:213 (1965)

E. S. Lee: *Ind. Eng. Chem. Fundamentals*, **3**:373 (1964)

F. Horn and U. Troltenier: *Chem. Ing. Tech.*, **35**:11 (1963)

The example in Sec. 9.11 is from

J. M. Douglas and M. M. Denn: *Ind. Eng. Chem.*, **57**(11):18 (1965)

while that in Secs. 9.12 and 9.13 is from the paper by Denn and Aris cited above. Other examples of steep descent and further references are contained in these papers and the reviews. An interesting use of penalty functions is described in

L. S. Lasdon, A. D. Waren, and R. K. Rice: *Preprints 1967 Joint Autom. Cont. Conf.*, Philadelphia, p. 538

Section 9.14: The min-H approach was suggested by Kelley, in the article cited above, and by

S. Katz: *Ind. Eng. Chem. Fundamentals*, **1**:226 (1962)

For implementation, including the extension to systems with final constraints, see

R. G. Gottlieb: *AIAA J.*, **5**:(1967)

R. T. Stancil: *AIAA J.*, **2**:1365 (1964)

The examination of convergence was in

M. M. Denn: *Ind. Eng. Chem. Fundamentals*, **4**:231 (1965)

The example is from

R. D. Megee, III: Computational Techniques in the Theory of Optimal Processes, B. S. thesis, University of Delaware, Newark, Del., 1965

Sections 9.15 and 9.16: The basic development of a second-variation procedure is in

H. J. Kelley, R. E. Kopp, and H. G. Moyer: *Progr. Astron. Aero.*, **14**:559 (1964)

C. W. Merriam: "Optimization Theory and the Design of Feedback Control Systems," McGraw-Hill Book Company, New York, 1964

The second variation is studied numerically in several of the reviews cited above; see also

D. Isaacs, C. T. Leondes, and R. A. Niemann: *Preprints 1966 Joint Autom. Contr. Conf.*, Seattle, p. 158

S. R. McReynolds and A. E. Bryson: *Preprints 1965 Joint Autom. Contr. Conf.,
Troy, N.Y.,* p. 551

S. K. Mitter: *Automatica,* **3**:135 (1966)

A. R. M. Noton, P. Dyer, and C. A. Markland: *Preprints 1966 Joint Autom. Contr.
Conf., Seattle,* p. 193

The equations for discrete systems have been obtained in

F. A. Fine and S. G. Bankoff: *Ind. Eng. Chem. Fundamentals,* **6**:293 (1967)

D. Mayne: *Intern. J. Contr.,* **3**:85 (1966)

*Efficient implementation of the second-variation technique requires an algorithm for solv-
ing the Riccati equation, such as the computer code in*

R. E. Kalman and T. S. Englar: "A User's Manual for the Automatic Synthesis
Program," *NASA Contractor Rept.* 475, June, 1966, available from the Clearinghouse for Federal Scientific and Technical Information, Springfield, Va. 22151.

PROBLEMS

9.1. Solve the following problem by each of the methods of this chapter.   If a computer
is not available, carry the formulation to the point where a skilled programmer with
no knowledge of optimization theory could code the program.   Include a detailed
logical flow sheet.

$$\dot{x}_1 = \frac{2}{\pi} \arctan u - x_1$$
$$\dot{x}_2 = x_1 - x_2$$
$$\dot{x}_3 = x_2 - x_3$$
$$x_1(0) = 0 \qquad x_2(0) = -0.4 \qquad x_3(0) = 1.5$$
$$\min \varepsilon = \int_0^3 [(2x_2)^{2\alpha} + x_3{}^2 + 0.01u^2] \, dt$$

Take $\alpha = 2$ and 10.   The latter case represents a penalty function approximation
for $|x_2| \leq \frac{1}{2}$.   The arctangent is an approximation to a term linear in u with the
constraint $|u| \leq 1$.   (This problem is due to Noton, and numerical results for some
cases may be found in his book.)   Repeat for the same system but with

$$\dot{x}_1 = u - x_1$$
$$|u| \leq 1$$

9.2. Solve Prob. 5.8 numerically for parameter values

$$k_1 = k_3 = 1 \qquad k_2 = 10$$
$$\theta = 0.4 \qquad \text{and} \qquad \theta = 1.0$$

Compare with the analytical solution.

9.3. Solve Prob. 7.2 using appropriate methods of this chapter.   (Boundary iteration
is difficult for this problem.   Why?)

9.6. Develop an extension of each of the algorithms of this chapter to the case in
which θ is specified only implicitly by a final constraint of the form

$$\psi[\mathbf{x}(\theta)] = 0$$

Note that it might sometimes be helpful to employ a duality of the type described in
Sec. 3.5.

10
Nonserial Processes

10.1 INTRODUCTION

A large number of industrial processes have a structure in which the flow of material and energy does not occur in a single direction because of the presence of bypass and recycle streams. Hence, decisions made at one point in a process can affect the behavior at a previous point. Though we have included the spatial dependence of decisions in processes such as the plug-flow reactor, our analyses thus far have been couched in the language and concepts of systems which evolve in time. For such systems the principle of causality prevents future actions from influencing the present, and in order to deal with the effect of feedback interactions in spatially complex systems we shall have to modify our previous analyses slightly.

The optimization of systems with complex structure involves only a single minor generalization over the treatment of simple systems in that the Green's functions satisfy a different set of boundary conditions. There has been some confusion in the engineering literature over this problem, however, and we shall proceed slowly and from several different points of view. The results of the analysis will include not only the optimization of systems with spatial interactions but also, because of mathematical similarities, the optimal operation of certain unsteady-state processes.

10.2 RECYCLE PROCESSES

We can begin our analysis of complex processes by a consideration of the recycle system shown in Fig. 10.1. A continuous process, such as a tubular chemical reactor, is described by the equations

$$\dot{x}_i = f_i(\mathbf{x}, u) \qquad 0 < t < \theta \tag{1}$$

where the independent variable t represents length or residence time. The initial state $\mathbf{x}(0)$ is made up of a mixture of a specified feed $\mathbf{x}_f$ and the effluent $\mathbf{x}(\theta)$ in the form

$$x_i(0) = G_i[\mathbf{x}_f, \mathbf{x}(\theta)] \tag{2}$$

The goal is to choose $u(t)$, $0 \le t \le \theta$, in order to minimize a function of the effluent, $\mathcal{E}[\mathbf{x}(\theta)]$.

We carry out an analysis here identical to that used in previous chapters. A function $\bar{u}(t)$ is specified, and Eqs. (1) and (2) are solved for the corresponding $\bar{\mathbf{x}}(t)$. We now change $u(t)$ by a small amount $\delta u(t)$ and obtain first-order variational equations for $\boldsymbol{\delta x}$

$$\delta \dot{x}_i = \sum_{j=1}^{S} \frac{\partial f_i}{\partial x_j} \delta x_j + \frac{\partial f_i}{\partial u} \delta u \tag{3}$$

The corresponding first-order change in the mixing boundary condition, Eq. (2), is

$$\delta x_i(0) = \sum_{j=1}^{S} \frac{\partial G_i}{\partial x_j(\theta)} \delta x_j(\theta) \tag{4}$$

Green's identity can be written for the linear equations (3) as

$$\sum_{i=1}^{S} \gamma_i(\theta)\,\delta x_i(\theta) = \sum_{i=1}^{S} \gamma_i(0)\,\delta x_i(0) + \int_0^\theta \sum_{i=1}^{S} \gamma_i \frac{\partial f_i}{\partial u}\,\delta u\,dt \tag{5}$$

Fig. 10.1 Schematic of a continuous recycle process.

where the Green's functions satisfy the equations

$$\dot{\gamma}_i = - \sum_{j=1}^{S} \gamma_j \frac{\partial f_j}{\partial x_i} \qquad 0 < t < \theta \tag{6}$$

Following substitution of Eq. (4), Green's identity, Eq. (5), becomes

$$\sum_{i=1}^{S} \gamma_i(\theta)\, \delta x_i(\theta) = \sum_{i=1}^{S} \gamma_i(0) \left[\sum_{j=1}^{S} \frac{\partial G_i}{\partial x_j(\theta)}\, \delta x_j(\theta) \right] + \int_0^\theta \sum_{i=1}^{S} \gamma_i \frac{\partial f_i}{\partial u}\, \delta u\, dt \tag{7}$$

or

$$\sum_{i=1}^{S} \left[\gamma_i(\theta) - \sum_{j=1}^{S} \gamma_j(0) \frac{\partial G_j}{\partial x_i(\theta)} \right] \delta x_i(\theta) = \int_0^\theta \frac{\partial H}{\partial u}\, \delta u\, dt \tag{8}$$

In Eq. (8) we have substituted the usual hamiltonian notation

$$H = \sum_{i=1}^{S} \gamma_i f_i \tag{9}$$

The first-order change in the objective, $\mathcal{E}[\mathbf{x}(\theta)]$, as a result of the change δu in decision, may be written

$$\delta \mathcal{E} = \sum_{i=1}^{S} \frac{\partial \mathcal{E}}{\partial x_i}\, \delta x_i(\theta) \tag{10}$$

This can be related explicitly to δu by means of Eq. (8), for if we write

$$\gamma_i(\theta) - \sum_{j=1}^{S} \gamma_j(0) \frac{\partial G_j}{\partial x_i(\theta)} = \frac{\partial \mathcal{E}}{\partial x_i} \tag{11}$$

Eq. (8) becomes

$$\delta \mathcal{E} = \int_0^\theta \frac{\partial H}{\partial u}\, \delta u\, dt \tag{12}$$

If $\bar{u}$ is the optimal decision, the weak minimum principle follows immediately from Eq. (12) by a proof identical to those used previously. The only change resulting from the recycle is that in the pair of boundary conditions, Eqs. (2) and (11).

An identical analysis can be carried out for a sequence of staged operations with recycle of the type shown in Fig. 10.2. The state is described by difference equations

$$x_i^n = f_i^n(\mathbf{x}^{n-1}, u^n) \qquad n = 1, 2, \ldots, N \tag{13}$$

where the input to the first stage is related to the effluent $\mathbf{x}^N$ and feed

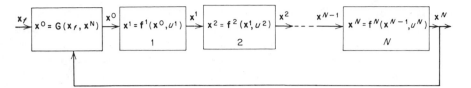

Fig. 10.2 Schematic of a staged recycle process.

$\mathbf{x}_f$ by the equation

$$x_i{}^0 = G_i(\mathbf{x}_f, \mathbf{x}^N) \tag{14}$$

and the objective is the minimization of $\mathcal{E}(\mathbf{x}^N)$.   The stage hamiltonian is defined by

$$H^n = \sum_{i=1}^{S} \gamma_i{}^n f_i{}^n \tag{15}$$

with Green's functions satisfying

$$\gamma_i{}^{n-1} = \sum_{j=1}^{S} \gamma_j{}^n \frac{\partial f_j{}^n}{\partial x_i{}^{n-1}} \qquad n = 1, 2, \ldots, N \tag{16}$$

$$\gamma_i{}^N - \sum_{j=1}^{S} \gamma_j{}^0 \frac{\partial G_j}{\partial x_i{}^N} = \frac{\partial \mathcal{E}}{\partial x_i{}^N} \tag{17}$$

As previously, only a weak minimum principle is generally satisfied. For continuous recycle systems a strong principle is proved in the usual way.

10.3 CHEMICAL REACTION WITH RECYCLE

We can apply the minimum principle to an elementary recycle problem by considering the process shown in Fig. 10.3.   The irreversible reaction

$$X \to Y$$

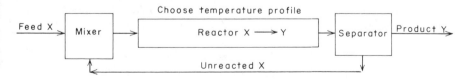

Fig. 10.3 Schematic of a reactor with recycle of unreacted feed.

is carried out in a tubular reactor, after which unreacted X is separated and recycled. We wish to find the temperature profile in the reactor which will maximize conversion less the cost of operation.

The concentration of X is denoted by x_1 and the temperature by u. With residence time as independent variable the governing equation for the reactor is then

$$\dot{x}_1 = -k(u)F(x_1) \qquad 0 < t \leq \theta \tag{1}$$

The mixing condition is

$$x_1(0) = (1 - \rho)x_f + \rho x_1(\theta) \tag{2}$$

where ρ is the fraction of the total volumetric flow which is recycled. The cost of operation is taken as the heating, which may be approximated as proportional to the integral of a function of temperature. Thus, we seek to maximize

$$\mathcal{P} = x_1(0) - x_1(\theta) - \int_0^\theta g[u(t)]\, dt \tag{3}$$

or, equivalently,

$$\mathcal{P} = \int_0^\theta [k(u)F(x_1) - g(u)]\, dt \tag{4}$$

This is put into the required form by defining a variable x_2,

$$\dot{x}_2 = -k(u)F(x_1) + g(u) \qquad x_2(0) = 0 \tag{5}$$

Then we wish to minimize

$$\mathcal{E} = x_2(\theta) \tag{6}$$

The hamiltonian is

$$H = -\gamma_1 k(u)F(x_1) + \gamma_2[-k(u)F(x_1) + g(u)] \tag{7}$$

with multiplier equations

$$\dot{\gamma}_1 = -\frac{\partial H}{\partial x_1} = (\gamma_1 + \gamma_2)k(u)F'(x_1) \tag{8a}$$

$$\dot{\gamma}_2 = -\frac{\partial H}{\partial x_2} = 0 \tag{8b}$$

The prime denotes differentiation with respect to the argument. The boundary conditions, from Eq. (17) of the preceding section, are

$$\gamma_1(\theta) = \rho\gamma_1(0) \tag{9a}$$
$$\gamma_2(\theta) = 1 \tag{9b}$$

Equations (8b) and (9b) imply that $\gamma_2 = 1$. If the optimum is taken to be interior, then

$$\frac{\partial H}{\partial u} = 0 = -(\gamma_1 + 1)F(x_1)k'(u) + g'(u) \tag{10}$$

or

$$\frac{g'(u)}{k'(u)} = (\gamma_1 + 1)F(x_1) \tag{11}$$

It is convenient to differentiate both sides of Eq. (11) with respect to t to obtain

$$\begin{aligned}
\frac{d}{dt}\frac{g'(u)}{k'(u)} &= \dot{\gamma}_1 F(x_1) + (\gamma_1 + 1)F'(x_1)\dot{x}_1 \\
&= (\gamma_1 + 1)k(u)F'(x_1)F(x_1) + (\gamma_1 + 1)F'(x_1)[-k(u)F(x_1)] \\
&= 0
\end{aligned} \tag{12}$$

Thus, $g'(u)/k'(u)$, a function only of u, is a constant, so that if the equation $g'(u)/k'(u) = $ const has a unique solution, the optimal temperature profile must be a constant. The problem then reduces to a one-dimensional search for the single constant u which maximizes $\mathcal{P}$.

For definiteness we take $F(x_1) = x_1$. Then Eqs. (1) and (2) can readily be solved for constant u and substituted into Eq. (3) to obtain

$$\mathcal{P} = \frac{x_f(1 - \rho)(1 - e^{-k(u)\theta})}{1 - \rho e^{-k(u)\theta}} - g(u)\theta \tag{13}$$

The maximization of $\mathcal{P}$ with respect to u can be carried out for given values of the parameters and functions by the methods of Chap. 2.

10.4 AN EQUIVALENT FORMULATION

The essential similarity between the results for recycle systems obtained in Sec. 10.2 and those for simple straight-chain systems obtained previously in Chaps. 4 to 7 suggests that the former might be directly derivable from the earlier results by finding an equivalent formulation of the problem in the spirit of Sec. 6.6. This can be done, though it is awkward for more complex situations. We shall carry out the manipulations for the continuous recycle system for demonstration purposes, but it is obvious that equivalent operations could be applied to staged processes.

The system satisfies the equations and recycle boundary condition

$$\dot{x}_i = f_i(\mathbf{x}, u) \qquad 0 < t \le \theta \tag{1}$$
$$x_i(0) = G_i[\mathbf{x}_f, \mathbf{x}(\theta)] \tag{2}$$

and the objective is the minimization of $\mathcal{E}[\mathbf{x}(\theta)]$. We shall define S

new variables, $y_1, y_2, \ldots, y_s$, as follows:

$$\dot{y}_i = 0 \tag{3}$$
$$y_i(0) - x_i(0) = 0 \tag{4}$$
$$y_i(\theta) - G_i[\mathbf{x}_f, \mathbf{x}(\theta)] = 0 \tag{5}$$

Equations (1) and (3) to (5) define a system of $2S$ equations with S initial constraints and S final constraints but with the recycle condition formally removed. The transversality conditions, Eqs. (12) and (13) of Sec. 6.4, can then be applied to the multipliers.

For convenience we denote the Green's functions corresponding to the x_i as γ_i and to the y_i as Γ_i. The γ_i satisfy the equations

$$\dot{\gamma}_i = - \sum_{j=1}^{S} \gamma_j \frac{\partial f_j}{\partial x_i} \qquad 0 \leq t < \theta \tag{6}$$

The transversality conditions resulting from the constraint equations (4) and (5) are, respectively,

$$\gamma_i(0) = -\eta_i \tag{7}$$
$$\gamma_i(\theta) = \frac{\partial \mathcal{E}}{\partial x_i} - \sum_{j=1}^{S} \nu_j \frac{\partial G_j}{\partial x_i} \tag{8}$$

where the η_i and ν_j are undetermined multipliers. The functions Γ_i satisfy the following equations and transversality conditions:

$$\dot{\Gamma}_i = 0 \qquad \Gamma_i = \text{const} \tag{9}$$
$$\Gamma_i(0) = \eta_i \tag{10}$$
$$\Gamma_i(\theta) = \nu_i \tag{11}$$

Combining Eqs. (7) to (11), we are led to the mixed boundary condition for the γ_i obtained previously

$$\gamma_i(\theta) - \sum_{j=1}^{S} \gamma_j(0) \frac{\partial G_j}{\partial x_i} = \frac{\partial \mathcal{E}}{\partial x_i} \tag{12}$$

The hamiltonian and strong minimum principle, of course, carry over directly to this equivalent formulation, but for the staged problem we obtain only the weak minimum principle.

10.5 LAGRANGE MULTIPLIERS

The Lagrange multiplier rule, first applied in Chap. 1 to the study of staged systems, is a particularly convenient tool for a limited class of problems. One of the tragedies of the minimum-principle–oriented analyses of recent years has been the failure to retain historical perspective through the relation to Lagrange multipliers as, for example, in Sec. 7.5. We shall repeat that analysis here for the staged recycle problem as a final demonstration of an alternative method of obtaining the proper multiplier boundary conditions.

The stage difference equations and recycle mixing condition may be written as

$$-x_i^n + f_i^n(\mathbf{x}^{n-1}, u^n) = 0 \qquad n = 1, 2, \ldots, N \tag{1}$$
$$-x_i^0 + G_i(\mathbf{x}_f, \mathbf{x}^N) = 0 \tag{2}$$

For the minimization of $\mathcal{E}(\mathbf{x}^N)$ the lagrangian is then written

$$\mathcal{L} = \mathcal{E} + \sum_{n=1}^{N} \sum_{i=1}^{S} \lambda_i^n[-x_i^n + f_i^n(\mathbf{x}^{n-1}, u^n)]$$

$$+ \sum_{i=1}^{S} \lambda_i^0[-x_i^0 + G_i(\mathbf{x}_f, \mathbf{x}^N)] \tag{3}$$

Setting partial derivatives with respect to each of the variables u^1, u^2, $\ldots$, u^N, x_i^0, x_i^1, $\ldots$, x_i^N to zero, we obtain

$$\frac{\partial \mathcal{L}}{\partial u^n} = \sum_{i=1}^{S} \lambda_i^n \frac{\partial f_i^n}{\partial u^n} = 0 \tag{4}$$

$$\frac{\partial \mathcal{L}}{\partial x_i^n} = -\lambda_i^n + \sum_{j=1}^{S} \lambda_j^{n+1} \frac{\partial f_j^{n+1}}{\partial x_i^n} = 0 \qquad n = 0, 1, 2, \ldots, N-1 \tag{5}$$

$$\frac{\partial \mathcal{L}}{\partial x_i^N} = \frac{\partial \mathcal{E}}{\partial x_i^N} - \lambda_i^N + \sum_{j=1}^{S} \lambda_j^0 \frac{\partial G_j}{\partial x_i^N} = 0 \tag{6}$$

Equations (4) and (5) are the usual weak-minimum-principle equations for an unconstrained decision, while Eq. (6) is the multiplier boundary condition obtained previously for the recycle problem.

10.6 THE GENERAL ANALYSIS

We can now generalize the recycle analysis to include plants with an arbitrary structure of interactions. The essence of the analysis is the observation that any complex structure can be broken down into a set of serial, or straight-chain, structures of the type shown in Fig. 10.4, with

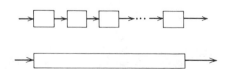

Fig. 10.4 Serial subsystems.

mixing conditions expressing the interactions between serial subsystems.

For notational purposes we shall denote the particular subsystem of interest by a parenthetical superscript. Thus $\mathbf{x}^{(k)}$ refers to the state of the kth serial subsystem. If it is continuous, the dependence on the continuous independent variable is expressed as $\mathbf{x}^{(k)}(t)$, where t ranges from zero (inlet) to $\theta^{(k)}$ (outlet). In a staged serial subsystem we would indicate stage number in the usual way, $\mathbf{x}^{(k)n}$, where n ranges from zero to $N^{(k)}$. It is convenient to express the inlet and outlet of continuous subsystems in the same way as the discrete, so that instead of $\mathbf{x}^{(k)}(0)$ we shall write $\mathbf{x}^{(k)0}$, and $\mathbf{x}^{(k)N}$ for $\mathbf{x}^{(k)}(\theta^{(k)})$.

The state of the kth serial subsystem is described by the difference or differential equations

$$x_i^{(k)n} = f_i^{(k)n}(\mathbf{x}^{(k)n-1}, u^{(k)n}) \qquad \begin{aligned} n &= 1, 2, \ldots, N^{(k)} \\ i &= 1, 2, \ldots, S^{(k)} \end{aligned} \qquad (1a)$$

$$\dot{x}_i^{(k)} = f_i^{(k)}(\mathbf{x}^{(k)}, u^{(k)}) \qquad \begin{aligned} 0 &< t < \theta^{(k)} \\ i &= 1, 2, \ldots, S^{(k)} \end{aligned} \qquad (1b)$$

where $S^{(k)}$ is the number of state variables required in the kth subsystem and a single decision is assumed for convenience. The subsystems are related to one another by mixing equations of the form

$$\mathbf{x}^{(k)0} = \mathbf{G}^{(k)}(\{\mathbf{x}_f\}, \{\mathbf{x}^{(l)N}\}) \qquad (2)$$

Equation (2) is a statement that the input to subsystem k depends in some known manner on the set of all external feeds to the system $\{\mathbf{x}_f\}$ and the set of outflow streams from all serial subsystems $\{\mathbf{x}^{(l)N}\}$. The simple recycle mixing condition, Eqs. (2) and (14) of Sec. 10.2, is a special case of Eq. (2). The objective is presumed to depend on all outflow streams

$$\mathcal{E} = \mathcal{E}\{\mathbf{x}^{(l)N}\} \qquad (3)$$

and for simplicity the outflows are presumed unconstrained.

Taking the usual variational approach, we suppose that a sequence of decisions $\{\bar{u}^{(k)}\}$ has been made for all parts of all subprocesses and that the system equations (1) have been solved subject to the mixing boundary conditions, Eq. (2). A small change in decisions then leads to the linear variational equations

$$\delta x_i^{(k)n} = \sum_{j=1}^{S^{(k)}} \frac{\partial f_i^{(k)n}}{\partial x_j^{(k)n-1}} \delta x_j^{(k)n-1} + \frac{\partial f_i^{(k)n}}{\partial u^{(k)n}} \delta u^{(k)n}$$

$$\begin{aligned} n &= 1, 2, \ldots, N^{(k)} \\ i &= 1, 2, \ldots, S^{(k)} \end{aligned} \qquad (4a)$$

$$\delta \dot{x}_i^{(k)} = \sum_{j=1}^{S^{(k)}} \frac{\partial f_i^{(k)}}{\partial x_j^{(k)}} \delta x_j^{(k)} + \frac{\partial f_i^{(k)}}{\partial u^{(k)}} \delta u^{(k)} \qquad \begin{aligned} 0 &< t < \theta^{(k)} \\ i &= 1, 2, \ldots, S^{(k)} \end{aligned} \qquad (4b)$$

The corresponding first-order changes in the mixing conditions and objective are

$$\delta x_i^{(k)0} = \sum_l \sum_{j=1}^{S^{(l)}} \frac{\partial G_i^{(k)}}{\partial x_j^{(l)N}} \delta x_j^{(l)N} \tag{5}$$

$$\delta \mathcal{E} = \sum_l \sum_{j=1}^{S^{(l)}} \frac{\partial \mathcal{E}}{\partial x_j^{(l)N}} \delta x_j^{(l)N} \tag{6}$$

The summations over l indicate summation over all subsystems.

The linear variational equations have associated Green's functions which satisfy difference or differential equations as follows:

$$\gamma_i^{(k)n-1} = \sum_{j=1}^{S^{(k)}} \gamma_j^{(k)n} \frac{\partial f_j^{(k)n}}{\partial x_i^{(k)n-1}} \qquad \begin{array}{l} n = 1, 2, \ldots, N^{(k)} \\ i = 1, 2, \ldots, S^{(k)} \end{array} \tag{7a}$$

$$\dot{\gamma}_i^{(k)} = -\sum_{j=1}^{S^{(k)}} \gamma_j^{(k)} \frac{\partial f_j^{(k)}}{\partial x_i^{(k)}} \qquad \begin{array}{l} 0 < t < \theta^{(k)} \\ i = 1, 2, \ldots, S^{(k)} \end{array} \tag{7b}$$

Green's identity for each subsystem is then

$$\sum_{i=1}^{S^{(k)}} \gamma_i^{(k)N} \delta x_i^{(k)N} = \sum_{i=1}^{S^{(k)}} \gamma_i^{(k)0} \delta x_i^{(k)0} + \sum_{n=1}^{N^{(k)}} \sum_{i=1}^{S^{(k)}} \gamma_i^{(k)n} \frac{\partial f_i^{(k)n}}{\partial u^{(k)n}} \delta u^{(k)n} \tag{8a}$$

$$\sum_{i=1}^{S^{(k)}} \gamma_i^{(k)N} \delta x_i^{(k)N} = \sum_{i=1}^{S^{(k)}} \gamma_i^{(k)0} \delta x_i^{(k)0} + \int_0^{\theta^{(k)}} \sum_{i=1}^{S^{(k)}} \gamma_i^{(k)} \frac{\partial f_i^{(k)}}{\partial u^{(k)}} \delta u^{(k)}(t) \, dt$$
$$\tag{8b}$$

The notation is made more compact by introducing the hamiltonian,

$$H^{(k)n} = \sum_{i=1}^{S^{(k)}} \gamma_i^{(k)n} f_i^{(k)n} \qquad \text{staged subsystem} \tag{9a}$$

$$H^{(k)} = \sum_{i=1}^{S} \gamma_i^{(k)} f_i^{(k)} \qquad \text{continuous subsystem} \tag{9b}$$

and the summation operator $\mathcal{S}_{(k)}$, defined by

$$\mathcal{S}_{(k)} \frac{\partial H^{(k)}}{\partial u^{(k)}} \delta u^{(k)} = \begin{cases} \displaystyle\sum_{n=1}^{N^{(k)}} \frac{\partial H^{(k)n}}{\partial u^{(k)n}} \delta u^{(k)n} & \text{staged subsystem} \\[3ex] \displaystyle\int_0^{\theta^{(k)}} \frac{\partial H^{(k)}}{\partial u^{(k)}} \delta u^{(k)}(t) \, dt & \text{continuous subsystem} \end{cases}$$
$$\tag{10}$$

Green's identity can then be written conveniently for staged or continuous subsystems as

$$\sum_{i=1}^{S^{(k)}} \gamma_i^{(k)N} \, \delta x_i^{(k)N} = \sum_{i=1}^{S^{(k)}} \gamma_i^{(k)0} \, \delta x_i^{(k)0} + \mathcal{S}_{(k)} \frac{\partial H^{(k)}}{\partial u^{(k)}} \, \delta u^{(k)} \tag{11}$$

We now substitute Eq. (5) for $\delta x_i^{(k)0}$ into Eq. (11), which is written as

$$\sum_{i=1}^{S^{(k)}} \gamma_i^{(k)N} \, \delta x_i^{(k)N} - \sum_{i=1}^{S^{(k)}} \gamma_i^{(k)0} \sum_l \sum_{j=1}^{S^{(l)}} \frac{\partial G_i^{(k)}}{\partial x_j^{(l)N}} \, \delta x_j^{(l)N} = \mathcal{S}_{(k)} \frac{\partial H^{(k)}}{\partial u^{(k)}} \, \delta u^{(k)} \tag{12}$$

Equation (12) is then summed over all subsystems

$$\sum_k \sum_{i=1}^{S^{(k)}} \gamma_i^{(k)N} \, \delta x_i^{(k)N} - \sum_k \sum_{i=1}^{S^{(k)}} \gamma_i^{(k)0} \sum_l \sum_{j=1}^{S^{(l)}} \frac{\partial G_i^{(k)}}{\partial x_j^{(l)N}} \, \delta x_j^{(l)N}$$
$$= \sum_k \mathcal{S}_{(k)} \frac{\partial H^{(k)}}{\partial u^{(k)}} \, \delta u^{(k)} \tag{13}$$

In the first term the dummy indices for summation, i and k, may be replaced by j and l, respectively, and order of finite summation interchanged in the second such that Eq. (13) is rewritten

$$\sum_l \sum_{j=1}^{S^{(l)}} \gamma_j^{(l)N} \, \delta x_j^{(l)N} - \sum_l \sum_{j=1}^{S^{(l)}} \left(\sum_k \sum_{i=1}^{S^{(k)}} \gamma_i^{(k)0} \frac{\partial G_i^{(k)}}{\partial x_j^{(l)N}} \right) \delta x_j^{(l)N}$$
$$= \sum_k \mathcal{S}_{(k)} \frac{\partial H^{(k)}}{\partial u^{(k)}} \, \delta u^{(k)} \tag{14}$$

or, finally,

$$\sum_l \sum_{j=1}^{S^{(l)}} \left(\gamma_j^{(l)N} - \sum_k \sum_{i=1}^{S^{(k)}} \gamma_i^{(k)0} \frac{\partial G_i^{(k)}}{\partial x_j^{(l)N}} \right) \delta x_j^{(l)N} = \sum_k \mathcal{S}_{(k)} \frac{\partial H^{(k)}}{\partial u^{(k)}} \, \delta u^{(k)} \tag{15}$$

Comparison with Eq. (6) for $\delta \mathcal{E}$ dictates the boundary conditions for the Green's functions

$$\gamma_j^{(l)N} - \sum_k \sum_{i=1}^{S^{(k)}} \gamma_i^{(k)0} \frac{\partial G_i^{(k)}}{\partial x_j^{(l)N}} = \frac{\partial \mathcal{E}}{\partial x_j^{(l)N}} \tag{16}$$

This is a generalization of Eqs. (11) and (17) of Sec. 10.2. The first-order variation in the objective can then be written explicitly in terms of the variations in decisions by combining Eqs. (6), (15), and (16)

$$\delta \mathcal{E} = \sum_k \mathcal{S}_{(k)} \frac{\partial H^{(k)}}{\partial u^{(k)}} \, \delta u^{(k)} \tag{17}$$

We may now adopt either of the two points of view which we have developed for variational problems. If we are examining variations about a presumed optimum, $\delta\mathcal{E}$ must be nonnegative and analyses identical to those in Secs. 6.5 and 7.4 lead to the weak minimum principle:

The Hamiltonian $H^{(k)n}$ or $H^{(k)}$ in each subprocess is made stationary by interior optimal decisions and a minimum (or stationary) by optimal decisions at a boundary or nondifferentiable point.

For continuous subsystems $H^{(k)}$ is a constant, and by a more refined analysis of the type in either Sec. 6.8 or 6.9, a strong minimum principle for continuous subsystems can be established. The necessary conditions for an optimum, then, require the simultaneous solution of Eqs. (1) and (7) with mixing conditions defined by Eqs. (2) and (16) and the optimum decision determined by means of the weak minimum principle.

Should we choose to develop a direct computational method based upon steep descent, as in Sec. 9.9, we wish to choose $\delta u^{(k)}(t)$ and $\delta u^{(k)n}$ to make $\delta\mathcal{E}$ nonpositive in order to drive $\mathcal{E}$ to a minimum. It follows from Eq. (17), then, that the changes in decision should be of the form

$$\delta u^{(k)n} = -w^{(k)n} \frac{\partial H^{(k)n}}{\partial u^{(k)n}} \tag{18a}$$

$$\delta u^{(k)}(t) = -w^{(k)}(t) \frac{\partial H^{(k)}}{\partial u^{(k)}} \tag{18b}$$

where $w^{(k)n}$ and $w^{(k)}(t)$ are nonnegative. A geometric analysis leads to the normalization analogous to Eqs. (12) and (23) of Sec. 9.9.

$$\delta u^{(k)n} = -\frac{\Delta G^{(k)n} \, \partial H^{(k)n}/\partial u^{(k)n}}{\left[\sum_k \mathcal{S}_{(k)} G^{(k)} \left(\frac{\partial H^{(k)}}{\partial u^{(k)}}\right)^2\right]^{1/2}} \tag{19a}$$

$$\delta u^{(k)}(t) = -\frac{\Delta G^{(k)}(t) \, \partial H^{(k)}/\partial u^{(k)}}{\left[\sum_k \mathcal{S}_{(k)} G^{(k)} \left(\frac{\partial H^{(k)}}{\partial u^{(k)}}\right)^2\right]^{1/2}} \tag{19b}$$

where Δ is a step size and $G^{(k)n}$, $G^{(k)}(t)$ are nonnegative weighting functions.

10.7 REACTION, EXTRACTION, AND RECYCLE

The results of the preceding section can be applied to the optimization of the process shown in Fig. 10.5. The reaction

$$X_1 \rightarrow X_2 \rightarrow \text{products}$$

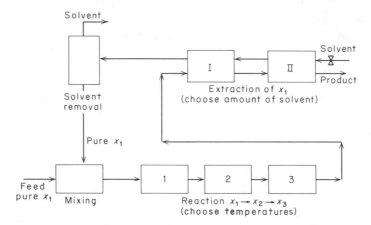

Fig. 10.5 Schematic of a reactor system with countercurrent extraction and recycle of unreacted feed. [*From M. M. Denn and R. Aris, Ind. Eng. Chem. Fundamentals,* **4**:248 (1965). *Copyright 1965 by the American Chemical Society. Reprinted by permission of the copyright owner.*]

is carried out in a sequence of three stirred-tank reactors. The product stream is passed through a two-staged countercurrent extraction unit, where the immiscible solvent extracts pure X_1, which is recycled to the feed to the first reactor. The reactor system will be denoted by arabic stage numbers and the extractor by roman numerals in order to avoid the use of parenthetical superscripts.

The kinetics of the reactor sequence are taken as second and first order, in which case the reactor material-balance equations are those used in Sec. 9.12

$$0 = x_1^{n-1} - x_1^n - \theta k_{10} e^{-E_1'/u^n}(x_1^n)^2 \qquad n = 1, 2, 3 \tag{1a}$$

$$0 = x_2^{n-1} - x_2^n + \theta k_{10} e^{-E_1'/u^n}(x_1^n)^2$$
$$\qquad\qquad - \theta k_{20} e^{-E_2'/u^n} x_2^n \qquad n = 1, 2, 3 \tag{1b}$$

The material-balance equations about the first and second stage of the extractor are, respectively,

$$x_1^I + u^{II}[\psi(x_1^I) - \psi(x_1^{II})] - x_1^3 = 0 \tag{2}$$

$$x_1^{II} + u^{II}\psi(x_1^{II}) - x_1^I = 0 \tag{3}$$

Here u^{II} is the ratio of volumetric flow rates of solvent to product stream and $\psi(x_1)$ is the equilibrium distribution between the concentration of x_1 in solvent and reactant stream. Because the solvent extracts X_1 only, there is no material-balance equation needed for x_2 about the extractor, for $x_2^3 = x_2^{II}$. The external feed is pure X_1, so that the feed to reactor 1 is

$$x_1^0 = x_{1f} + u^{II}\psi(x_1^I) \tag{4a}$$
$$x_2^0 = 0 \tag{4b}$$

We wish to maximize the production of X_2 while allowing for costs of raw material and extraction, and hence we seek to minimize

$$\mathcal{E} = -x_2^3 - cx_1^{II} + \sigma u^{II} \tag{5}$$

The temperatures u^1, u^2, u^3 and solvent ratio u^{II} are to be chosen subject to constraints

$$u_* \leq u^1, u^2, u^3 \leq u^* \tag{6a}$$
$$0 \leq u^{II} \tag{6b}$$

Equations (1) to (5) are not in the form required for application of the theory, for, though we need not have done so, we have restricted the analysis to situations in which a decision appears only in one stage transformation and not in mixing conditions or objective. This is easily rectified by defining a variable† x_3 with

$$x_3^I = x_3^z \tag{7}$$
$$x_3^{II} = x_3^I + u^{II} \tag{8}$$

Equations (2) and (3) are then rewritten, after some manipulation,

$$\psi(x_1^I) - x_1^z = 0 \tag{9}$$
$$x_1^{II} + u^{II}\psi(x_1^{II}) - x_1^I = 0 \tag{10}$$

where the mixing boundary conditions are now rearranged as

$$x_1^0 = x_{1f} + x_1^3 - x_1^{II} \tag{11a}$$
$$x_2^0 = 0 \tag{11b}$$
$$x_1^z = \frac{x_1^3 - x_1^{II}}{x_3^{II}} \tag{11c}$$
$$x_3^z = 0 \tag{11d}$$

The system is then defined by Eqs. (1) and (7) to (11), and the objective is rewritten

$$\mathcal{E} = -x_2^3 - x_1^{II} + \sigma x_3^{II} \tag{12}$$

The structure defining the interactions represented by Eqs. (11) is shown in Fig. 10.6.

The equations for the Green's functions are defined by Eq. (7) of

---

† The superscript z (for zero) will denote the input to the first stage for the roman-numeraled variables.

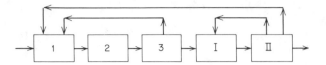

Fig. 10.6 Structure of the interactions in the reaction-extraction-recycle system. [*From M. M. Denn and R. Aris, Ind. Eng. Chem. Fundamentals,* **4:**248 (1965). *Copyright 1965 by the American Chemical Society. Reprinted by permission of the copyright owner.*]

Sec. 10.6 and, using implicit differentiation as in Sec. 7.6, are

$$\gamma_1{}^{n-1} = \frac{\gamma_1{}^n}{1 + 2\theta k_{10}e^{-E_1'/u^n}x_1{}^n}$$
$$+ \frac{2\theta\gamma_2{}^n k_{10}e^{-E_1'/u^n}x_1{}^n}{(1 + 2\theta k_{10}e^{-E_1'/u^n}x_1{}^n)(1 + \theta k_{20}e^{-E_2'/u^n})} \qquad n = 1, 2, 3 \quad (13a)$$

$$\gamma_2{}^{n-1} = \frac{\gamma_2{}^n}{1 + k\theta_{20}e^{-E_2'/u^n}} \qquad n = 1, 2, 3 \tag{13b}$$

$$\gamma_1{}^z = \frac{\gamma_1{}^I}{\psi'(x_1{}^I)} \tag{14a}$$

$$\gamma_1{}^I = \frac{\gamma_1{}^{II}}{1 + u^{II}\psi'(x_1{}^I)} \tag{14b}$$

$$\gamma_3{}^{n-1} = \gamma_3{}^n \qquad n = I, II \tag{14c}$$

The prime denotes differentiation with respect to the argument. The boundary conditions, obtained from Eqs. (11) and (12) by means of the defining equation (16) of the preceding section, are

$$\gamma_1{}^{II} + \frac{\gamma_1{}^z}{x_3{}^{II}} + \gamma_1{}^0 = -c \tag{15a}$$

$$\gamma_1{}^3 - \frac{\gamma_1{}^z}{x_3{}^{II}} - \gamma_1{}^0 = 0 \tag{15b}$$

$$\gamma_2{}^3 = -1 \tag{15c}$$

$$\gamma_3{}^{II} + \frac{\gamma_1{}^z}{x_3{}^{II}}\,\psi\,(x_1{}^I) = \sigma \tag{15d}$$

Finally, the partial derivatives of the hamiltonians with respect to the decisions are

$$\frac{\partial H^n}{\partial u^n} = \frac{\theta}{(u^n)^2}\frac{\begin{array}{l}[\gamma_2{}^n - \gamma_1{}^n(1 + \theta k_{20}e^{-E_2'/u^n})]E'k_{10}e^{-E_1'/u^n}(x_1{}^n)^2 \\ - \gamma_2{}^n E_2' k_{20}e^{-E_2'/u^n}x_2{}^n(1 + 2\theta k_{10}e^{-E_1'/u^n}x_1{}^n)\end{array}}{(1 + \theta k_{20}e^{-E_2'/u^n})(1 + 2\theta k_{10}e^{-E_1'/u^n}x_1{}^n)}$$
$$n = 1, 2, 3 \quad (16a)$$

$$\frac{\partial H^{II}}{\partial u^{II}} = \gamma_1{}^{II}\frac{\partial f_1{}^{II}}{\partial u^{II}} + \gamma_3{}^{II}\frac{\partial f_3{}^{II}}{\partial u^{II}}$$

$$= -\frac{\gamma_1{}^{II}\psi(x_1{}^{II})}{1 + u^{II}\psi'(x_1{}^{II})} + \gamma_3{}^{II} \tag{16b}$$

or, substituting Eq. (15d) into (16b),

$$\frac{\partial H^{II}}{\partial u^{II}} = \sigma - \frac{\gamma_1{}^z}{u^{II}}\,\psi(x_1{}^I) - \frac{\gamma_1{}^{II}\psi(x_1{}^{II})}{1 + u^{II}\psi'(x_1{}^{II})} \tag{16c}$$

It is evident that the artificial variables $x_3{}^I$, $x_3{}^{II}$ and $\gamma_3{}^I$, $\gamma_3{}^{II}$ are never needed in actual computation.

The simultaneous solution of the material-balance relations, Eqs. (1) to (4), and Green's function, Eqs. (13) to (15), with the optimal temperatures and solvent ratio determined from the weak minimum principle by means of Eqs. (5) and (16), is a difficult task. Application of the indirect methods developed in Chap. 9 would require iterative solution of both the state and multiplier equations for any given set of boundary conditions because of the mixing conditions for both sets of variables, and even with stable computational methods this would be a time-consuming operation. Steep descent, on the other hand, is quite attractive, for although the material-balance equations must be solved iteratively for each assumed set of decisions, the multiplier equations are then linear with fixed coefficients and can be solved by superposition.

To illustrate this last point let us suppose that the decisions and resulting state variables have been obtained. Addition of Eqs. (15a) and (15b) leads to

$$\gamma_1{}^{II} + \gamma_1{}^3 = -c \tag{17}$$

and so it is evident that a knowledge of $\gamma_1{}^3$ is sufficient to solve Eqs. (13a), (14), (15a), (15b), and (15d). We assume a value $\hat{\gamma}_1{}^3$ and compute the set $\{\hat{\gamma}_1{}^n\}$, $n = 1, 2, 3, z, I, II$. The correct values are denoted by $\{\hat{\gamma}_1{}^n + \delta\gamma_1{}^n\}$. From Eq. (17)

$$\delta\gamma_1{}^{II} = -\delta\gamma_1{}^3 \tag{18}$$

in which case it follows from Eqs. (14a) and (14b) that

$$\delta\gamma_1{}^z = \frac{\delta\gamma_1{}^3}{\psi'(x_1{}^I)[1 + u^{II}\psi'(x_1{}^I)]} \tag{19}$$

From Eq. (13a),

$$\delta\gamma_1{}^0 = \delta\gamma_1{}^3 \sum_{m=1}^{3} \frac{1}{1 + 2\theta x_1{}^m k_{20} e^{-E_1{}'/u^m}} \tag{20}$$

But it follows from Eq. (15b) that

$$\delta\gamma_1{}^3 - \frac{\delta\gamma_1{}^z}{u^{II}} - \delta\gamma_1{}^0 = -\hat{\gamma}_1{}^3 + \frac{\hat{\gamma}_1{}^z}{u^{II}} + \hat{\gamma}_1{}^0 \tag{21}$$

and combining Eqs. (19) to (21), it follows that the correct value of $\gamma_1{}^3$ is

$$\gamma_1{}^3 = \hat{\gamma}_1{}^3 + \cfrac{\hat{\gamma}_1{}^0 - \hat{\gamma}_1{}^3 + \hat{\gamma}_1{}^z/u^{II}}{1 + \cfrac{1}{u^{II}\psi'(x_1{}^I)[1 + u^{II}\psi'(x^I)]} - \prod_{m=1}^{3} \cfrac{1}{1 + 2\theta x_1{}^m k_{10}e^{-E_1'/u^m}}} \tag{22}$$

Thus, once the temperatures and solvent ratio have been specified and the material-balance equations solved iteratively, the equations for the Green's functions need be solved only twice with one application of Eq. (22).

The corrections to the values of u^1, u^2, u^3, u^{II} are calculated using Eqs. (16a) and (16c) from the relations

$$\delta u^n = -\Delta \frac{G^n \, \partial H^n/\partial u^n}{\left[\sum_{m=1}^{3} G^m \left(\dfrac{\partial H^m}{\partial u^m}\right)^2 + G^{II}\left(\dfrac{\partial H^{II}}{\partial u^{II}}\right)^2\right]^{\frac{1}{2}}} \tag{23a}$$

$$\delta u^{II} = -\Delta \frac{G^{II} \, \partial H^{II}/\partial u^{II}}{\left[\sum_{m=1}^{3} G^m \left(\dfrac{\partial H^m}{\partial u^m}\right)^2 + G^{II}\left(\dfrac{\partial H^{II}}{\partial u^{II}}\right)^2\right]^{\frac{1}{2}}} \tag{23b}$$

The physical parameters used in the calculations in Secs. 9.3 and 9.12 were used here, with $\theta = 2$ and a total reactor residence time of 6. The function $\psi(x_1)$ is shown in Fig. 10.7 and has the analytical form

$$\psi(x_1) = \begin{cases} 2.5x_1 - 2(x_1)^2 & 0 \leq x_1 \leq 0.6 \\ 1.08 - 1.1x_1 + (x_1)^2 & 0.6 < x_1 < 0.9 \end{cases} \tag{24}$$

The cost of extraction σ was allowed to range over all values, and G^1, G^2, G^3 were set equal to unity. Following some preliminary experimen-

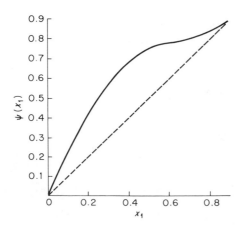

Fig. 10.7 Equilibrium distribution function of feed between solvent and reactant streams. [*From M. M. Denn and R. Aris, Ind. Eng. Chem. Fundamentals,* **4:**248 (1965). *Copyright 1965 by the American Chemical Society. Reprinted by permission of the copyright owner.*]

tation, G^{II} was taken as 0.005. As in Sec. 9.12 the initial step size Δ was set equal to 2, with the step size halved for each move not resulting in a decrease in the value of ε. The criterion for convergence was taken to be a step size less than 10^{-3}. In all calculations the material-balance equations were solved by a one-dimensional direct search. Figure 10.8 shows a typical convergence sequence, in this case for $\sigma = 0.30$.

The profit $-\varepsilon$ is plotted in Fig. 10.9 as a function of extraction cost. The horizontal line corresponds to the optimal nonrecycle solution found in Sec. 9.12, and in the neighborhood of the intersection two locally optimal solutions were found. Figure 10.10, for example, shows another set of calculations for $\sigma = 0.30$ starting at the same initial temperature policy

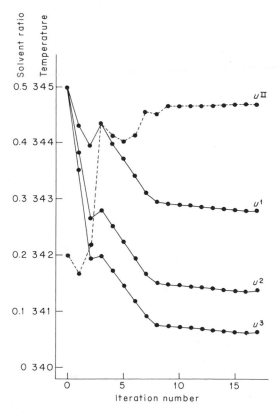

Fig. 10.8 Successive approximations to the optimal temperature sequence and solvent ratio using steep descent. [*From M. M. Denn and R. Aris, Ind. Eng. Chem. Fundamentals*, **4**:248 (1965). *Copyright 1965 by the American Chemical Society. Reprinted by permission of the copyright owner.*]

as in Fig. 10.8 but at a different solvent ratio, and the resulting temperatures are those for the optimal three-stage serial process. The optimal temperature and solvent policies as functions of σ are shown in Figs. 10.11 and 10.12. For sufficiently inexpensive separation the temperatures go to u_*, indicating low conversion and large recycle, while for sufficiently costly separation the optimal policy is nonrecycle. Multiple solutions are shown by dashed lines. The discontinuous nature of the extraction process with separation cost has obvious economic implications if capital investment costs have not yet been taken into account in deriving the cost factors for the process.

A mixed continuous staged process with the structure shown in Figs. 10.5 and 10.6 is obtained by replacing the three stirred-tank reactors with a plug-flow tubular reactor of residence time $\theta = 6$. The material-balance equations for the continuous serial process are then

$$\dot{x}_1 = -k_{10}e^{-E_1'/u}(x_1)^2 \qquad 0 < t < 6 \tag{25a}$$
$$\dot{x}_2 = k_{10}e^{-E_1'/u}(x_1)^2 - k_{20}e^{-E_2'/u}x_2 \qquad 0 < t < 6 \tag{25b}$$

with corresponding Green's functions

$$\dot{\gamma}_1 = 2k_{10}e^{-E_1'/u}x_1(\gamma_1 - \gamma_2) \qquad 0 < t < 6 \tag{26a}$$
$$\dot{\gamma}_2 = k_{20}e^{-E_2'/u}\gamma_2 \qquad 0 < t < 6 \tag{26b}$$

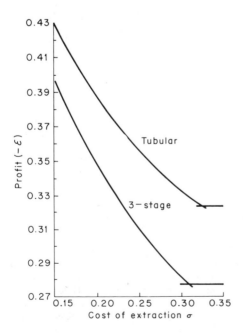

Fig. 10.9 Profit as a function of cost of extraction. [*From M. M. Denn and R. Aris, Ind. Eng. Chem. Fundamentals,* **4**:248 (1965). *Copyright* 1965 *by the American Chemical Society. Reprinted by permission of the copyright owner.*]

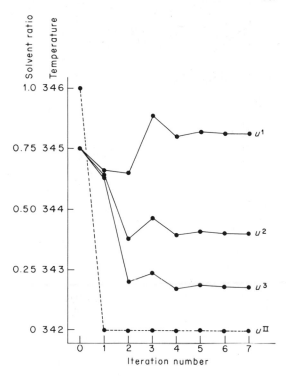

Fig. 10.10 Successive approximations to the optimal temperature sequence and solvent ratio using steep descent. [*From M. M. Denn and R. Aris, Ind. Eng. Chem. Fundamentals,* **4**:248 (1965). *Copyright* 1965 *by the American Chemical Society. Reprinted by permission of the copyright owner.*]

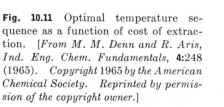

Fig. 10.11 Optimal temperature sequence as a function of cost of extraction. [*From M. M. Denn and R. Aris, Ind. Eng. Chem. Fundamentals,* **4**:248 (1965). *Copyright* 1965 *by the American Chemical Society. Reprinted by permission of the copyright owner.*]

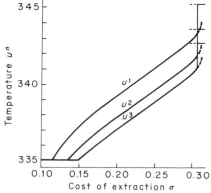

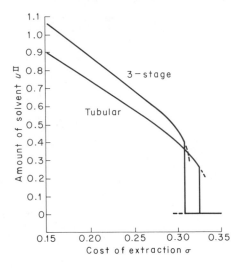

Fig. 10.12 Optimal solvent ratio as a function of cost of extraction. [*From M. M. Denn and R. Aris, Ind. Eng. Chem. Fundamentals,* **4**:248 (1965). *Copyright 1965 by the American Chemical Society. Reprinted by permission of the copyright owner.*]

The partial derivative of the continuous hamiltonian with respect to the decision is

$$\frac{\partial H}{\partial u} = \frac{1}{u^2}\left[x_1{}^2 k_{10} e^{-E_1'/u} E_1'(\gamma_1 - \gamma_2) - x_2 k_{20} e^{-E_2'/u} E_2' \gamma_2\right] \tag{27}$$

If we denote the effluent at $t = 6$ by the superscript $n = 3$, Eqs. (2) to (4), (7) to (11), (14), (15), and (16c) remain valid. The Green's function equations can still be solved with a single correction, but instead of Eq. (22) we must use

$$\gamma_1(6) = \hat{\gamma}_1(6)$$
$$+ \frac{\hat{\gamma}_1(0) - \hat{\gamma}_1(6) + \hat{\gamma}_1{}^z/u^{\mathrm{II}}}{1 + \dfrac{1}{u^{\mathrm{II}}\psi'(x_1{}^{\mathrm{I}})[1 + u^{\mathrm{II}}\psi'(x_1{}^{\mathrm{I}})]} - \exp\left(-\displaystyle\int_0^6 2x_1 k_{10} e^{-E_1'/u}\, dt\right)} \tag{28}$$

The corrections in temperature and solvent ratio are

$$\delta u(t) = -\Delta \frac{G(t)\, \partial H/\partial u}{\left[\displaystyle\int_0^6 G(t)\left(\frac{\partial H}{\partial u}\right)^2 dt + G^{\mathrm{II}}\left(\frac{\partial H^{\mathrm{II}}}{\partial u^{\mathrm{II}}}\right)^2\right]^{\frac{1}{2}}} \tag{29a}$$

$$\delta u^{\mathrm{II}} = -\Delta \frac{G^{\mathrm{II}}\, \partial H^{\mathrm{II}}/\partial u^{\mathrm{II}}}{\left[\displaystyle\int_0^6 G(t)\left(\frac{\partial H}{\partial u}\right)^2 dt + G^{\mathrm{II}}\left(\frac{\partial H^{\mathrm{II}}}{\partial u^{\mathrm{II}}}\right)^2\right]^{\frac{1}{2}}} \tag{29b}$$

Following the results of Sec. 9.11, the continuous variables were stored at intervals of 0.1 with linear interpolation, and an initial step size of 8 was taken. $G(t)$ was set equal to unity and G^{II} to 0.0005. A typi-

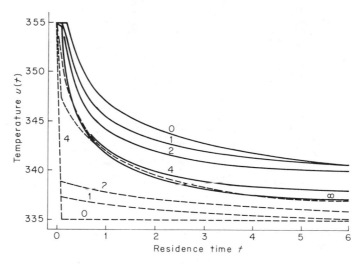

Fig. 10.13 Successive approximations to the optimal temperature profile in a continuous reactor using steep descent. [*From M. M. Denn and R. Aris, Ind. Eng. Chem. Fundamentals,* **4:**248 (1965). *Copyright 1965 by the American Chemical Society. Reprinted by permission of the copyright owner.*]

cal convergence sequence is shown in Figs. 10.13 and 10.14 for $\sigma = 0.25$. The solid starting-temperature profile is the optimal nonrecycle solution found in Sec. 9.3, while the shape of the dashed starring curve was dictated by the fact that the unconstrained solution for the temperature profile can be shown from the necessary conditions to require an infinite

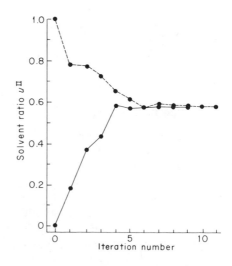

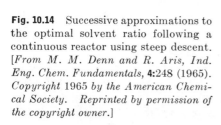

Fig. 10.14 Successive approximations to the optimal solvent ratio following a continuous reactor using steep descent. [*From M. M. Denn and R. Aris, Ind. Eng. Chem. Fundamentals,* **4:**248 (1965). *Copyright 1965 by the American Chemical Society. Reprinted by permission of the copyright owner.*]

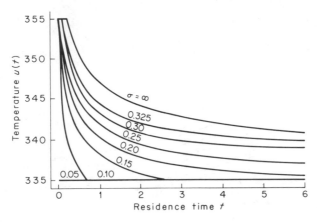

Fig. 10.15 Optimal temperature profile as a function of cost of extraction. [*From M. M. Denn and R. Aris, Ind. Eng. Chem. Fundamentals,* **4**:248 (1965). *Copyright* 1965 *by the American Chemical Society. Reprinted by permission of the copyright owner.*]

slope at $t = 0$ (compare Sec. 4.12). Convergence was obtained in 9 and 11 iterations from the solid and dashed curves, respectively, with some difference in the ultimate profiles. The apparent discontinuity in slope at $t = 0.1$ is a consequence of the use of a finite number of storage locations and linear interpolation.

The profit is shown as a function of σ in Fig. 10.9, the solvent allocation in Fig. 10.12, and the optimal temperature profiles in Fig. 10.15. As in the staged system there is a region in which multiple solutions were found. For $\sigma = 0.325$, for example, the same profit was obtained by the curve shown in Fig. 10.15 and solvent in Fig. 10.12 and by the nonrecycle solution, shown as $\sigma = \infty$. It is evident that multiple solutions must be expected and sought in complex systems in which the amount of recycle or bypass is itself a decision which has monetary value.

10.8 PERIODIC PROCESSES

The observation that the efficiency of some separation and reaction systems can be enhanced by requiring the system to operate in the unsteady state, as demonstrated, for example, in Sec. 6.11, has led to substantial interest in the properties of *periodic* processes. These are processes in which a regular cyclic behavior is established and therefore, in terms of time-averaged behavior, allows the overall operation to be considered from a production point of view in steady-state terms. As we shall see, the fact that a decision made during one cycle has an effect on an earlier part of the successive cycle is equivalent to feedback of information, and such processes are formally equivalent to recycle processes.

We consider a process which evolves in time according to the differential equations

$$\dot{x}_i = f_i(\mathbf{x},u) \qquad \begin{array}{l} 0 < t < \theta \\ i = 1, 2, \ldots, S \end{array} \tag{1}$$

The process is to be operated periodically, so that we have boundary conditions

$$x_i(0) = x_i(\theta) \qquad i = 1, 2, \ldots, S \tag{2}$$

This is clearly a special case of the recycle boundary conditions. For simplicity we shall assume that θ is specified. The decision function $u(t)$ is to be chosen to minimize some time-averaged performance criterion

$$\mathcal{E} = \frac{1}{\theta} \int_0^\theta \mathcal{F}(\mathbf{x},u) \, dt \tag{3}$$

This can be put in the form of a function of the state at $t = \theta$ by defining a new variable

$$\dot{x}_{s+1} = \frac{1}{\theta} \mathcal{F}(\mathbf{x},u) \qquad 0 < t \leq \theta \tag{4}$$

$$x_{s+1}(0) = 0 \tag{5}$$

In that case

$$\mathcal{E}[\mathbf{x}(\theta)] = x_{s+1}(\theta) \tag{6}$$

The hamiltonian for this problem is

$$H = \gamma_{s+1} \frac{1}{\theta} \mathcal{F} + \sum_{i=1}^{S} \gamma_i f_i \tag{7}$$

where the multiplier equations are

$$\dot{\gamma}_i = -\gamma_{s+1} \frac{1}{\theta} \frac{\partial \mathcal{F}}{\partial x_i} - \sum_{j=1}^{S} \gamma_j \frac{\partial f_j}{\partial x_i} \qquad i = 1, 2, \ldots, S \tag{8a}$$

$$\dot{\gamma}_{s+1} = 0 \tag{8b}$$

The boundary conditions, following Eq. (11) of Sec. 10.2, are

$$\gamma_i(\theta) - \gamma_i(0) = 0 \qquad i = 1, 2, \ldots, S \tag{9a}$$
$$\gamma_{s+1}(\theta) = 1 \tag{9b}$$

Thus we obtain, finally,

$$H = \frac{1}{\theta}\mathcal{F} + \sum_{i=1}^{S} \gamma_i f_i \tag{10}$$

$$\dot{\gamma}_i = -\frac{1}{\theta}\frac{\partial \mathcal{F}}{\partial x_i} - \sum_{j=1}^{S} \gamma_j \frac{\partial f_j}{\partial x_i} \qquad i = 1, 2, \ldots, S \tag{11}$$

$$\gamma_i(0) = \gamma_i(\theta) \qquad i = 1, 2, \ldots, S \tag{12}$$

The partial derivative of the hamiltonian with respect to the decision function is, of course,

$$\frac{\partial H}{\partial u} = \frac{1}{\theta}\frac{\partial \mathcal{F}}{\partial u} + \sum_{i=1}^{S} \gamma_i \frac{\partial f_i}{\partial u} \tag{13}$$

The calculation of the periodic control function $u(t)$ for optimal periodic operation is easily carried out by steep descent by exploiting the periodic boundary conditions on $\mathbf{x}$ and $\boldsymbol{\gamma}$. An initial periodic decision function is specified and Eqs. (1) integrated from some initial state until periodic response is approached. Next, Eqs. (11) for $\boldsymbol{\gamma}$ are integrated in reverse time from some starting value until a periodic response results. Reverse-time integration is used because of the stability problems discussed in Sec. 9.8. Finally, the new decision is chosen by the usual relation

$$\delta u = -w(t)\frac{\partial H}{\partial u} \qquad 0 \le t \le \theta \tag{14}$$

and the process is repeated. Convergence to periodic response might sometimes be slow using this simulation procedure, and a Newton-Raphson scheme with second-order convergence properties can be developed.

We have already seen for the special case in Sec. 6.11 that $\partial H/\partial u$ vanishes at an interior optimal steady state, necessitating the use of the strong minimum principle in the analysis of optimality. This is, in fact, true in general, and it then follows that a steady state cannot be used as the starting value for a steep-descent calculation, for then δu in Eq. (14) would not lead to an improvement beyond the optimal steady state. We prove the general validity of this property of the steady state by noting that in the steady state we seek to minimize $\mathcal{F}(\mathbf{x},u)$ subject to the restrictions

$$f_i(\mathbf{x},u) = 0 \qquad i = 1, 2, \ldots, S \tag{15}$$

The lagrangian is then

$$\mathcal{L} = \mathcal{F}(\mathbf{x},u) + \sum_{i=1}^{S} \lambda_i f_i(\mathbf{x},u) \tag{16}$$

and it is stationary at the solutions of the following equations:

$$\frac{\partial \mathcal{L}}{\partial u} = \frac{\partial \mathcal{F}}{\partial u} + \sum_{i=1}^{S} \lambda_i \frac{\partial f_i}{\partial u} = 0 \tag{17}$$

$$\frac{\partial \mathcal{L}}{\partial x_i} = \frac{\partial \mathcal{F}}{\partial x_i} + \sum_{j=1}^{S} \lambda_j \frac{\partial f_j}{\partial x_i} = 0 \qquad i = 1, 2, \ldots, S \tag{18}$$

Identifying λ_i with $\theta \gamma_i$, Eq. (17) is equivalent to the vanishing of $\partial H / \partial u$ in Eq. (13), while Eqs. (15) and (17) are the steady-state equivalents of Eqs. (1) and (11), whose solutions trivially satisfy the periodicity boundary conditions.

As a computational example of the development of an optimal periodic operating policy using steep descent we shall again consider the reactor example of Horn and Lin introduced in Sec. 6.11. Parallel reactions are carried out in a stirred-tank reactor with material-balance equations

$$\dot{x}_1 = -ux_1{}^n - au^r x_1 - x_1 + 1 \tag{19a}$$
$$\dot{x}_2 = ux_1{}^n - x_2 \tag{19b}$$

x_1 and x_2 are periodic over the interval θ. The temperature-dependent rate coefficient $u(t)$ is to be chosen periodically, subject to constraints

$$u_* \leq u \leq u^* \tag{20}$$

to maximize the time-average conversion of X_2, that is, to minimize

$$\mathcal{E} = -\frac{1}{\theta} \int_0^\theta x_2(t) \, dt \tag{21}$$

Then

$$\mathcal{F} = -\frac{1}{\theta} x_2 \tag{22}$$

The periodic Green's functions satisfy the differential equations

$$\dot{\gamma}_1 = \gamma_1 (nux_1{}^{n-1} + au^r + 1) - n\gamma_2 ux_1{}^{n-1} \tag{23a}$$

$$\dot{\gamma}_2 = \frac{1}{\theta} + \gamma_2 \tag{23b}$$

The hamiltonian is

$$H = -\frac{1}{\theta} x_2 - \gamma_1 (ux_1{}^n + au^r x_1 + x_1 - 1) + \gamma_2 (ux_1{}^n - x_2) \tag{24}$$

with a partial derivative with respect to u

$$\frac{\partial H}{\partial u} = -(\gamma_1 x_1{}^n + a\gamma_1 r u^{r-1} x_1 + \gamma_2 x_1{}^n) \tag{25}$$

It is shown in Sec. 6.1 that when

$$nr - 1 > 0 \qquad r < 1 \tag{26}$$

improvement can be obtained over the best steady-state solution.

For this particular problem some simplification results.   Equation (19a) involves only u and x_1 and can be driven to a periodic solution for periodic $u(t)$ in the manner described above.   Equations (19b), (23a), and (23b) can then be solved in terms of u and x_1 with periodic behavior by quadrature, as follows:

$$x_2(t) = x_2(0)e^{-t} + \int_0^t e^{-(t-\tau)}u(\tau)x_1{}^n(\tau)\, d\tau \tag{27a}$$

$$x_2(0) = \frac{1}{1 - e^{-\theta}} \int_0^\theta e^{-(\theta-\tau)}u(\tau)x_1{}^n(\tau)\, d\tau \tag{27b}$$

$$\gamma_2(t) = \frac{1}{\theta} = \text{const} \tag{28}$$

$$\gamma_1(t) = \gamma_1(0) \exp\left[\int_0^t (nux_1{}^{n-1} + au^r + 1)\, d\sigma\right]$$

$$+ \frac{n}{\theta} \int_0^t \exp\left[\int_t^\tau (nux_1{}^{n-1} + au^r + 1)\, d\sigma\right] u(\tau)x_1{}^n(\tau)\, d\tau \tag{29a}$$

$$\gamma_1(0) = \frac{\dfrac{n}{\theta} \int_0^\theta \exp\left[\int_\theta^\tau (nux_1{}^{n-1} + au^r + 1)\, d\sigma\right] u(\tau)x_1{}^n(\tau)\, d\tau}{1 - \exp\left[\int_0^\theta (nux_1{}^{n-1} + au^r + 1)\, d\sigma\right]} \tag{29b}$$

The adjustment in $u(t)$ is then computed from Eq. (14) using $\partial H/\partial u$ in Eq. (25).

The numerical values used are as follows:

$$\begin{aligned}
n &= 2 \qquad r = 0.75 \\
a &= 1 \qquad \theta = 0.1 \\
u_* &= 1 \qquad u^* = 5
\end{aligned}$$

The optimal steady state, computed from Eq. (15) of Sec. 6.11, is $u = 2.5198$, with corresponding values

$$\begin{aligned}
x_1 &= 0.27144 \\
x_2(&= -\mathcal{E}) = 0.18567 \\
\gamma_1 &= -3.1319
\end{aligned}$$

The starting value of $u(t)$ for the steep-descent calculation was taken as

$$u(t) = 2.5198 + 0.40 \sin 20\pi t \tag{30}$$

The weighting factor $w(t)$ for steep descent was based on the normalized form and taken as

$$w(t) = \frac{G(t)}{\left[\int_0^\theta \left(\dfrac{\partial H}{\partial u}\right)^2 dt\right]^{\frac{1}{2}}} \tag{31}$$

with $G(t)$ equal to 0.1 or the maximum required to reach a bound and halved in case of no improvement in $\mathcal{E}$.

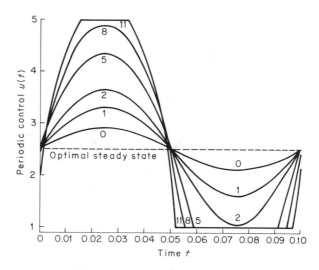

Fig. 10.16 Successive approximations to the optimal periodic control using steep descent.

A response for $x_1(t)$ which was periodic to within 1 part in 1,000 was reached in no more than eight cycles for each iteration, using the value of $x_1(\theta)$ from the previous iteration as the starting value. This error generally corresponded to less than 1 percent of the amplitude of the oscillation. Successive values of the periodic temperature function $u(t)$ are shown in Fig. 10.16 and corresponding values of the objective in Fig. 10.17. The dynamic behavior of x_1 and x_2 is shown in Figs. 10.18

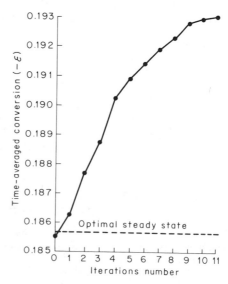

Fig. 10.17 Time-averaged conversion on successive iterations.

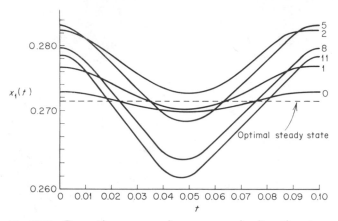

Fig. 10.18 Dynamic response of x_1 on successive iterations to periodic forcing.

and 10.19, respectively. Note that following the first iteration the entire time response of $x_2(t)$ lies above the optimal steady-state value. The course of the iterations is clearly influenced by the sinusoidal starting function and appears to be approaching a bang-bang policy, though this was not obtained for these calculations. Horn and Lin report that the optimum forcing function for this problem is in fact one of infinitely rapid switching between bounds or, equivalently, one for which the length of the period goes to zero. The steep-descent calculation evidently cannot suggest this result for a fixed value of θ.

In realistic applications the decision function will normally not influence the system response directly, as the temperature does in this

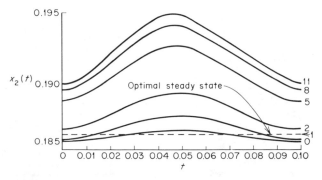

Fig. 10.19 Dynamic response of x_2 on successive iterations to periodic forcing.

example, but will do so through a mechanism with which damping will be associated. Coolant flow-rate variations, for example, must be transmitted to the output conversion through the energy equation for temperature, and high-frequency oscillations will be masked and therefore have the same effect as steady-state operation. An optimal period will therefore exist, and can most easily be found by a one-dimensional search.

10.9 DECOMPOSITION

The computational procedure which we have developed for the solution of variational problems in complex structures has depended upon adjusting the boundary conditions of the Green's function in order to incorporate the mixing boundary conditions. There is an alternative approach that has sometimes been suggested which we can briefly illustrate by means of the examples used in this chapter.

We consider first the reactor problem from Sec. 10.3. The problem is formulated in terms of two variables

$$\dot{x}_1 = -k(u)F(x_1) \tag{1a}$$
$$\dot{x}_2 = -k(u)F(x_1) + g(u) \tag{1b}$$

The boundary conditions are

$$x_1(0) = (1 - \rho)x_f + \rho x_1(\theta) \tag{2a}$$
$$x_2(0) = 0 \tag{2b}$$

with objective

$$\mathcal{E} = x_2(\theta) \tag{3}$$

If we knew the value of the input $x_1(0)$, this would specify the output $x_1(\theta)$. For any fixed input, say x_1^*, we can find the function $u(t)$ which minimizes $\mathcal{E}$ subject to Eqs. (1) and (2b) and

$$x_1(0) = x_1^* \tag{4a}$$
$$x_1(\theta) = \frac{1}{\rho} x_1^* - \frac{1 - \rho}{\rho} x_f \tag{4b}$$

This is a problem for a serial process with constrained output. We call the optimum $\mathcal{E}(x_1^*)$. For some physically realizable value of x_1^*, $\mathcal{E}$ takes on its minimum, and this must then represent the solution to the original recycle problem. The sequence of operations is then:

1. Choose x_1^*. Solve a constrained-output minimization problem to find $\mathcal{E}(x_1^*)$.
2. Search over values of x_1^* to find the minimum of $\mathcal{E}(x_1^*)$ using, for example, the Fibonacci method of Sec. 2.4.

In this case the ultimate computation by either this decomposition method or the use of recycle boundary conditions for the Green's functions is identical, and no saving occurs by use of one method or the other.

A somewhat more revealing example is the plant studied in Sec. 10.7. The reactor equations are

$$0 = x_1^{n-1} - x_1^n - \theta k_{10} e^{-E_1'/u^n}(x_1^n)^2 \qquad n = 1, 2, 3 \tag{5a}$$

$$0 = x_2^{n-1} - x_2^n + \theta k_{10} e^{-E_1'/u^n}(x_1^n)^2 - \theta k_{20} e^{-E_2'/u^n} x_2^n \qquad n = 1, 2, 3 \tag{5b}$$

The extractor equations and the boundary conditions are

$$x_1^I + u^{II}[\psi(x_1^I) - \psi(x_1^{II})] - x_1^3 = 0 \tag{6}$$

$$x_1^{II} + u^{II}\psi(x_1^{II}) - x_1^I = 0 \tag{7}$$

$$x_1^0 = x_{1f} + u^{II}\psi(x_1^I) \tag{8a}$$

$$x_2^0 = 0 \tag{8b}$$

and the objective is

$$\mathcal{E} = -x_2^3 - cx^{II} + \sigma u^{II} \tag{9}$$

The key to the decomposition of this problem is the observation that [Eq. (11a) of Sec. 10.7]

$$x_1^0 = x_{1f} + x_1^3 - x_1^{II} \tag{10}$$

Thus, by setting x_1^3 and x_1^{II} to fixed values x_1^{3*} and x_1^{II*} we completely fix the feed and effluent of X_1 from the reactor system. We have already seen in Sec. 9.13 how to solve the problem of minimizing $-x_2^3$ subject to fixed inputs and outputs of X_1. Call this minimum $-x_2^{3*}(x_1^{3*}, x_1^{II*})$. Furthermore, Eqs. (6) and (7) can be solved for u^{II} in terms of x_1^{3*} and x_1^{II*}. Call this value $u^{II*}(x_1^{3*}, x_1^{II*})$. After carrying out these two operations we then seek the proper values of x_1^3 and x_1^{II} by the operation

$$\min_{\substack{x_1^{3*}, x_1^{II*} \\ x_1^{3*} \geq x_1^{II*}}} [-x_2^{3*}(x_1^{3*}, x_1^{II*}) - cx_1^{II*} + \sigma u^{II*}(x_1^{3*}, x_1^{II*})] \tag{11}$$

The decomposition approach appears to offer no advantage in this latter case. The computation of a single optimum temperature sequence for a constrained output in Sec. 9.13 requires nearly as extensive a set of calculations as the complete analysis in Sec. 10.7. The subminimization would have to be carried out for each pair of values x_1^3, x_1^{II} in the search for the optimum pair using, for example, an approximation to steep descent such as that in Sec. 2.9. The approach of breaking a large minimization problem down into a repeated sequence of smaller problems has been applied usefully in other areas, however, particularly certain types of linear programming problems, and might find application in some problems of the type treated here.

BIBLIOGRAPHICAL NOTES

Section 10.2: The development here was contained in

M. M. Denn and R. Aris: *AIChE J.*, **11**:367 (1965)

—— and ——: *Ind. Eng. Chem. Fundamentals*, **4**:7 (1965)

A proof of a strong minimum principle for continuous systems is in

M. M. Denn and R. Aris: *Chem. Eng. Sci.*, **20**:373 (1965)

A first incorrect attempt to extend results for straight-chain processes to recycle processes was by

D. F. Rudd and E. D. Blum: *Chem. Eng. Sci.*, **17**:277 (1962)

The error was demonstrated by counterexample in

R. Jackson: *Chem. Eng. Sci.*, **18**:215 (1963)

A correct development of the multiplier boundary conditions for recycle was also obtained by

L. T. Fan and C. S. Wang: *Chem. Eng. Sci.*, **19**:86 (1964)

Section 10.3: The example was introduced by

G. S. G. Beveridge and R. S. Schechter: *Ind. Eng. Chem. Fundamentals*, **4**:257 (1965)

Section 10.4: The reformulation to permit use of the straight-chain process result is pointed out in

F. J. M. Horn and R. C. Lin: *Ind. Eng. Chem. Process Design Develop.*, **6**:21 (1967)

Section 10.5: The lagrangian approach was taken by Jackson to obtain results applicable to the more general problem of the following section in

R. Jackson: *Chem. Eng. Sci.*, **19**:19, 253 (1964)

Sections 10.6 and 10.7: The general development and the example follow

M. M. Denn and R. Aris: *Ind. Eng. Chem. Fundamentals*, **4**:248 (1965)

Further examples of the optimization of nonserial processes may be found in

L. T. Fan and C. S. Wang: "The Discrete Maximum Principle," John Wiley & Sons, Inc., New York, 1964

—— and associates: "The Continuous Maximum Principle," John Wiley & Sons, Inc., New York, 1966

J. D. Paynter and S. G. Bankoff: *Can. J. Chem. Eng.*, **44**:340 (1966); **45**:226 (1967)

A particularly interesting use of the Green's functions to investigate the effects on reaction systems of local mixing and global mixing by recycle and bypass is in

F. J. M. Horn and M. J. Tsai: *J. Optimization Theory Appl.*, **1**:131 (1967)

R. Jackson: *J. Optimization Theory Appl.*, **2**:240 (1968)

Section 10.8: This section is based on the paper by Horn and Lin cited above. The paper was presented at a symposium on periodic processes at the 151st meeting of the American Chemical Society in Pittsburgh, March, 1966, and many of the other papers were also published in the same issue of Industrial and Engineering Chemistry Process Design and Development. Another symposium was held at the 60th annual meeting of the AIChE in New York, November, 1967. An enlightening introduction to the notion of periodic process operation is

J. M. Douglas and D. W. T. Rippin: *Chem. Eng. Sci.*, **21**:305 (1966)

A practical demonstration of periodic operation is described in

R. H. Wilhelm, A. Rice, and A. Bendelius: *Ind. Eng. Chem. Fundamentals*, **5**:141 (1966)

Section 10.9: The basic paper on decomposition of recycle processes is

L. G. Mitten and G. L. Nemhauser: *Chem. Eng. Progr.*, **59**:52 (1963)

A formalism is developed in

R. Aris, G. L. Nemhauser, and D. J. Wilde: *AIChE J.*, **10**:913 (1964)

and an extensive discussion of strategies is contained in

D. J. Wilde and C. S. Beightler: "Foundations of Optimization," Prentice-Hall, Inc., Englewood-Cliffs, N.J., 1967

See also

D. F. Rudd and C. C. Watson: "Strategy of Process Engineering," John Wiley & Sons, Inc., New York, 1968

PROBLEMS

10.1. The autocatalytic reaction $X_1 + X_2 \rightleftharpoons 2X_2$ in a tubular reactor is described by the equations

$$\dot{x}_1 = k_{10} \exp\left(\frac{-E_1'}{u}\right) x_1 x_2 - k_{20} \exp\left(\frac{-E_2'}{u}\right) x_2{}^2$$

$$x_1 + x_2 = \text{const}$$

The reaction is to be carried out in a reactor of length L with product recycle such that

$$x_1(0) = (1 - \rho)x_1^* + \rho x_1(L)$$
$$x_2(0) = (1 - \rho)x_2^* + \rho x_2(L)$$

Develop an equation for the maximum conversion to $x_2(L)$ when

$$\frac{E_2}{E_1} = 2 \qquad x_1^* = 1 - \epsilon \qquad x_2^* = \epsilon$$

(The problem is due to Fan and associates.)

10.2. A chemical reaction system in a tubular reactor is described by the differential equations

$$\dot{x}_i = f_i(\mathbf{x})$$
$$x_i(0) = x_{i0}$$

with an objective $\mathcal{O}[\mathbf{x}(\theta)]$. Show that an improvement in $\mathcal{O}$ can be obtained by removing a small stream at a point t_1 and returning it at t_2 only if

$$\sum_i [\gamma_i(t_2) - \gamma_i(t_1)]x_i(t_1) > 0$$

Obtain the equivalent result for recycle of a stream from t_2 to t_1. (The problem is due to Jackson and Horn and Tsai.)

11
Distributed-parameter Systems

11.1 INTRODUCTION

A great many physical processes must be described by functional equations of a more general nature than the ordinary differential and difference equations we have considered thus far. The two final examples in Chap. 3 are illustrative of such situations. The procedures we have developed in Chaps. 6, 7, and 9 for obtaining necessary conditions and computational algorithms, however, based upon the construction of Green's function for the linear variational equations, may be readily extended to these more general processes. We shall illustrate the essentially straightforward nature of the extension in this chapter, where we focus attention on distributed-parameter systems describable by partial differential equations in two independent variables.

11.2 A DIFFUSION PROCESS

As an introduction to the study of distributed-parameter systems we shall consider the generalization of the slab-heating problem of Sec.

3.12, for the results which we shall obtain will be applicable not only to a more realistic version of that problem but to problems in chemical reaction analysis as well. The state of the system is described† by the (generally nonlinear) diffusion-type partial differential equation

$$\frac{\partial x}{\partial t} = f\left(x, \frac{\partial x}{\partial z}, \frac{\partial^2 x}{\partial z^2}, z, t\right) \qquad \begin{array}{l} 0 < z < 1 \\ 0 < t \leq \theta \end{array} \tag{1}$$

with a symmetry condition at $z = 1$

$$\frac{\partial x}{\partial z} = 0 \qquad \text{at } z = 1 \text{ for all } t \tag{2}$$

and a "cooling" condition at $z = 0$,

$$\frac{\partial x}{\partial z} = g(x,v) \qquad \text{at } z = 0 \text{ for all } t \tag{3}$$

where $v(t)$ represents the conditions in the surrounding environment. The initial distribution is given

$$x = \phi(z) \qquad \text{at } t = 0 \tag{4}$$

The environmental condition $v(t)$ may be the decision variable, or it may be related to the decision variable through the first-order ordinary differential equation

$$\tau \frac{dv}{dt} = h(v,u) \tag{5}$$

The former case (*lagless control*) is included by taking the time constant τ to be zero and h as $u - v$. It is assumed that the purpose of control is to minimize some function of the state at time θ, together with the possibility of a cost-of-control term; i.e., we seek an admissible function $u(t)$, $0 \leq t \leq \theta$, in order to minimize

$$\mathcal{E}[u] = \int_0^1 E[x(\theta,z);z]\, dz + \int_0^\theta C[u(t)]\, dt \tag{6}$$

11.3 VARIATIONAL EQUATIONS

We shall construct the equations describing changes in the objective as a result of changes in the decision in the usual manner. We assume that we have available functions $\bar{u}(t)$, $\bar{v}(t)$, and $\bar{x}(t,z)$ which satisfy Eqs. (1) to (5) of the preceding section. If we suppose that in some way we have caused an infinitesimal change $\delta v(t)$ in $v(t)$ over all or part of the range $0 \leq t \leq \theta$, this change is transmitted through the boundary conditions to the function $x(t,z)$ and we can write the variational equations to

---

† Note that the state at any time t^* is represented by an entire *function* of z, $x(t^*,z)$, $0 < z < 1$.

first order as†

$$(\delta x)_t = f_x \, \delta x + f_{x_z}(\delta x)_z + f_{x_{zz}}(\delta x)_{zz} \tag{1}$$
$$(\delta x)_z = g_x \, \delta x + g_v \, \delta v \qquad \text{at } z = 0 \text{ for all } t \tag{2}$$
$$(\delta x)_z = 0 \qquad \text{at } z = 1 \text{ for all } t \tag{3}$$
$$\delta x = 0 \qquad \text{when } t = 0 \text{ for all } z \tag{4}$$

We construct the Green's function $\gamma(t,z)$ for this linear partial differential equation as in Secs. 6.2 and 7.2. We consider first the product $\gamma \, \delta x$ and write

$$
\begin{aligned}
(\gamma \, \delta x)_t &= \gamma_t \, \delta x + \gamma(\delta x)_t \\
&= \gamma_t \, \delta x + \gamma f_x \, \delta x + \gamma f_{x_z}(\delta x)_z + \gamma f_{x_{zz}}(\delta x)_{zz}
\end{aligned}
\tag{5}
$$

or, integrating with respect to z,

$$\frac{\partial}{\partial t} \int_0^1 (\gamma \, \delta x) \, dz = \int_0^1 [\gamma_t \, \delta x + \gamma f_x \, \delta x + \gamma f_{x_z}(\delta x)_z + \gamma f_{x_{zz}}(\delta x)_{zz}] \, dz \tag{6}$$

The last two terms can be integrated by parts

$$\int_0^1 \gamma f_{x_z}(\delta x)_z \, dz = - \int_0^1 (\gamma f_{x_z})_z \, \delta x \, dz + \gamma f_{x_z} \, \delta x \, \Big|_0^1 \tag{7a}$$
$$\int_0^1 \gamma f_{x_{zz}}(\delta x)_{zz} \, dz = \int_0^1 (\gamma f_{x_{zz}})_{zz} \, \delta x \, dz + [\gamma f_{x_{zz}}(\delta x)_z - (\gamma f_{x_{zz}})_z \, \delta x] \, \Big|_0^1 \tag{7b}$$

so that Eq. (6) becomes

$$
\begin{aligned}
\frac{\partial}{\partial t} \int_0^1 (\gamma \, \delta x) \, dz = \int_0^1 [\gamma_t + \gamma f_x - (\gamma f_{x_z})_z + (\gamma f_{x_{zz}})_{zz}] \, \delta x \, dz \\
+ \{ \delta x[\gamma f_{x_z} - (\gamma f_{x_{zz}})_z] + (\delta x)_z \gamma f_{x_{zz}} \} \, \Big|_0^1 \quad (8)
\end{aligned}
$$

Integrating with respect to t from 0 to θ and using Eqs. (2) to (4), we then obtain

$$
\begin{aligned}
\int_0^1 \gamma(\theta,z) \, \delta x(\theta,z) \, dz = \int_0^\theta \int_0^1 [\gamma_t + \gamma f_x - (\gamma f_{x_z})_z \\
+ (\gamma f_{x_{zz}})_{zz}] \, \delta x \, dz \, dt + \int_0^\theta \delta x[\gamma f_{x_z} - (\gamma f_{x_{zz}})_z] \, dt \, \Big|_{z=1} \\
- \int_0^\theta \delta x[\gamma f_{x_z} + \gamma f_{x_{zz}} g_x - (\gamma f_{x_{zz}})_z] \, dt \, \Big|_{z=0} \\
- \int_0^\theta \gamma(t,0) f_{x_{zz}}(t,0) g_v(t) \, \delta v(t) \, dt \quad (9)
\end{aligned}
$$

† In accordance with common practice for partial differential equations and in order to avoid some cumbersome notation we shall use subscripts to denote partial differentiation in Secs. 11.3 to 11.7. Thus,

$$x_t \equiv \frac{\partial x}{\partial t} \qquad x_z \equiv \frac{\partial x}{\partial z} \qquad f_{x_{zz}} \equiv \frac{\partial f}{\partial (\partial^2 x / \partial z^2)}$$

and so forth. Since we are dealing here with a single state variable and a single decision variable, and hence have no other need for subscripts, there should be no confusion.

As before, we shall define the Green's function so as to remove explicit dependence upon δx from the right-hand side of Eq. (9). Thus, we require $\gamma(t,z)$ to satisfy the equation

$$\gamma_t = -\gamma f_x + (\gamma f_{x_z})_z - (\gamma f_{x_{zz}})_{zz} \qquad \begin{array}{l} 0 < z < 1 \\ 0 \leq t < \theta \end{array} \qquad (10)$$

$$\gamma(f_{x_z} + f_{x_z}g_x) - (\gamma f_{x_{zz}})_z = 0 \qquad \text{at } z = 0 \text{ for all } t \qquad (11)$$

$$\gamma f_{x_z} - (\gamma f_{x_{zz}})_z = 0 \qquad \text{at } z = 1 \text{ for all } t \qquad (12)$$

Equation (9) then becomes

$$\int_0^1 \gamma(\theta,z)\, \delta x(\theta,z)\, dz = -\int_0^\theta \gamma(t,0) f_{x_{zz}}(t,0) g_v(t)\, \delta v(t)\, dt \qquad (13)$$

Finally, we note that the variation in the first term of the objective [Eq. (6), Sec. 11.2] as a result of the infinitesimal change $\delta v(t)$ is

$$\int_0^\theta E_x[x(\theta,z);z]\, \delta x(\theta,z)\, dz$$

If we complete the definition of γ with

$$\gamma = E_x \qquad \text{at } t = \theta \text{ for all } z \qquad (14)$$

then we may write

$$\int_0^1 E_x[x(\theta,z);z]\, \delta x(\theta,z)\, dz = -\int_0^\theta \gamma(t,0) f_{x_{zz}}(t,0) g_v(t)\, \delta v(t)\, dt \qquad (15)$$

11.4 THE MINIMUM PRINCIPLE

In order to establish a minimum principle analogous to that obtained in Sec. 6.8 we must consider the effects on the objective of a finite perturbation in the optimal decision function over a very small interval. If $\bar{u}(t)$ is the optimal decision, we choose the particular variation

$$\delta u(t) = 0 \qquad \begin{array}{l} 0 \leq t < t_1 \\ t_1 + \Delta < t \leq \theta \end{array} \qquad (1)$$

$$\delta u(t) = \text{finite} \qquad t_1 \leq t \leq t_1 + \Delta$$

where Δ is infinitesimal. Thus, during the interval $0 \leq t < t_1$ the value of $v(t)$ is unchanged, and $\delta v(t_1) = 0$.

During the interval $t_1 \leq t \leq t_1 + \Delta$ it follows from Eq. (5) of Sec. 11.2 that δv must satisfy the equation

$$\tau(\delta v)_t = h(\bar{v} + \delta v, \bar{u} + \delta u) - h(\bar{v},\bar{u}) \qquad (2)$$

or

$$\delta v(t) = \frac{1}{\tau} \int_{t_1}^t [h(\bar{v} + \delta v, \bar{u} + \delta u) - h(\bar{v},\bar{u})]\, dt \qquad (3)$$

The integral may be expressed to within $o(\Delta)$ as the integrand evaluated at t_1 multiplied by the integration interval. Thus, using the fact that

$\delta v(t_1)$ is zero we may write

$$\delta v(t) = \frac{t - t_1}{\tau} \{h[\bar{v}(t_1),\ \bar{u}(t_1) + \delta u(t_1)] - h[\bar{v}(t_1),\bar{u}(t_1)]\} + o(\Delta)$$

$$t_1 \leq t \leq t_1 + \Delta \quad (4)$$

and, in particular,

$$\delta v(t_1 + \Delta) = \frac{\Delta}{\tau} \{h[\bar{v}(t_1),\ \bar{u}(t_1) + \delta u(t_1)] - h[\bar{v}(t_1),\bar{u}(t_1)]\} + o(\Delta)$$

$$(5)$$

Because Δ is infinitesimal this change in $v(t)$ resulting from the finite change δu is infinitesimal.

Following time $t_1 + \Delta$ the variation δu is again zero. The variational equation describing the infinitesimal change in v is then, to within $o(\Delta)$,

$$\tau(\delta v)_t = h_v(\bar{v},\bar{u})\ \delta v \qquad t_1 + \Delta < t \leq \theta \quad (6)$$

which has the solution, subject to Eq. (5),

$$\delta v(t) = \frac{\Delta}{\tau} [h(\bar{v},u) - h(\bar{v},\bar{u})] \exp \left\{\frac{1}{\tau} \int_{t_1+\Delta}^{t} h_v[\bar{v}(\xi),\bar{u}(\xi)]\ d\xi\right\} + o(\Delta)$$

$$t_1 + \Delta \leq t \leq \theta \quad (7)$$

where the term in brackets preceding the exponential is evaluated at $t = t_1$.

Combining Eq. (7) with Eqs. (6) and (15) of Secs. 11.2 and 11.3, respectively, we obtain the expression for the change in $\mathcal{E}$ resulting from the finite change in u

$$\delta \mathcal{E} = -\frac{\Delta}{\tau} [h(\bar{v},u) - h(\bar{v},\bar{u})] \int_{t_1+\Delta}^{\theta} \gamma(s,0)f_{x_{zz}}(s,0)g_v(s)$$

$$\exp \left[\frac{1}{\tau} \int_{t_1+\Delta}^{s} h_v(\xi)\ d\xi\right] ds + \Delta[C(u) - C(\bar{u})] + o(\Delta) \geq 0 \quad (8)$$

where the inequality follows from the fact that $\bar{u}(t)$ is the function which minimizes $\mathcal{E}$. Dividing by Δ and then taking the limit as $\Delta \to 0$ and noting that t_1 is completely arbitrary, we obtain the inequality

$$C(u) - \frac{h(\bar{v},u)}{\tau} \int_{t}^{\theta} \gamma(s,0)f_{x_{zz}}(s,0)g_v(s)\ \exp \left[\frac{1}{\tau}\int_{t}^{s} h_v(\xi)\ d\xi\right] ds$$

$$\geq C(\bar{u}) - \frac{h(\bar{v},\bar{u})}{\tau} \int_{t}^{\theta} \gamma(s,0)f_{x_{zz}}(s,0)g_v(s)\ \exp \left[\frac{1}{\tau}\int_{t}^{s} h_v(\xi)\ d\xi\right] ds \quad (9)$$

That is, it is necessary that the optimal function $\bar{u}(t)$ minimize

$$\min_{u(t)} C(u) - \frac{h(v,u)}{\tau} \int_{t}^{\theta} \gamma(s,0)f_{x_{zz}}(s,0)g_v(s) \exp \left[\frac{1}{\tau}\int_{t}^{s} h_v(\xi)\ d\xi\right] ds$$

$$(10)$$

everywhere except possibly at a set of discrete points.   If the final time θ is not specified, we obtain a further necessary condition by differentiating Eq. (6) of Sec. 11.2 with respect to θ and equating the result to zero to obtain

$$\theta \text{ unspecified:} \qquad \int_0^1 \gamma(\theta,z) f(\theta,z) \, dz + C[u(\theta)] = 0 \qquad (11)$$

In the important case of the lagless controller, where

$$h(v,u) = u - v$$

and τ goes to zero, we can utilize the fact that for small τ the integrand in Eq. (10) is negligibly small for all time not close to t.   Taking the limit as $\tau \to 0$ then leads to the result

$$u = v: \min_{u(t)} C(u) - u\gamma(t,0) f_{x_{zz}}(t,0) g_v(t) \qquad (12)$$

with v equal to the minimizing value of u.

11.5 LINEAR HEAT CONDUCTION

The linear heat-conduction system examined in Sec. 3.12 with a quadratic objective has the property that γ and x can be obtained explicitly in terms of u.   If we consider

$$x_t = x_{zz} \qquad \begin{matrix} 0 < z < 1 \\ 0 < t \le \theta \end{matrix} \qquad (1)$$

$$x_z = 0 \qquad \text{at } z = 1 \text{ for all } t \qquad (2)$$

$$x_z = \rho(x - v) \qquad \text{at } z = 0 \text{ for all } t \qquad (3)$$

$$x = 0 \qquad \text{at } t = 0 \text{ for all } z \qquad (4)$$

$$\tau v_t = u - v \qquad 0 < t \le \theta \qquad (5)$$

$$\min \mathcal{E} = \int_0^1 [x^*(z) - x(\theta,z)]^2 \, dz + \int_0^\theta C[u(t)] \, dt \qquad (6)$$

where θ is specified, η a constant, and $x^*(z)$ a specified function, Eqs. (10) to (12) and (14) of Sec. 11.3 for γ then become

$$\gamma_t = -\gamma_{zz} \qquad \begin{matrix} 0 < z < 1 \\ 0 \le t < \theta \end{matrix} \qquad (7)$$

$$\rho\gamma - \gamma_z = 0 \qquad \text{at } z = 0 \text{ for all } t \qquad (8)$$

$$\gamma_z = 0 \qquad \text{at } z = 1 \text{ for all } t \qquad (9)$$

$$\gamma = -2[x^*(z) - x(\theta,z)] \qquad \text{at } t = \theta \text{ for all } z \qquad (10)$$

Equations (1) to (5) and (7) to (10) are solved in terms of $u(t)$ by elementary methods such that the minimum principle, Eq. (10) of the preceding section, becomes

$$\min_{u(t)} C[u(t)] + u(t) \left[\int_0^\theta G(t,s) u(s) \, dx - \psi(t) \right] \qquad (11)$$

where, as in Sec. 3.12,

$$\psi(t) = \int_0^1 x^*(z) K(\theta - t, z) \, dz \tag{12}$$

$$G(t,s) = \int_0^1 K(\theta - t, z) K(\theta - s, z) \, dz \tag{13}$$

and

$$K(t,z) = \frac{\alpha^2 \cos \alpha(1 - z)}{\cos \alpha - \alpha/\eta \sin \alpha} e^{-\alpha^2 t}$$

$$+ 2\alpha^2 \sum_{i=1}^{\infty} \frac{\cos (1 - z)\beta_i}{(\alpha^2 - \beta_i^2)[1/\rho + (1 + \rho)/\beta_i^2] \cos \beta_i} e^{-\beta_i^2 t} \tag{14}$$

with $\alpha = 1/\sqrt{\tau}$ and β_i the real roots of

$$\beta \tan \beta = \rho \tag{15}$$

Thus, for example, if $C(u) = \frac{1}{2}c^2 u^2$ and $u(t)$ is unconstrained, the minimum of Eq. (11) is obtained by setting the partial derivative with respect to $u(t)$ to zero. The optimal control is then found as the solution of the Fredholm integral equation of the second kind

$$\int_0^\theta G(t,s) u(s) \, ds + c^2 u(t) = \psi(t) \tag{16}$$

If, on the other hand, $C(u) = 0$ and $u(t)$ is constrained by limits $u_* \leq u \leq u^*$, Eq. (11) is linear in u and the optimal control is at a limit defined by

$$u(t) = \begin{cases} u^* & \int_0^\theta G(t,s) u(s) \, ds - \psi(t) < 0 \\ u_* & \int_0^\theta G(t,s) u(s) \, ds - \psi(t) > 0 \end{cases} \tag{17}$$

An intermediate (singular) solution is possible as the solution of the Fredholm integral equation of the first kind

$$\int_0^\theta G(t,s) u(s) \, ds = \psi(t) \tag{18}$$

Equation (18) is the result obtained as a solution to the simplest problem in the calculus of variations in Sec. 3.12.

11.6 STEEP DESCENT

For determination of the optimum by means of a steep-descent calculation we first choose a trial function $\bar{u}(t)$ and then seek a small change $\delta u(t)$ over all t which leads to an improved value of the objective $\mathcal{E}$. We need, then, the effect on $\mathcal{E}$ of a continuous small variation in u, rather than the finite sharply localized variation used in the development of the minimum principle.

From Eqs. (6) and (15) of Secs. 11.2 and 11.3, respectively, we have, for any small change δu, the first-order expression for the corresponding change in $\mathcal{E}$,

$$\delta\mathcal{E} = \int_0^\theta C_u(t)\, \delta u(t)\, dt - \int_0^\theta \gamma(t,0)f_{x_{zz}}(t,0)g_v(t)\, \delta v(t)\, dt \tag{1}$$

Here, δv can be related to a continuous small change δu by the variational equation derived from Eq. (4) of Sec. 11.2

$$\tau\, \delta v_t = h_v\, \delta v + h_u\, \delta u \tag{2}$$

or

$$\delta v(t) = \frac{1}{\tau} \int_0^t h_u(s) \exp\left[\frac{1}{\tau} \int_s^t h_v(\xi)\, d\xi\right] \delta u(s)\, ds \tag{3}$$

Thus, by substituting for δv in Eq. (1) and changing the order of integration we obtain the expression for the variation in the objective

$$\delta\mathcal{E} = \int_0^\theta \left\{ C_u \right.$$
$$\left. - \frac{1}{\tau} h_u \int_t^\theta \gamma(s,0)f_{x_{zz}}(s,0)g_v(s) \exp\left[\frac{1}{\tau} \int_t^s h_v(\xi)\, d\xi\right] \right\} \delta u(t)\, dt \tag{4}$$

The choice of $\delta u(t)$ which leads to a decrease in $\mathcal{E}$ is then

$$\delta u(t) = -w(t) \left\{ C_u \right.$$
$$\left. - \frac{1}{\tau} h_u \int_t^\theta \gamma(s,0)f_{x_{zz}}(s,0)g_v(s) \exp\left[\frac{1}{\tau} \int_t^s h_v(\xi)\, d\xi\right] ds \right\} \tag{5}$$

where $w(t) \geq 0$ reflects the geometry of the space and is chosen sufficiently small to ensure that a constraint is not violated. For $\tau = 0$, with u equal to v, we obtain

$$\delta u(t) = -w(t)[C_u - \gamma(t,0)f_{x_{zz}}(t,0)g_v(t)] \tag{6}$$

11.7 COMPUTATION FOR LINEAR HEAT CONDUCTION

As a first application of the computational procedure to a distributed system we return to the linear heat-conduction problem described in Sec. 11.5 without the cost-of-control term. The system equations are

$$x_t = x_{zz} \qquad \begin{array}{c} 0 < z < 1 \\ 0 < t \leq \theta \end{array} \tag{1}$$

$$x_z = 0 \qquad \text{at } z = 1 \text{ for all } t \tag{2}$$

$$x_z = \rho(x - v) \qquad \text{at } z = 0 \text{ for all } t \tag{3}$$

$$x = 0 \qquad \text{at } t = 0 \text{ for all } z \tag{4}$$

$$\tau v_t = u - v \qquad 0 < t \leq \theta \tag{5}$$

$$u_* \leq u \leq u^* \tag{6}$$
$$E = \frac{1}{2}[x^*(z) - x(\theta,z)]^2 \tag{7}$$
$$C = 0 \tag{8}$$
$$\gamma_t = -\gamma_{zz} \qquad \begin{matrix} 0 < z < 1 \\ 0 \leq t < \theta \end{matrix} \tag{9}$$
$$\rho\gamma - \gamma_t = 0 \qquad \text{at } z = 0 \text{ for all } t \tag{10}$$
$$\gamma_z = 0 \qquad \text{at } z = 1 \text{ for all } t \tag{11}$$
$$\gamma = -2[x^*(z) - x(\theta,z)] \qquad \text{at } t = \theta \text{ for all } z \tag{12}$$

The necessary conditions for an optimum are given in Sec. 11.5 by Eqs. (17) and (18), while the direction of steep descent is determined by

$$\delta u(t) = -w(t)\frac{\rho}{\tau}\int_t^\theta \gamma(s,0)e^{(t-s)/\tau}\,ds \tag{13}$$

The linearity of the system makes it possible to obtain and use an analytical expression for $\gamma(s,0)$.

The parameters used in the numerical solution are

$$\rho = 10 \qquad \tau = 0.04$$
$$x^*(z) = \text{constant} = 0.2$$
$$u_* = 0 \leq u \leq 1 = u^*$$

Two cases are considered, $\theta = 0.2$ and $\theta = 0.4$, with the starting functions $u(t)$ taken as constant values of $u(t) = 0, 0.5,$ and 1.0 in each case. Numerical integration was carried out using Simpson's rule, with both z and t coordinates divided into 50 increments. $w(t)$ was arbitrarily set as 10, or the maximum less than 10 which would carry u to a boundary of the admissible region, and halved whenever a decrease in the objective was not obtained.

Figure 11.1 shows several of the successive control policies calculated from Eq. (13) for $\theta = 0.2$ with a constant starting policy of $u = 0.5$, no improvement being found after 10 iterations. The policy is seen to approach a bang-bang controller, with the exception of a small time increment at the end, and the final policies from the other two starting values were essentially the same. The successive values of the objective are shown in Fig. 11.2, with curves I, II, and III corresponding, respectively, to starting values $u(t) = 0, 0.5,$ and 1. The three temperature profiles $x(\theta,z)$ at $t = \theta$ corresponding to the starting control policies, together with the final profile, are shown in Fig. 11.3, where it can be observed from curve III that no control policy bounded by unity can raise the temperature at $z = 1$ to a value of 0.2 in a time period $\theta = 0.2$.

Figure 11.4 shows several of the iterations for the optimal control with $\theta = 0.4$, starting with the constant policy $u(t) = 0.5$, with no improvement possible after 11 descents. The curve for the second itera-

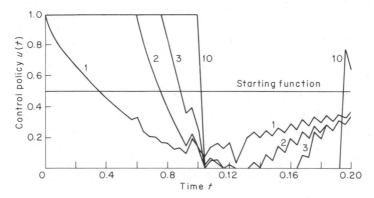

Fig. 11.1 Successive approximations to the optimal control policy using steep descent, $\theta = 0.2$. [*From M. M. Denn, Intern. J. Control,* **4**:167 (1966). *Copyright* 1966 *by Taylor and Francis, Ltd. Reprinted by permission of the copyright owner.*]

tion is not shown beyond $t = 0.25$ because it differs in general by too little from the optimal curve to be visible on this scale. The sharp oscillations in the optimal controller at alternate points in the spatial grid were felt to have no physical basis, and a crude smoothed approximation to the optimum was then used as a starting function for the steep-descent program. This converged to the dashed curve shown in Fig. 11.4, which has very nearly the same value of the objective as the oscillating result.

Figure 11.5 shows the reduction in the objective for steep-descent

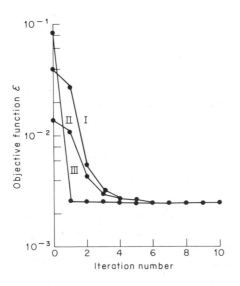

Fig. 11.2 Reduction in the objective on successive iterations, $\theta = 0.2$. [*From M. M. Denn, Intern. J. Control,* **4**:167 (1966). *Copyright* 1966 *by Taylor and Francis, Ltd. Reprinted by permission of the copyright owner.*]

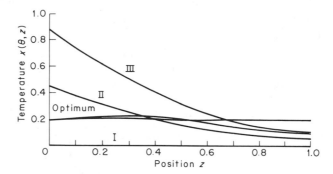

Fig. 11.3 Temperature profiles at $t = \theta$ using initial and
optimal control policies, $\theta = 0.2$. [*From M. M. Denn,
Intern. J. Control*, **4**:167 (1966). *Copyright 1966 by Taylor
and Francis, Ltd. Reprinted by permission of the copyright
owner.*]

calculations starting from constant control policies of $u(t) = 0, 0.5$, and 1,
corresponding to curves I, II, and III, respectively, with the temperature
profile at $t = \theta$ for each of these starting policies shown in Fig. 11.6.   The
optimum is indistinguishable from 0.2 on the scale used.   The three
unsmoothed policies are shown in Fig. 11.7.   They indicate substantial
latitude in choosing the control, but with the exception of the oscillations
in the later stages, which can be smoothed by the same computer program,
all are smooth, intermediate policies, corresponding to approximations
to the singular solution defined by the integral equation (18) of Sec. 11.5.

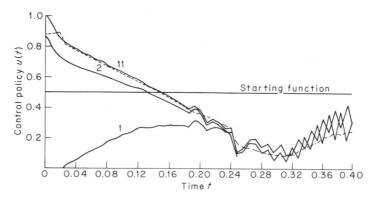

Fig. 11.4 Successive approximations to the optimal control policy
using steep descent, $\theta = 0.4$. [*From M. M. Denn, Intern. J. Control*,
4:167 (1966). *Copyright 1966 by Taylor and Francis, Ltd. Reprinted by permission of the copyright owner.*]

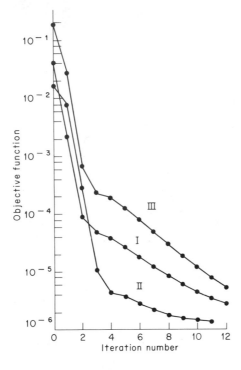

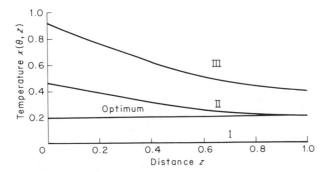

Fig. 11.5 Reduction in the objective on successive iterations, $\theta = 0.4$. [*From M. M. Denn, Intern. J. Control,* **4**:167 (1966). *Copyright* 1966 *by Taylor and Francis, Ltd. Reprinted by permission of the copyright owner.*]

It is interesting to note the change in the optimal control strategy from bang-bang to intermediate as more time is available. It has sometimes been suggested in the control literature that steep descent cannot be used to obtain singular solutions to variational problems, but it is clear from this example that such is not the case.

Fig. 11.6 Temperature profiles at $t = \theta$ using initial and optimal control policies, $\theta = 0.4$. [*From M. M. Denn, Intern. J. Control,* **4**:167 (1966). *Copyright* 1966 *by Taylor and Francis, Ltd. Reprinted by permission of the copyright owner.*]

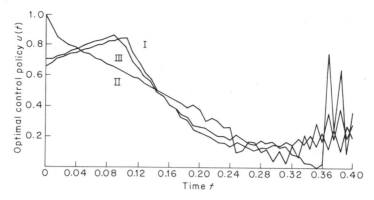

Fig. 11.7   Unsmoothed intermediate optimal control policies for three starting functions, $\theta = 0.4$. [*From M. M. Denn, Intern. J. Control,* **4:**167 (1966). *Copyright 1966 by Taylor and Francis, Ltd. Reprinted by permission of the copyright owner.*]

11.8   CHEMICAL REACTION WITH RADIAL DIFFUSION

We have considered the optimal operation of a batch or pipeline chemical reactor in Sec. 3.5 with respect to the choice of operating temperature and in Sec. 5.11 with respect to the optimal heat flux.   In any realistic consideration of a packed tubular reactor it will generally be necessary to take radial variations of temperature and concentration into account, requiring even for the steady state a description in terms of two independent variables.   The study of optimal heat-removal rates in such a reactor provides an interesting extension of the results of the previous sections and illustrates as well how an optimization study can be used in arriving at a practical engineering design.

In describing the two-dimensional reactor it is convenient to define $x(t,z)$ as the degree of conversion and $y(t,z)$ as the true temperature divided by the feed temperature.   The coordinates t and z, the axial residence time (position/velocity) and radial coordinate, respectively, are normalized to vary from zero to unity.   With suitable physical approximations the system is then described by the equations

$$\frac{\partial x}{\partial t} = \frac{\Lambda}{Pe}\left[\frac{1}{z}\frac{\partial}{\partial z}\left(z\frac{\partial x}{\partial z}\right)\right] + Da_{\mathrm{I}}r(x,y) \qquad \begin{matrix} 0 < t < 1 \\ 0 < z < 1 \end{matrix} \qquad (1)$$

$$\frac{\partial y}{\partial t} = \frac{\Lambda}{Pe'}\left[\frac{1}{z}\frac{\partial}{\partial z}\left(z\frac{\partial y}{\partial z}\right)\right] + Da_{\mathrm{III}}r(x,y) \qquad \begin{matrix} 0 < t < 1 \\ 0 < z < 1 \end{matrix} \qquad (2)$$

Here Λ is the ratio of length to radius, Pe and Pe' the radial Peclet numbers for mass and heat transfer, respectively, Da_{I} and Da_{III} the first and third Damkohler numbers, and $r(x,y)$ the dimensionless reaction rate.

The boundary conditions are

$$x = 0 \qquad \text{at } t = 0 \text{ for all } z \tag{3a}$$
$$y = 1 \qquad \text{at } t = 0 \text{ for all } z \tag{3b}$$

$$\frac{\partial x}{\partial z} = 0 \qquad \text{at } z = 0 \text{ for all } t \tag{3c}$$

$$\frac{\partial y}{\partial z} = 0 \qquad \text{at } z = 0 \text{ for all } t \tag{3d}$$

$$\frac{\partial x}{\partial z} = 0 \qquad \text{at } z = 1 \text{ for all } t \tag{3e}$$

$$\frac{\partial y}{\partial z} = -u(t) \qquad \text{at } z = 1 \text{ for all } t \tag{3f}$$

The dimensionless wall heat flux $u(t)$ is to be chosen subject to bounds

$$u_* \leq u(t) \leq u^* \tag{4}$$

in order to maximize the total conversion in the effluent

$$\bar{x}(1) = 2 \int_0^1 zx(1,z)\, dz \tag{5}$$

The generalization of the analysis of Secs. 11.2 to 11.4 and 11.6 to include two dependent variables is quite straightforward, and we shall simply use the results. We shall denote partial derivatives explicitly here and use subscripts only to distinguish between components γ_1 and γ_2 of the Green's vector. The multiplier equations corresponding to Eq. (10) of Sec. 11.3 are

$$\frac{\partial \gamma_1}{\partial t} = -(Da_I \gamma_I + Da_{III} \gamma_2)\frac{\partial r}{\partial x} + \frac{\Lambda}{Pe}\left(\frac{\partial}{\partial z}\frac{\gamma_1}{z} - \frac{\partial^2 \gamma_1}{\partial z^2}\right) \tag{6}$$

$$\frac{\partial \gamma_2}{\partial t} = -(Da_I \gamma_1 + Da_{III} \gamma_2)\frac{\partial r}{\partial y} + \frac{\Lambda}{Pe'}\left(\frac{\partial}{\partial z}\frac{\gamma_2}{z} - \frac{\partial^2 \gamma_2}{\partial z^2}\right) \tag{7}$$

while the boundary conditions corresponding to Eqs. (11), (12), and (14) are

$$\gamma_1 = -2z \qquad \text{at } t = 1 \text{ for all } z \tag{8a}$$
$$\gamma_2 = 0 \qquad \text{at } t = 1 \text{ for all } z \tag{8b}$$
$$\gamma_1 = \gamma_2 = 0 \qquad \text{at } z = 0 \text{ for all } t \tag{8c}$$

$$\gamma_1 - \frac{\partial \gamma_1}{\partial z} = \gamma_2 - \frac{\partial \gamma_2}{\partial z} = 0 \qquad \text{at } z = 1 \text{ for all } t \tag{8d}$$

The minimum principle corresponding to Eq. (12) of Sec. 11.4 is

$$\min_{u(t)} -\frac{\Lambda}{Pe'}\gamma_2(t,1)u(t) \tag{9}$$

The linearity of the minimum criterion requires that the optimal cooling function have the form

$$u(t) = \begin{cases} u^* & \gamma_2(t,1) > 0 \\ u_* & \gamma_2(t,1) < 0 \end{cases} \tag{10}$$

with intermediate cooling possible only if $\gamma_2(t,1)$ vanishes for some finite interval. This is the structure found for the optimal operation of the reactor without radial effects in Sec. 5.11. Indeed, the similarity can be extended further by noting that if Eq. (8d) is taken to imply that γ_2 is approximately proportional to z near $z = 1$, then the second parenthesis on the right side of Eq. (7) must be close to zero near $z = 1$. Thus $\gamma_2(t,1)$ can vanish for a finite interval only if the reaction rate near the wall is close to a maximum; i.e., the singular solution implies

$$\frac{\partial r}{\partial y} \approx 0 \qquad z = 1 \tag{11}$$

which must be identically true without radial effects. If the parameters are chosen such that $u_* = 0$ and $\gamma_2(t,1)$ is initially zero, then since when $u = 0$ initially there are no radial effects, the switch to intermediate cooling should be close to the switch for the one-dimensional reactor.

Computations for this system were carried out using steep descent, the direction corresponding to Eq. (6) of Sec. 11.6 being

$$\delta u(t) = w(t)\gamma_2(t,1) \tag{12}$$

For the computations the reaction was taken to be first order, in which case the dimensionless reaction rate has the form

$$r(x,y) = e^{E_1'/T_0}[(1 - x)e^{-E_1'/yT_0} - kxe^{-E_2'/yT_0}] \tag{13}$$

where T_0 is the temperature of the feed. The physical parameters were chosen as follows:

$$
\begin{aligned}
&Pe = 110 && Pe' = 84.5 \\
&E_1' = 12{,}000 && E_2' = 25{,}000 \\
&\Lambda = 50 && k = 3.585 \times 10^9 \\
&Da_{\mathrm{I}} = 0.25 && Da_{\mathrm{III}} = 0.50 \\
&T_0 = 620 && u_* = 0
\end{aligned}
$$

u^* was not needed for the parameters used here. The solution for the case without radial effects, following the analysis of Sec. 5.11, gives an initially adiabatic policy ($u = 0$) with a switch to an intermediate policy at $t = 0.095$.

Solutions of the nonlinear partial differential equations were obtained using an implicit Crank-Nicholson procedure with various grid sizes. Figure 11.8 shows successive values of the heat flux starting from

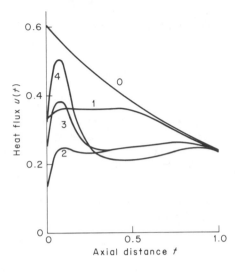

Fig. 11.8 Successive approximations to the optimal heat flux using steep descent. [*From M. M. Denn, R. D. Gray, Jr., and J. R. Ferron, Ind. Eng. Chem. Fundamentals,* **5**:59 (1966). *Copyright 1966 by the American Chemical Society. Reprinted by permission of the copyright owner.*]

a nominal curve arbitrarily chosen so that a reactor without diffusion would operate at a dimensionless temperature y of 1.076. $w(t)$ was taken initially as $0.1/\int_0^1 [\gamma_2(t,1)]^2 \, dt$ and halved when no improvement was obtained. For these computations a coarse integration grid was used initially, with a fine grid for later computations. No improvement could be obtained after the fourth iteration, and the technique was evidently unable to approach a discontinuous profile, although the sharp peak at approximately $t = 0.075$ is suggestive of the possibility of such a solution. The successive improvements in the conversion are shown by curve I in Fig. 11.9.

The next series of calculations was carried out by assuming that the optimal policy would indeed be discontinuous and fixing the point at which a switch occurs from adiabatic to intermediate operation. The steep-descent calculations were then carried out for the intermediate section only. Figures 11.10 and 11.11 show such calculations for the switch at $t = 0.095$, the location in the optimal reactor without radial variations. The starting policy in Fig. 11.10 is such that the section beyond $t = 0.095$ in a reactor without radial effects would remain at a constant value of y of 1.076 [this is the value of y at $t = 0.095$ for $u(t) = 0$, $t < 0.095$]. The values of the objective are shown in Fig. 11.9 as curve II, with no improvement after two iterations. The starting policy in Fig. 11.11 is the optimum for a reactor without radial effects, with successive values of the conversion shown as curve III in Fig. 11.9. It is significant that an optimal design which neglects radial effects when they should be included can result in a conversion far from the best possible. The maximum conversion was found to occur for a switch to intermediate

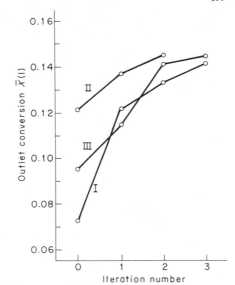

Fig. 11.9 Improvement in outlet conversion on successive iterations. [*From M. M. Denn, R. D. Gray, Jr., and J. R. Ferron, Ind. Eng. Chem. Fundamentals,* **5:**59 (1966). *Copyright* 1966 *by the American Chemical Society. Reprinted by permission of the copyright owner.*]

operation at $t = 0.086$, but the improvement beyond the values shown here for a switch at $t = 0.095$ is within the probable error of the computational scheme.

One interesting feature of these calculations was the discovery that further small improvements could be obtained following the convergence of the steep-descent algorithm by seeking changes in $u(t)$ in the direction opposite that predicted by Eq. (12). This indicates that the cumulative

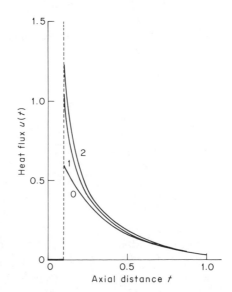

Fig. 11.10 Successive approximations to the optimal heat flux using steep descent with switch point specified at $t = 0.095$. [*From M. M. Denn, R. D. Gray, Jr., and J. R. Ferron, Ind. Eng. Chem. Fundamentals,* **5:**59 (1966). *Copyright* 1966 *by the American Chemical Society. Reprinted by permission of the copyright owner.*]

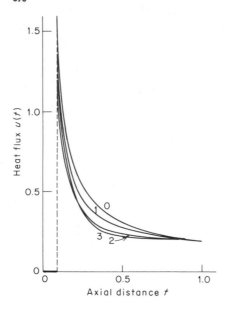

Fig. 11.11 Successive approximations to the optimal heat flux using steep descent with switch point specified at $t = 0.095$. [*From M. M. Denn, R. D. Gray, Jr., and J. R. Ferron, Ind. Eng. Chem. Fundamentals,* **5**:59 (1966). *Copyright* 1966 *by the American Chemical Society. Reprinted by permission of the copyright owner.*]

error associated with solving Eqs. (1) and (2) followed by Eqs. (6) and (7) was sufficiently large to give an incorrect sign to the small quantity $\gamma_2(t,1)$ in the region of the optimum. Such numerical difficulties must be expected in the neighborhood of the optimum in the solution of variational problems for nonlinear distributed systems, and precise results will not be obtainable.

From an engineering point of view the results obtained thus far represent a goal to be sought by a practical heat-exchanger design. The coolant temperature required to produce the calculated heat-removal rate $u(t)$ can be determined by the relation

$$\frac{\partial y(t,1)}{\partial z} = -u(t) = \eta[y_c(t) - y(t,1)] \tag{14}$$

where y_c is the reduced coolant temperature and η a dimensionless overall heat-transfer coefficient times surface area. Using a value of $\eta = 10$, the functions $y_c(t)$ were calculated for the values of $u(t)$ obtained from the steep-descent calculation. The variation of y_c as a function of t was generally found to be small, and a final series of calculations was carried out to find the best *constant* value of y_c, since an isothermal coolant is easily obtained in practice. Here the best switch from adiabatic operation was found at $t = 0.075$ and $y_c = 0.927$, but the results were essentially independent of the switch point in the range studied. For a switch at $t = 0.095$, corresponding to the calculations discussed here, the best value of y_c was found to be 0.928. The corresponding function $u(t)$, calculated from Eq. (14), is plotted as the dashed line in Fig. 11.12, together with the results from the steep-descent calculations. The correspondence with the results from the rigorous optimization procedure is striking, indicating for this case the true optimality of the conventional heat-exchanger design, and, in fact, as a consequence of the numerical difficulties in the more complicated steep-descent procedure, the conversion for the best constant coolant is slightly above that obtained from the steep-descent computation.

This example is a useful demonstration of several of the practical engineering aspects of the use of optimization theory. An optimal design based on an incomplete physical model, such as neglect of important radial effects, might give extremely poor results in practice. A careful optimization study, however, can be used to justify a practical engineering design by providing a theoretical guideline for required performance. Finally, in complicated systems the application of the theory is limited by the sophistication and accuracy of the numerical analysis, and all results in such situations must be taken as approximate.

11.9 LINEAR FEEDFORWARD–FEEDBACK CONTROL

The two examples of optimization problems in distributed-parameter systems which we have examined thus far have both been situations in which the decision function has appeared in the boundary conditions. In a number of important applications the decision function enters the partial differential equation directly. The simplest such situation, which

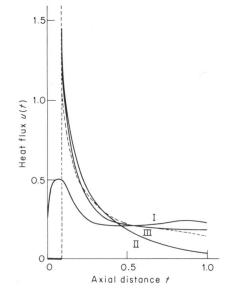

Fig. 11.12 Heat-flux profiles obtained using steep descent from three starting functions. The dashed line corresponds to an isothermal coolant. [*After M. M. Denn, R. D. Gray, Jr., and J. R. Ferron, Ind. Eng. Chem. Fundamentals,* **5**:59 (1966). *Copyright 1966 by the American Chemical Society. Reprinted by permission of the copyright owner.*]

includes as a special case the regulation of outflow temperature in a cross-flow heat exchanger, would be described by the single linear hyperbolic equation

$$\frac{\partial x}{\partial t} + V \frac{\partial x}{\partial z} = Ax + Bu \qquad \begin{matrix} 0 < t \leq \theta \\ 0 < z \leq 1 \end{matrix} \tag{1}$$

where the coefficients V, A, and B are constants and u is a function only of t. We shall restrict attention to this case, although the generalization to several equations and spatially varying coefficients and control functions, as well as to higher-order spatial differential operators, is direct.

The system is presumed to be in some known state at time $t = 0$. We shall suppose that disturbances enter with the feed stream at $z = 0$, so that the boundary condition has the form

$$x(t,0) = d(t) \tag{2}$$

and that the object of control $u(t)$ is to maintain a weighted position average of x^2 as small as possible, together with a cost-of-control term. That is, we seek the function $u(t)$ which minimizes the cumulative error

$$\mathcal{E} = \frac{1}{2} \int_0^\theta \left[\int_0^1 C(z)x^2(t,z) \, dz + u^2(t) \right] dt \tag{3}$$

This formulation includes the special case in which we wish only to regulate x at $z = 1$, for the special choice $C(z) = \bar{C}\delta(1 - z)$, where $\delta(\xi)$ is the Dirac delta, leads to

$$\mathcal{E} = \frac{1}{2} \int_0^\theta [\bar{C}x^2(t,1) + u^2(t) \, dt] \tag{4}$$

The necessary conditions for optimality are obtained in the usual manner, by constructing the Green's function for the variational equations and examining the effect of a finite change in $u(t)$ over an infinitesimal interval. The details, which are sufficiently familiar to be left as a problem, lead to a result expressible in terms of a hamiltonian,

$$H = \frac{1}{2}Cx^2 + \frac{1}{2}u^2 + \gamma(Ax + Bu) \tag{5}$$

as

$$\frac{\partial x}{\partial t} + V \frac{\partial x}{\partial z} = \frac{\partial H}{\partial \gamma} = Ax + Bu \tag{6}$$

$$\frac{\partial \gamma}{\partial t} + V \frac{\partial \gamma}{\partial z} = -\frac{\partial H}{\partial x} = -Cx - A\gamma \tag{7}$$

$$\gamma(\theta,z) = \gamma(t,1) = 0 \tag{8}$$

$$\min_{u(t)} \int_0^1 H \, dz \tag{9}$$

The minimum condition in turn implies

$$u(t) = -B \int_0^1 \gamma(t,\xi) \, d\xi \tag{10}$$

with a minimum ensured by requiring $C(z) \geq 0$.

The system of Eqs. (6), (7), and (10) are analogous to the lumped-parameter system studied in Sec. 8.2, where we found that the Green's functions could be expressed as a linear combination of the state and disturbance variables, provided the disturbances could be approximated as constant for intervals long with respect to the system response time. We shall make the same assumption here concerning $d(t)$, in which case, interpreting x and γ at each value of z as corresponding to one component of the vectors $\mathbf{x}$ and $\boldsymbol{\gamma}$ in the lumped system, the relation analogous to Eq. (7) of Sec. 8.2 is

$$\gamma(t,z) = \int_0^1 M(t,z,\xi) x(t,\xi) \, d\xi + D(t,z) \, d(t) \tag{11}$$

The boundary conditions of Eq. (8) require

$$M(t,1,\xi) = M(\theta,z,\xi) = D(t,1) = D(\theta,z) = 0 \tag{12}$$

It follows from the results for the lumped system (and will be established independently below) that M is symmetric in its spatial arguments

$$M(t,z,\xi) = M(t,\xi,z) \tag{13}$$

Substitution of Eq. (11) into the partial differential equation (7) yields

$$\frac{\partial \gamma}{\partial t} + V \frac{\partial \gamma}{\partial z} = -C(z) x(z) - A \int_0^1 M(z,\xi) x(\xi) \, d\xi - A D(z) d \tag{14a}$$

Here and henceforth we shall suppress any explicit dependence on t, but it will frequently be necessary to denote the spatial dependence. It is convenient to rewrite Eq. (14a) in the equivalent form

$$\frac{\partial \gamma}{\partial t} + V \frac{\partial \gamma}{\partial z} = -\int_0^1 [C(\xi)\delta(z - \xi) + A M(z,\xi)] x(\xi) \, d\xi - A D(z) d \tag{14b}$$

On the other hand, Eq. (11) requires that

$$\frac{\partial \gamma}{\partial t} + V \frac{\partial \gamma}{\partial z} = \int_0^1 \frac{\partial M(z,\xi)}{\partial t} x(\xi) \, d\xi + \int_0^1 M(z,\xi) \frac{\partial x(\xi)}{\partial t} \, d\xi$$
$$+ \frac{\partial D}{\partial t} d + \int_0^1 V \frac{\partial}{\partial z} M(z,\xi) x(\xi) \, d\xi + V \frac{\partial D}{\partial z} d \tag{15}$$

which becomes, following an integration by parts and substitution of

Eqs. (2), (6), (7), (11), and (13),

$$\frac{\partial \gamma}{\partial t} + V \frac{\partial \gamma}{\partial z} = \int_0^1 \left\{ \frac{\partial M(z,\xi)}{\partial t} + V \frac{\partial M(z,\xi)}{\partial z} + V \frac{\partial M(z,\xi)}{\partial \xi} + A M(z,\xi) \right.$$

$$\left. - B^2 \left[\int_0^1 M(z,\sigma) \, d\sigma \right] \left[\int_0^1 M(\varsigma,\xi) \, d\varsigma \right] \right\} x(\xi) \, d\xi + \left\{ \frac{\partial D}{\partial t} + V \frac{\partial D}{\partial z} \right.$$

$$\left. - B^2 \left[\int_0^1 M(z,\sigma) \, d\sigma \right] \left[\int_0^1 D(\xi) \, d\xi \right] + V M(z,0) \right\} d \quad (16)$$

Equations (14) and (16) represent the same quantity, and thus the coefficients of x and d must be identical. We therefore obtain the two equations

$$\frac{\partial M}{\partial t} + V \left(\frac{\partial M}{\partial z} + \frac{\partial M}{\partial \xi} \right) + 2AM$$

$$- B^2 \left[\int_0^1 M(z,\sigma) \, d\sigma \right] \left[\int_0^1 M(\varsigma,\xi) \, d\varsigma \right] + C(\xi)\delta(z - \xi) = 0 \quad (17)$$

$$\frac{\partial D}{\partial t} + V \frac{\partial D}{\partial z} + AD - B^2 \left[\int_0^1 M(z,\sigma) \, d\sigma \right] \left[\int_0^1 D(\xi) \, d\xi \right]$$

$$+ V M(z,0) = 0 \quad (18)$$

Any solution of Eq. (17) clearly satisfies the symmetry condition, Eq. (13). In the special case $V = 0$ Eq. (17) reduces with two integrations to the usual Riccati equation for lumped systems, Eq. (10) of Sec. 8.2. For the industrially important case of $\theta \to \infty$ Eqs. (17) and (18), as in the lumped-parameter analogs, tend to solutions which are independent of time. We shall consider regulation only of the exit stream, so that $C(\xi) = \bar{C}\delta(1 - \xi)$, and we let $\theta \to \infty$. The optimal control is computed from Eqs. (10) and (11) as

$$u(t) = \int_0^1 G^{\mathrm{FB}}(z) x(t,z) \, dz + G^{\mathrm{FF}} d(t) \quad (19)$$

where the feedback (FB) and feedforward (FF) gains are written in terms of the solutions of Eqs. (17) and (18) as

$$G^{\mathrm{FB}}(z) = -B \int_0^1 M(z,\xi) \, d\xi \quad (20)$$

$$G^{\mathrm{FF}} = -B \int_0^1 D(\xi) \, d\xi \quad (21)$$

Because of the term $\delta(1 - \xi)$ the steady-state solution to Eq. (17) is discontinuous at the values $r = 1$, $\xi = 1$. By using the method of characteristics a solution valid everywhere except at these points can be obtained in the implicit form

$$M(z,\xi) = -\frac{1}{V} \int_z^{1-(\xi-z)S(\xi-z)} e^{-(2A/V)(z-\eta)} G^{\mathrm{FB}}(\eta) G^{\mathrm{FB}}(\eta - z + \xi) \, d\eta$$

$$+ \frac{\bar{C}}{V} e^{(2A/V)(1-\xi)} \delta(\xi - z) \quad (22)$$

Here the Heaviside step function is defined as

$$S(\xi - z) = \begin{cases} 0 & \xi < z \\ 1 & \xi \geq z \end{cases} \tag{23}$$

Integration of Eq. (22) then leads to an equation for the feedback gain

$$G^{\mathrm{FB}}(z) = -\frac{\bar{C}B}{V} e^{(2A/V)(1-z)}$$
$$+ B \int_z^1 G^{\mathrm{FB}}(\eta) e^{-(2A/V)(z-\eta)} \, d\eta \int_{\eta-z}^1 G^{\mathrm{FB}}(\xi) \, d\xi \tag{24}$$

Equation (24) is in a form which is amenable to iterative solution for the optimal gain. Equation (18) in the steady state is simply a linear first-order equation which can be solved explicitly for the feedforward gain

$$G^{\mathrm{FF}} = \frac{B \int_0^1 dz \int_z^1 e^{-(A/V)(z-\xi)} \, d\xi \int_\xi^1 e^{-(2A/V)(\xi-\eta)} G^{\mathrm{FB}}(\eta) G^{\mathrm{FB}}(\eta - \xi) \, d\eta}{V - B \int_0^1 dz \int_z^1 e^{-(A/V)(z-\xi)} G^{\mathrm{FB}}(\xi) \, d\xi} \tag{25}$$

This result can be easily generalized to include nonconstant disturbances which are random with known statistical properties, in which case the feedforward gain G^{FF} depends upon the statistics of the disturbance. In particular, if $d(t)$ is uncorrelated ("white") noise, G^{FF} is zero and the optimal control is entirely feedback. The feedback section of the optimal control requires the value of x at every position z, which is clearly impossible. The practical design problem then becomes one of locating a minimum number of sensing elements in order to obtain an adequate approximation to the integral in Eq. (19) for a wide range of disturbance states $x(t, \xi)$.

11.10 OPTIMAL FEED DISTRIBUTION IN PARAMETRIC PUMPING

Several separation processes developed in recent years operate periodically in order to obtain improved mass transfer. One such process, developed by Wilhelm and his associates and termed *parametric pumping*, exploits the temperature dependence of adsorption equilibria to achieve separation, as shown schematically in Fig. 11.13. A stream containing dissolved solute is alternately fed to the top and bottom of the column. When the feed is to the top of the column, which we denote by the time interval τ_1, the stream is heated, while during feed to the bottom in the interval τ_3 the stream is cooled. There may be intervals τ_2 and τ_4 in which no feed enters. A temperature gradient is therefore established in the column. During τ_1 cold product lean in the solute is removed from the bottom, and during τ_3 hot product rich in

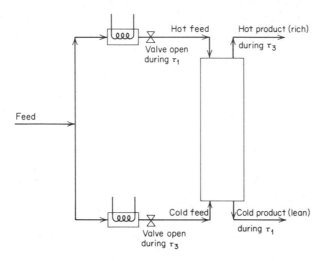

Fig. 11.13 Schematic of a parametric-pumping separation process.

solute is removed from the top. This physical process leads to an interesting variational problem when we seek the distribution of hot and cold feed over a cycle of periodic operation which maximizes the separation.

Denoting normalized temperatures by T and concentrations by c, with subscripts f and s referring to fluid and solid phases, respectively, the mass and energy balances inside the column ($0 < z < 1$) are

$$-\psi \frac{\partial^2 T_f}{\partial z^2} + u(t)\frac{\partial T_f}{\partial z} + \frac{\partial T_f}{\partial t} + \beta\kappa\frac{\partial T_s}{\partial t} = 0 \tag{1a}$$

$$\frac{\partial T_s}{\partial t} + \gamma(T_s - T_f) = 0 \tag{1b}$$

$$-\eta \frac{\partial^2 c_f}{\partial z^2} + u(t)\frac{\partial c_f}{\partial z} + \frac{\partial c_f}{\partial t} + \kappa\frac{\partial c_s}{\partial t} = 0 \tag{1c}$$

$$\frac{\partial c_s}{\partial t} + \lambda(c^* - c_s) = 0 \tag{1d}$$

Here, c^* is the equilibrium concentration of solute which, for the salt solutions studied by Rice, is related to solution concentration and temperature by the empirical equation

$$c^* = 111\left(\frac{c_s}{88}\right)^{3.16-0.44T_s} \tag{2}$$

$u(t)$ is the fluid velocity, which is negative during τ_1 (downflow, hot feed), positive during τ_3 (upflow, cold feed), and zero during τ_2 and τ_4 (periods of no flow).   The values used for the other physical parameters are

$$\beta\kappa = 1.10 \qquad \kappa = 1.38$$
$$\gamma = 200 \qquad \lambda = 0.3$$
$$\psi, \eta \approx 10^{-3}$$

In order to simplify the model for this preliminary study two approximations were made based on these values.   First, γ was taken as approaching infinity, implying instantaneous temperature equilibration.   In that case $T_s = T_f = T$.   Secondly, dispersion of heat and mass were neglected compared to convection and ψ and η set equal to zero. This latter approximation changes the order of the equations and was retained only during the development of the optimization conditions. All calculations were carried out including the dispersion terms.   The simplified equations describing the process behavior are then

$$(1 + \beta\kappa) \frac{\partial T}{\partial t} = -u \frac{\partial T}{\partial z} \tag{3a}$$

$$\frac{\partial c_f}{\partial t} = -u \frac{\partial c_f}{\partial z} - \kappa \frac{\partial c_s}{\partial t} \tag{3b}$$

$$\frac{\partial c_s}{\partial t} = \lambda(c_f - c^*) \tag{3c}$$

with boundary conditions

$$\tau_1, u < 0: \qquad T = 1 \qquad c_f = 1 \qquad \text{at } z = 1 \tag{4a}$$
$$\tau_3, u > 0: \qquad T = 0 \qquad c_f = 1 \qquad \text{at } z = 0 \tag{4b}$$

All variables are periodic in time with period θ.   An immediate consequence of this simplification is that Eq. (3a) can be formally solved to yield

$$T\left[z - \frac{1}{1 + \beta\kappa} \int_0^t u(\tau) \, d\tau\right] = \text{const} \tag{5}$$

Periodicity then requires

$$\int_0^\theta u(\tau) \, d\tau = 0 \tag{6}$$

The amount of dissolved solute obtained in each stream during one cycle is

Rich product stream: $\qquad \int_{\tau_3} u(t) c_f(t,1) \, dt \tag{7a}$

Lean product stream: $\qquad -\int_{\tau_1} u(t) c_f(t,0) \, dt \tag{7b}$

Thus, the net separation can be expressed as

$$\mathcal{P} = \int_{\tau_3} u(t) c_f(t,1) \, dt + \int_{\tau_1} u(t) c_f(t,0) \, dt \tag{8}$$

The total feed during one cycle is

$$V = \int_0^\theta |u(t)| \, dt = \int_{\tau_3} u(t) \, dt - \int_{\tau_1} u(t) \, dt \tag{9}$$

The optimization problem is that of choosing $u(t)$ such that $\mathcal{O}$ is maximized ($-\mathcal{O}$ minimized) for a fixed value of V. To complete the specification of parameters we take $\theta = 2$, $V = 2$.

The linearization and construction of Green's functions proceeds in the usual manner, though it is necessary here to show that terms resulting from changes in the ranges of τ_1 and τ_3 are of second order and that Eqs. (6) and (9) combine to require

$$\int_{\tau_1} \delta u(t) \, dt = \int_{\tau_3} \delta u(t) \, dt = 0 \tag{10}$$

The Green's functions are then found to satisfy the equations

$$(1 + \beta\kappa)\frac{\partial\gamma_1}{\partial t} + u\frac{\partial\gamma_1}{\partial z} - \lambda\gamma_3\frac{\partial c^*}{\partial T} = 0 \tag{11a}$$

$$\frac{\partial\gamma_2}{\partial t} + u\frac{\partial\gamma_2}{\partial z} + \lambda\gamma_3 = 0 \tag{11b}$$

$$\frac{\partial\gamma_3}{\partial t} - \lambda\gamma_3\frac{\partial c^*}{\partial c_s} + \kappa\frac{\partial\gamma_2}{\partial t} = 0 \tag{11c}$$

$$\tau_1: \qquad \gamma_1(t,0) = 0 \qquad \gamma_2(t,0) = -1 \tag{12a}$$
$$\tau_3: \qquad \gamma_1(t,1) = 0 \qquad \gamma_2(t,1) = +1 \tag{12b}$$

All three functions are periodic over θ. The direction of steep descent for maximizing the separation $\mathcal{O}$ is

$$\delta u(t) = \begin{cases} w(t)\left[\gamma_{41} + c_f(t,0) - \int_0^1\left(\gamma_1\frac{\partial T}{\partial z} + \gamma_2\frac{\partial c_f}{\partial z}\right)dz\right] & \text{in } \tau_1 \\ w(t)\left[\gamma_{43} + c_f(t,1) - \int_0^1\left(\gamma_1\frac{\partial T}{\partial z} + \gamma_2\frac{\partial c_f}{\partial z}\right)dz\right] & \text{in } \tau_3 \end{cases} \tag{13}$$

Here $w(t) \geq 0$, and γ_{41} and γ_{43} are constants chosen to satisfy the restrictions of Eq. (10) on δu.

The initial choice for the feed distribution was taken to be sinusoidal

$$\bar{u}(t) = -0.5 \sin t \tag{14}$$

The periodic solution to the system model was obtained by an implicit finite difference method using the single approximation $T_s = T_f = T$ ($\gamma \to \infty$) and with dispersion terms retained. The multiplier equations were solved only approximately in order to save computational time, for an extremely small grid size appeared necessary for numerical stability.

First Eqs. (11*b*) and (11*c*) were combined to give a single equation in γ_2, and the following form was assumed for application of the Galerkin method introduced in Sec. 3.9:

$$\gamma_2 = (2z - 1) + C_1 z(1 - z) \cos t + C_2[\beta(1 - z) + (1 - \beta)z] \sin t$$
$$+ C_3 z(1 - z) \cos 2t + C_4[\beta(1 - z) + (1 - \beta)z] \sin 2t \quad (15)$$

where $\beta = +1$ when $u > 0$, $\beta = 0$ when $u < 0$. Evaluating the constants in the manner described in Sec. 3.9 shows them to be

$$C_1 = -5.54 \qquad C_2 = 0.461$$
$$C_3 = 0.056 \qquad C_4 = -0.046$$

indicating rapid convergence, and, in fact, the C_3 and C_4 terms were neglected for computation. γ_3 was then found by analytically integrating Eq. (11*c*), treating the equation as a linear first-order equation with variable coefficients. γ_1 was also estimated using the Galerkin method, leading to the solution

$$\gamma_1 = 0.0248\{z(1 - z) \cos t + [\beta(1 - z) + (1 - \beta)z] \sin t\} \quad (16)$$

The function $\delta u(t)$ was computed from Eq. (13) for a constant weighting function $w(t) = k$, and a function $w(t) = k|\sin t|$. These are shown normalized with respect to k in Fig. 11.14 as a dotted and solid line, respectively. The dashed line in the figure is the function $0.5\,\bar{u}(t)$. Both weighting functions indicate the same essential features when the

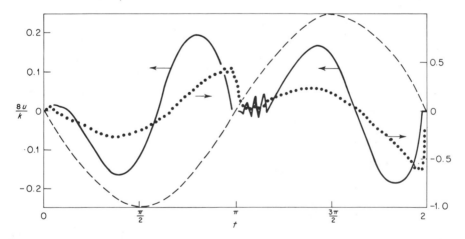

Fig. 11.14 Normalized correction to velocity. Solid line $w(t) = k|\sin t|$; dotted line $w(t) = k$; dashed line $0.5\bar{u}(t) = -0.25 \sin t$.

starting velocity function $\bar{u}(t)$ is compared with the change indicated
by the optimization theory in the form of δu.   Amplification of the flow
is required in some regions for improved performance and reduction in
others.   The value of the separation obtained for the initial velocity distribution ($k = 0$) was $\mathcal{P} = 0.150$.   For sinusoidal weighting and values
of $k = 0.2$, 0.4, and 0.8 the values of $\mathcal{P}$ were, respectively, 0.160, 0.170,
and 0.188.   For constant weighting the value of $\mathcal{P}$ for $k = 0.1$ was 0.160.
Further increase in k in either case would have required a new computer
program to solve the state equations, which did not appear to be a
fruitful path in a preliminary study of this kind.   The dashed sinusoid
in Fig. 11.15 is the initial velocity and the solid line the value for sinusoidal weighting and $k = 0.8$.

The trend suggested by the calculated function $\delta u(t)$ is clearly
towards regions of no flow followed by maximal flow.   Such a flow pattern was estimated by taking the sign changes of δu for sinusoidal weighting to define the onset and end of flow, leading to the rectangular wave
function shown in Fig. 11.15.   The separation is 0.234, an increase of
36 percent over the starting sinusoidal separation.

No further study was undertaken because the approximate nature

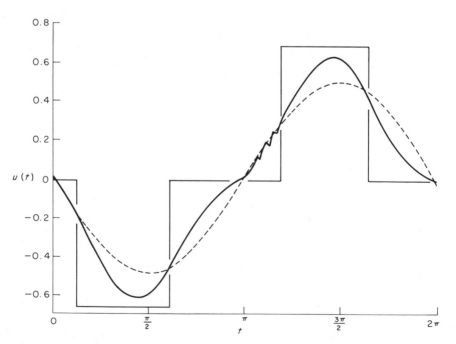

Fig. 11.15   Velocity functions.   Dashed line $\bar{u}(t) = -0.5 \sin t$; solid line sinusoidal
weighting with $k = 0.8$; rectangular wave final estimate of optimum.

of the model did not appear to justify the additional computational effort. The preliminary results obtained here do suggest a fruitful area of application of optimization theory when physical principles of the type considered here have been carried to the point of process implementation.

11.11 CONCLUDING REMARKS

This concludes our study of the use of Green's functions in the solution of variational problems. In addition to the examination of several examples which are of considerable practical interest in themselves, the purpose of this chapter has been to demonstrate the general applicability of the procedures first introduced in Chaps. 6, 7, and 9. The variational solution of optimization problems inevitably reduces to a consideration of linearized forms of the state equations, whether they are ordinary differential, difference, integral, partial differential, difference-differential, differential-integral, or any other form. The rational treatment of these linearized systems requires the use of Green's functions.

As we have attempted to indicate, in all situations the construction of the Green's function follows in a logical fashion from the application of the timelike operator to the inner product of Green's function and state function. It is then a straightforward matter to express the variation in objective explicitly in terms of finite or infinitesimal variations in the decisions, thus leading to both necessary conditions and computational procedures. As long as this application of standard procedures in the theory of linear systems is followed, there is no inherent difficulty in applying variational methods to classes of optimization problems not examined in this book.

BIBLIOGRAPHICAL NOTES

Sections 11.2 to 11.7: These sections follow quite closely

M. M. Denn: *Intern. J. Contr.*, **4**:167 (1966)

The particular problem of temperature control has been considered in terms of a minimum principle for integral formulations by

A. G. Butkovskii: *Proc. 2d Intern. Congr. IFAC*, paper 513 (1963)

and by classical variational methods by

Y. Sakawa: *IEEE Trans. Autom. Cont.*, **AC9**:420 (1964); **AC11**:35 (1966)

Modal analysis, approximation by a lumped system, and a procedure for approximating the solution curves are used by

I. McCausland: *J. Electron. Contr.*, **14**:655 (1963)
———: *Proc. Inst. Elec. Engrs. (London)*, **112**:543 (1965)

Section 11.8: The essential results are obtained in the paper

M. M. Denn, R. D. Gray, Jr., and J. R. Ferron: *Ind. Eng. Chem. Fundamentals,*
   5:59 (1966)

Details of the construction of the model and numerical analysis are in

R. D. Gray, Jr.: Two-dimensional Effects in Optimal Tubular Reactor Design, Ph.D.
   thesis, University of Delaware, Newark, Del., 1965

The usefulness of an isothermal coolant and the effect of design parameters are investigated in

A. R. Hoge: Some Aspects of the Optimization of a Two-dimensional Tubular Reactor,
   B.S. thesis, University of Delaware, Newark, Del., 1965
J. D. Robinson: A Parametric Study of an Optimal Two-dimensional Tubular Reactor
   Design, M.Ch.E. thesis, University of Delaware, Newark, Del., 1966

Section 11.9: This section is based on

M. M. Denn: *Ind. Eng. Chem. Fundamentals,* **7**:410 (1968)

*where somewhat more general results are obtained, including control at the boundary z = 0.
   An equation for the feedback gain corresponding to Eq. (17) for more general spatial
   differential operators was first presented by Wang in his review paper*

P. K. C. Wang: in C. T. Leondes (ed.), "Advances in Control Systems," vol. 1,
   Academic Press, Inc., New York, 1964

A procedure for obtaining the feedback law using modal analysis is given by

D. M. Wiberg: *J. Basic Eng.,* **89D**:379 (1967)

*Wiberg's results require a discrete spectrum of eigenvalues and are not applicable to
   hyperbolic systems of the type studied here. An approach similar to that used here
   leading to an approximate but easily calculated form for the feedback gain is in*

L. B. Koppel, Y. P. Shih, and D. R. Coughanowr: *Ind. Eng. Chem. Fundamentals,*
   7:296 (1968)

*Section 11.10: This section follows a paper with A. K. Wagle presented at the 1969
   Joint Automatic Control Conference and published in the preprints of the meeting.
   More details are contained in*

A. K. Wagle: Optimal Periodic Separation Processes, M.Ch.E. Thesis, University of
   Delaware, Newark, Del., 1969

*Separation by parametric pumping was developed by Wilhelm and coworkers and reported
   in*

R. H. Wilhelm, A. W. Rice, and A. R. Bendelius: *Ind. Eng. Chem. Fundamentals,*
   5:141 (1966)
———, ———, R. W. Rolke, and N. H. Sweed: *Ind. Eng. Chem. Fundamentals,*
   7:337 (1968)

*The basic equations are developed and compared with experiment in the thesis of Rice,
   from which the parameters used here were taken:*

A. W. Rice: Some Aspects of Separation by Parametric Pumping, Ph.D. thesis, Princeton University, Princeton, N.J., 1966

Section 11.11: Variational methods for distributed-parameter systems have received attention in recent years. The paper by Wang noted above contains an extensive bibliography through 1964, although it omits a paper fundamental to all that we have done, namely,

S. Katz: *J. Electron. Contr.*, **16**:189 (1964)

More current bibliographies are contained in

E. B. Lee and L. Markus: "Foundations of Optimal Control Theory," John Wiley & Sons, Inc., New York, 1967
P. K. C. Wang: *Intern. J. Contr.*, **7**:101 (1968)

An excellent survey of Soviet publications in this area is

A. G. Butkovsky, A. I. Egorov, and K. A. Lurie: *SIAM J. Contr.*, **6**:437 (1968)

Many pertinent recent papers can be found in the SIAM Journal on Control; IEEE Transactions on Automatic Control; Automation and Remote Control; and Automatica; in the annual preprints of the Joint Automatic Control Conference; and in the University of Southern California conference proceedings:

A. V. Balakrishnan and L. W. Neustadt (eds.): "Mathematical Theory of Control," Academic Press, Inc., New York, 1967

The following are of particular interest in process applications:

R. Jackson: *Proc. Inst. Chem. Engrs.–AIChE Joint Meeting*, **4**:32 (1965)
———: *Intern. J. Contr.*, **4**:127, 585 (1966)
———: *Trans. Inst. Chem. Engrs. (London)*, **45**:T160 (1967)
K. A. Lurie: in G. Leitmann (ed.), "Topics in Optimization," Academic Press, Inc., New York, 1967

A minimum principle for systems described by integral equations is outlined in the paper by Butkovskii listed above.   Results for difference-differential equations, together with further references, are in

D. H. Chyung: *Preprints 1967 Joint Autom. Contr. Conf.*, p. 470
M. M. Denn and R. Aris: *Ind. Eng. Chem. Fundamentals*, **4**:213 (1965)
M. N. Oğüztöreli: "Time-lag Control Systems," Academic Press, Inc., New York, 1966
H. R. Sebesta and L. G. Clark: *Preprints 1967 Joint Autom. Contr. Conf.*, p. 326

Results due to Kharatishvili can be found in the book by Pontryagin and coworkers and in the collection above edited by Balakrishnan and Neustadt.

PROBLEMS

11.1. Obtain the minimum-principle and steep-descent direction used in Sec. 11.8.
11.2. Obtain necessary conditions and the direction of steep descent for the problem in Sec. 11.10.   What changes result when the objective is maximization of the separation

factor,

$$S = -\frac{\displaystyle\int_{\tau_3} u(t)c_f(t,1)\,dt}{\displaystyle\int_{\tau_1} u(t)c_f(t,0)\,dt}$$

11.3. Optimal control of a heat exchanger to a new set point by adjustment of wall temperature is approximated by the equation

$$\frac{\partial x}{\partial t} + \frac{\partial x}{\partial z} = P(u - x)$$

$$x(0,t) = 0$$

Obtain the optimal feedback gain as a function of P and C for the objective

$$\mathcal{E} = \frac{1}{2}\int_0^\infty [Cx^2(1,t) + u^2(t)]\,dt$$

Compare with the approximate solution of Koppel, Shih, and Coughanowr,

$$G^{\text{FB}}(z) = \begin{cases} -C\exp\left[-(2P + PK)(1 - z)\right] & 0 \le z < 1 \\ 0 & z = 1 \end{cases}$$

where K is the solution of a transcendental equation

$$K = \frac{C}{2 + K}[1 - \exp(-2P - PK)]$$

11.4. The reversible exothermic reaction $X \rightleftharpoons Y$ in a tubular catalytic reactor is approximated by the equations

$$\frac{\partial x_1}{\partial z} = x_2\tau k_0 e^{-E_2'/u}[(1 - y_0 - x_1)K_0 e^{(E_2' - E_1')/u} - (y_0 + x_1)] \qquad 0 < z < 1$$

$$\frac{\partial x_2}{\partial t} = -\alpha u x_2 \qquad 0 < \tau < 1$$

where x_1 is the extent of reaction, x_2 the catalyst efficiency, u the temperature, τ the reactor residence time, and y_0 the fraction of reaction product in the feed. Boundary and initial conditions are

$$x_1 = 0 \qquad \text{at } z = 0 \text{ for all } t$$
$$x_2 = 1 \qquad \text{at } t = 0 \text{ for all } z$$

The temperature profile $u(z)$ is to be chosen at each time t to maximize average conversion

$$\mathcal{O} = \int_0^1 x_1(1,t)\,dt$$

Obtain all the relevant equations for solution. Carry out a numerical solution if necessary. Parameters are

$$\tau k_0 = 3 \times 10^7 \qquad K_0 = 2.3 \times 10^{-5} \qquad y_0 = 0.06$$
$$E_2' = 10{,}000 \qquad E_1' = 5{,}000 \qquad \alpha = 4 \times 10^{-3}$$

(The problem is due to Jackson.)

11.5. Discuss the minimum time control of the function $x(z,t)$ to zero by adjusting surroundings temperature $u(t)$ in the following radiation problem:

$$\frac{\partial x}{\partial t} = \frac{\partial^2 x}{\partial z^2} \qquad \begin{array}{l} 0 < z < 1 \\ 0 < t \le \theta \end{array}$$

$$x = 0 \qquad \text{at } t = 0 \text{ for all } z$$

$$\frac{\partial x}{\partial z} = 0 \qquad \text{at } z = 0 \text{ for all } t$$

$$\frac{\partial x}{\partial z} = k[u^4(t) - x^4] \qquad \text{at } z = 1 \text{ for all } t$$

(The problem is due to Uzgiris and D'Souza.)

11.6. Obtain necessary conditions and the direction of steep descent for a system described by the difference-differential equations

$$\dot{x}_i(t) = f_i[\mathbf{x}(t), \mathbf{x}(t - \tau), u(t), u(t - \tau)]$$
$$x_i(t) = x_{i0}(t) \qquad -\tau \le t \le 0$$
$$\min \mathcal{E} = \int_0^\theta \mathcal{F}[\mathbf{x}(t), u(t)] \, dt$$

12
Dynamic Programming and Hamilton-Jacobi Theory

12.1 INTRODUCTION

The major part of this book has been concerned with the application of classical variational methods to sequential decision making, where the sequence was either continuous in a timelike variable or, as in Chap. 7, over discrete stages.   Simultaneously with the refinement of these methods over the past two decades an alternative approach to such problems has been developed and studied, primarily by Bellman and his coworkers. Known as *dynamic programming*, this approach has strong similarities to variational methods and, indeed, for the types of problems we have studied often leads to the same set of equations for ultimate solution. In this brief introduction to dynamic programming we shall first examine the computational aspects which differ from those previously developed and then demonstrate the essential equivalence of the two approaches to sequential optimization for many problems.

12.2 THE PRINCIPLE OF OPTIMALITY AND COMPUTATION

For illustrative purposes let us consider the process shown schematically in Fig. 12.1. The sequence of decisions u_1, u_2, . . . , u_N is to be made to minimize some function $\mathcal{E}(\mathbf{x}^N)$, where there is a known input-output relation at each stage, say

$$\mathbf{x}^n = \mathbf{f}^n(\mathbf{x}^{n-1}, u^n) \qquad n = 1, 2, \ldots, N \tag{1}$$

The dynamic programming approach is based on the *principle of optimality*, as formulated by Bellman:

> *An optimal policy has the property that whatever the initial state and initial decision are, the remaining decisions must constitute an optimal policy with regard to the state resulting from the first decision.*

That is, having chosen u^1 and thus determined $\mathbf{x}^1$, the remaining decisions u^2, u^3, . . . , u^N must be chosen so that $\mathcal{E}(\mathbf{x}^N)$ is minimized for that particular $\mathbf{x}^1$. Similarly, having chosen u^1, u^2, . . . , u^{N-1} and thus determined $\mathbf{x}^{N-1}$, the remaining decision u^N must be chosen so that $\mathcal{E}(\mathbf{x}^N)$ is a minimum for that $\mathbf{x}^{N-1}$. The proof of the principle of optimality is clear by contradiction, for if the choice u^1 happened to be the *optimal* first choice and the remaining decisions were not optimal with respect to that $\mathbf{x}^1$, we could always make $\mathcal{E}(\mathbf{x}^N)$ smaller by choosing a new set of remaining decisions.

The principle of optimality leads immediately to an interesting computational algorithm. If we suppose that we have somehow determined $\mathbf{x}^{N-1}$, the choice of the remaining decision simply involves the search over all values of u^N to minimize $\mathcal{E}(\mathbf{x}^N)$ or, substituting Eq. (1),

$$\min_{u^N} \mathcal{E}[\mathbf{f}^N(\mathbf{x}^{N-1}, u^N)] \tag{2}$$

Since we do not know what the proper value of $\mathbf{x}^{N-1}$ is, however, we must do this for *all* values of $\mathbf{x}^{N-1}$ or, more realistically, for a representative selection of values. We can then tabulate for each $\mathbf{x}^{N-1}$ the minimizing value of u^N and the corresponding minimum value of $\mathcal{E}$.

We now move back one stage and suppose that we have available

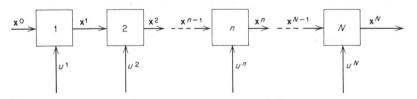

Fig. 12.1 Schematic of a sequential decision process.

$\mathbf{x}^{N-2}$, for which we must find the decisions u^{N-1} and u^N that minimize $\mathcal{E}(\mathbf{x}^N)$. A specification of u^{N-1} will determine $\mathbf{x}^{N-1}$, and for any given $\mathbf{x}^{N-1}$ we already have tabulated the optimal u^N and value of $\mathcal{E}$. Thus, we need simply search over u^{N-1} to find the *tabulated*† $\mathbf{x}^{N-1}$ that results in the minimum value of $\mathcal{E}$. That is, we search over u^{N-1} only, and not simultaneously over u^{N-1} and u^N. The dimensionality of the problem is thus reduced, but again we must carry out this procedure and tabulate the results for all (representative) values of $\mathbf{x}^{N-2}$, since we do not know which value is the optimal one.

We now repeat the process for $\mathbf{x}^{N-3}$, choosing u^{N-2} by means of the table for $\mathbf{x}^{N-2}$, etc., until finally we reach $\mathbf{x}^0$. Since $\mathbf{x}^0$ is known, we can then choose u^1 by means of the table for $\mathbf{x}^1$. This gives the optimal value for $\mathbf{x}^1$, so that u^2 can then be found from the table for $\mathbf{x}^2$, etc. In this way the optimal sequence $u^1, u^2, \ldots, u^N$ is constructed for a given $\mathbf{x}^0$ by means of a sequence of minimizations over a single variable. Note that we have made no assumptions concerning differentiability of the functions or bounds on the decisions. Thus, this algorithm can be used when those outlined in Chap. 9 might be difficult or inapplicable. We shall comment further on the computational efficiency following an example.

12.3 OPTIMAL TEMPERATURE SEQUENCES

As an example of the computational procedure we shall again consider the optimal temperature sequence for consecutive chemical reactions introduced in Secs. 1.12 and 7.6 and studied computationally in Chaps. 2 and 9. Taking the functions F and G as linear for simplicity, the state equations have the form

$$x^n = \frac{x^{n-1}}{1 + \theta k_{10} e^{-E_1'/u^n}} \tag{1a}$$

$$y^n = \frac{y^{n-1} + k_{10} e^{-E_1'/u^n}(y^{n-1} + x^{n-1})}{(1 + \theta k_{10} e^{-E_1'/u^n})(1 + \theta k_{20} e^{-E_2'/u^n})} \tag{1b}$$

where we wish to choose $u^1, u^2, \ldots, u^N$ in order to minimize

$$\mathcal{E} = -y^N - \rho x^N \tag{2}$$

We shall use the values of the parameters as follows:

$$k_{10} = 5.4 \times 10^{10} \qquad k_{20} = 4.6 \times 10^{17}$$
$$E_1' = 9 \times 10^3 \qquad E_2' = 15 \times 10^3$$
$$\theta = 5 \qquad \rho = 0.3$$
$$x^0 = 1 \qquad y^0 = 0$$

† Clearly some interpolation among the tabulated representative values of $\mathbf{x}^{N-1}$ will be required.

**Table 12.1   Optimal decisions and values of the objective at
stage N for various inputs resulting from stage $N - 1$**

x^{N-1} / y^{N-1}	0	0.1	0.2	0.3	0.4	0.5	0.6	0.7	0.8	0.9	1.0
0		340 0.056	340 0.112	340 0.168	340 0.224	340 0.281	340 0.337	340 0.393	340 0.449	340 0.505	340 0.561
0.1	330 0.096	334 0.145	338 0.199	340 0.255	340 0.310	340 0.367	340 0.423	340 0.479	340 0.535	340 0.591	340 0.647
0.2	330 0.192	330 0.240	334 0.291	336 0.344	338 0.399	338 0.454	340 0.509	340 0.565	340 0.621	340 0.677	340 0.733
0.3	330 0.288	330 0.336	332 0.385	334 0.436	336 0.489	338 0.543	338 0.598	338 0.654	338 0.709	340 0.764	340 0.820
0.4	330 0.384	330 0.432	330 0.481	332 0.530	334 0.582	336 0.634	336 0.688	338 0.742	338 0.798	338 0.853	338 0.908
0.5	330 0.480	330 0.528	330 0.577	330 0.625	332 0.675	334 0.727	336 0.779	336 0.833	336 0.887	338 0.942	338 0.997
0.6	330 0.576	330 0.624	330 0.673	330 0.721	332 0.770	334 0.820	334 0.872	334 0.925	336 0.978	336 1.032	338 1.086
0.7	330 0.672	330 0.720	330 0.769	330 0.817	330 0.866	332 0.913	332 0.966	334 1.018	334 1.070	336 1.123	336 1.177
0.8	330 0.768	330 0.816	330 0.865	330 0.913	330 0.962	330 1.010	332 1.060	334 1.111	334 1.163	334 1.215	336 1.268
0.9	330 0.864	330 0.912	330 0.961	330 1.009	330 1.058	330 1.106	332 1.155	332 1.205	332 1.255	334 1.309	334 1.361
1.0	330 0.960	330 1.008	330 1.057	330 1.105	330 1.154	330 1.202	330 1.250	330 1.329	332 1.354	334 1.402	334 1.454

Furthermore, we shall restrict u^n by the constraints

$$330 \leq u^n \leq 340 \tag{3}$$

and, to demonstrate a situation where techniques based on the calculus
are not applicable, we shall further restrict each u^n to values which are
even integers.   That is, we shall restrict the choice to the values 330,
332, 334, 336, 338, and 340.

We begin by considering the last (Nth) stage.   For each value of
x^{N-1}, y^{N-1} we can compute x^N and y^N by means of Eqs. (1) for the six
possible choices of u^N.   The particular value of u^N which minimizes ε
and the corresponding value of $-\varepsilon$ is recorded in Table 12.1, where it is

assumed that increments of 0.1 in x^{N-1} and y^{N-1} are sufficient for purposes of interpolation.

Next we consider the penultimate $[(N-1)\text{st}]$ stage. For example, for $x^{N-2} = 0.7$, $y^{N-2} = 0.2$, we obtain:

u^{N-1}	x^{N-1}	y^{N-1}	$-\varepsilon$
330	0.504	0.380	0.617
332	0.481	0.397	0.621
334	0.455	0.415	0.626
336	0.429	0.430	0.625
338	0.403	0.443	0.624
340	0.375	0.452	0.617

where the values of $-\varepsilon$ are obtained by linear interpolation in Table 12.1. Thus, for this pair of values x^{N-2}, y^{N-2} the optimal u^{N-1} is 334 with a corresponding value of ε of -0.626. In this way Table 12.2 is constructed for all values of x^{N-2}, y^{N-2}, where, for simplicity, the range has now been reduced.

In a similar manner we move on to the second from last stage and consider values of x^{N-3}, y^{N-3}, now finding the optimum by interpolation in Table 12.2. The only entry we note here is $x^{N-3} = 1$, $y^{N-3} = 0$, with $u^{N-2} = 336$ and $\varepsilon = -0.706$. When we have finally reached the first stage, we can reconstruct the optimal policy. For example, if $N = 1$, we find from Table 12.1 that for $x^0 = 1$, $y^0 = 0$ the optimum is -0.56 with $u^1 = 340$. For $N = 2$† we begin with Table 12.2, where we find for $x^0 = 1$, $y^0 = 0$ the optimum is -0.662 and $u^1 = 340$. Equations (1) then indicate that $x^1 = 0.536$, $y^1 = 0.400$, and Table 12.1 for the last (second) stage indicates the optimal decision $u^2 = 336$. For $N = 3$ our sole tabular entry recorded above indicates $u^1 = 336$ and $\varepsilon = -0.71$. Then, from Eqs. (1), $x^1 = 0.613$ and $y^1 = 0.354$, and Table 12.2 indicates that $u^2 = 334$. Applying Eqs. (1) again, $x^2 = 0.398$, $y^2 = 0.530$, and from Table 12.1 we find $u^3 = 332$. Finally, then, $x^3 = 0.274$, $y^3 = 0.624$. This process can be continued for N as large as we wish.

We can now evaluate the strong and weak points of this computational procedure. The existence of constraints increases efficiency, as does the restriction to discrete values of the decisions. The former is

---

† For this case ($N\theta = 10$) we found in Chap. 2 that $u^1 = 338$, $u^2 = 336.5$, and $\varepsilon = -0.655$ when there were no restrictions on u^1, u^2. Since the restricted optimum cannot be better than the unrestricted, we see some small error resulting from the interpolation.

Table 12.2 Optimal decisions at stage $N - 1$ and optimal values of the objective for various inputs resulting from stage $N - 2$

x^{N-2} / y^{N-2}	0.5	0.6	0.7	0.8	0.9	1.0
0	336 0.332	338 0.404	338 0.460	338 0.524	338 0.590	340 0.662
0.1	334 0.414	336 0.475	338 0.541	336 0.605	336 0.672	338 0.736
0.2	334 0.509	336 0.559	336 0.626	334 0.696	336 0.757	334 0.818
0.3	334 0.555	334 0.648	336 0.713	334 0.776	336 0.840	334 0.905
0.4	334 0.649	334 0.735	336 0.799	334 0.863	334 0.926	334 0.992
0.5	330 0.743	332 0.826	334 0.887	334 0.951	334 1.014	334 1.078

not a major limitation on procedures derivable from the calculus, but the latter effectively eliminates their use. Thus, if such a restriction is physically meaningful, an algorithm such as this one is essential, while if the restriction is used simply as a means of obtaining first estimates by use of a coarse grid, we might expect to obtain first estimates as easily using steep descent as outlined in Chap. 9. In reducing the N-dimensional search problem to a sequence of one-dimensional searches we have traded a single difficult problem for a large number of simpler ones, and indeed, we have simultaneously solved not only the problem of interest but related problems for a range of initial values. If we are truly interested in a range of initial values, this is quite useful, but if we care only about a single initial condition, the additional information is of little use and seemingly wasteful. Dynamic programming does not eliminate the two-point boundary-value problem which has occupied us to such a great extent, but rather solves it by considering a complete range of final values.

The fact that the optimum is automatically computed as a feedback policy, depending only upon the present state and number of remaining decisions, suggests utility in control applications. The storage problem, however, is a serious restriction, for with two state variables we have been able to tabulate data in a two-dimensional array, but three variables would require a three-dimensional array, four variables four dimensions, etc. Fast memory capabilities in present computers effectively limit direct storage to systems with three state variables, and approximation techniques must be used for higher-dimensional problems. Thus, the algorithm resulting from direct application of the principle of optimality will be of greatest use in systems with many stages but few state variables, particularly when variables are constrained or limited to discrete values.

12.4 THE HAMILTON-JACOBI-BELLMAN EQUATION

For analytical purposes it is helpful to develop the mathematical formalism which describes the tabulation procedure used in the previous section. At the last stage the minimum value of $\mathcal{E}(\mathbf{x}^N)$ depends only upon the value of $\mathbf{x}^{N-1}$. Thus, we define a function S^N of $\mathbf{x}^{N-1}$ as

$$S^N(\mathbf{x}^{N-1}) = \min_{u^N} \mathcal{E}[\mathbf{f}^N(\mathbf{x}^{N-1}, u^N)] \tag{1}$$

which is simply a symbolic representation of the construction of Table 12.1 in the example. Similarly with two stages to go we can define

$$S^{N-1}(\mathbf{x}^{N-2}) = \min_{u^{N-1}} \min_{u^N} \mathcal{E}(\mathbf{x}^N) \tag{2}$$

or, using Eq. (1),

$$S^{N-1}(\mathbf{x}^{N-2}) = \min_{u^{N-1}} S^N[\mathbf{f}^{N-1}(\mathbf{x}^{N-2}, u^{N-1})] \tag{3}$$

which is the symbolic representation of the construction of Table 12.2. In general, then, we obtain the recursive relation

$$S^n(\mathbf{x}^{n-1}) = \min_{u^n} S^{n+1}[\mathbf{f}^n(\mathbf{x}^{n-1}, u^n)] \tag{4}$$

Consistent with this definition we can define

$$S^{N+1}(\mathbf{x}^N) = \mathcal{E}(\mathbf{x}^N) \tag{5}$$

Equation (4) is a difference equation for the function S^n, with the boundary condition Eq. (5). It is somewhat distinct from the type of difference equation with which we normally deal, for it contains a minimization operation, but we can conceive of first carrying out the minimization and then using the minimizing value of u^n to convert Eq. (4) to a classical difference equation in the variable n, with S^n depending explicitly upon $\mathbf{x}^{n-1}$. By its definition the optimum must satisfy Eq. (4), and furthermore, when we have found the functions S^n, we have, by definition, found the optimum. Thus, solution of the difference equation (4) with boundary-condition equation (5) is a necessary and sufficient condition for an optimum. We shall call Eq. (4) a *Hamilton-Jacobi-*

Bellman equation, for it is closely related to the Hamilton-Jacobi equation of mechanics.

If we assume that the functions S^n are differentiable with respect to $\mathbf{x}^{n-1}$ and, further, that $\mathbf{f}^n$ is differentiable with respect to both $\mathbf{x}^{n-1}$ and u^n, we can relate the Hamilton-Jacobi theory to the results of Chap. 7. For simplicity we shall temporarily presume that u^n is unconstrained. Then, since S^{n+1} depends explicitly upon $\mathbf{x}^n$, the minimum of the right-hand side of Eq. (4) occurs at the solution of

$$\frac{\partial S^{n+1}}{\partial u^n} = \sum_i \frac{\partial S^{n+1}}{\partial x_i{}^n} \frac{\partial f_i{}^n}{\partial u^n} = 0 \tag{6}$$

The value of u^n satisfying this equation, $\bar{u}^n$, is, of course, a function of the particular value of $\mathbf{x}^{n-1}$, and Eq. (4) can now be written as

$$S^n(\mathbf{x}^{n-1}) = S^{n+1}[\mathbf{f}^n(\mathbf{x}^{n-1}, \bar{u}^n)] \tag{7}$$

Equation (7) is an identity, valid for all $\mathbf{x}^{n-1}$. The partial derivatives in Eq. (6) can then be evaluated by differentiating both sides of Eq. (7) with respect to each component of $\mathbf{x}^{n-1}$, giving

$$\frac{\partial S^n}{\partial x_i{}^{n-1}} = \sum_j \frac{\partial S^{n+1}}{\partial x_j{}^n} \frac{\partial f_j{}^n}{\partial x_i{}^{n-1}} + \sum_j \left(\frac{\partial S^{n+1}}{\partial x_j{}^n} \frac{\partial f_j{}^n}{\partial u^n} \right) \frac{\partial \bar{u}^n}{\partial x_i{}^{n-1}} \tag{8}$$

where the second term is required because of the dependence of $\bar{u}^n$ on $\mathbf{x}^{n-1}$. From Eq. (6), however, the quantity in parenthesis vanishes, and

$$\frac{\partial S^n}{\partial x_i{}^{n-1}} = \sum_j \frac{\partial S^{n+1}}{\partial x_j{}^n} \frac{\partial f_j{}^n}{\partial x_i{}^{n-1}} \tag{9}$$

From Eq. (5), if there are no relations among the components of $\mathbf{x}^N$,

$$\frac{\partial S^{N+1}}{\partial x_i{}^N} = \frac{\partial \mathcal{E}}{\partial x_i{}^N} \tag{10}$$

Now, it is convenient to define a vector $\boldsymbol{\gamma}^n$ as

$$\gamma_i{}^n = \frac{\partial S^{n+1}}{\partial x_i{}^n} \tag{11}$$

that is, as the rate of change of the optimum with respect to the state at the nth stage. Then Eqs. (9) and (10) may be written

$$\gamma_i{}^{n-1} = \sum_j \gamma_j{}^n \frac{\partial f_j{}^n}{\partial x_i{}^{n-1}} \tag{12}$$

$$\gamma_i{}^N = \frac{\partial \mathcal{E}}{\partial x_i{}^N} \tag{13}$$

and Eq. (6) for the optimum $\bar{u}^n$

$$\sum_i \gamma_i{}^n \frac{\partial f_i{}^n}{\partial u^n} = 0 \tag{14}$$

These are the equations for the weak minimum principle derived in Sec. 7.4, and they relate the Green's functions to the sensitivity interpretation of the Lagrange multiplier in Sec. 1.15.

Even if u^n is bounded, the minimization indicated in Eq. (4) can be carried out and the minimizing value $\bar{u}^n$ found as a function of $\mathbf{x}^{n-1}$. Equation (8) is still valid, and if the optimum occurs at an interior value, the quantity in parentheses will vanish. If, on the other hand, the optimal $\bar{u}^n$ lies at a constraint, small variations in the stage input will not cause any change and $\partial u^n/\partial x_i{}^{n-1}$ will be zero. Thus, Eqs. (9) and (10), or equivalently (12) and (13), are unchanged by constraints. The minimum in Eq. (4) at a bound is defined by

$$\frac{\partial S^{n+1}}{\partial u^n} = \sum_i \frac{\partial S^{n+1}}{\partial x_i{}^n} \frac{\partial f_i{}^n}{\partial u^n} \begin{cases} \geq 0 & u^n \text{ at lower bound} \\ \leq 0 & u^n \text{ at upper bound} \end{cases} \tag{15a}$$

or

$$\sum_i \frac{\partial S^{n+1}}{\partial x_i{}^n} f_i{}^n = \sum_i \gamma_i{}^n f_i{}^n = \min \qquad u^n \text{ at bound} \tag{15b}$$

Thus we have the complete weak minimum principle. We must reiterate, however, that while the Hamilton-Jacobi-Bellman equation is both necessary and sufficient, the operations following Eq. (5) merely define conditions that the solution of Eq. (4) must satisfy and are therefore only necessary.

12.5 A SOLUTION OF THE HAMILTON–JACOBI–BELLMAN EQUATION

It is sometimes possible to construct a solution to the Hamilton-Jacobi-Bellman equation directly, thus obtaining a result known to be sufficient for an optimum. We can demonstrate this procedure for the case of a system described by the linear separable equations

$$x_i{}^n = \sum_j A_{ij}{}^n x_j{}^{n-1} + b_i{}^n(u^n) \tag{1}$$

where we wish to minimize a linear function of the final state

$$\mathcal{E} = \sum_j c_j x_j{}^N \tag{2}$$

The Hamilton-Jacobi-Bellman equation, Eqs. (4) and (5) of the preceding section, is

$$S^n(\mathbf{x}^{n-1}) = \min_{u^n} S^{n+1}[\mathbf{f}^n(\mathbf{x}^{n-1}, u^n)] \tag{3}$$

$$S^{N+1}(\mathbf{x}^N) = \mathcal{E} = \sum_j c_j x_j^N \tag{4}$$

The linearity of $\mathbf{f}^n$ and $\mathcal{E}$ suggest a linear solution

$$S^n(\mathbf{x}^{n-1}) = \sum_j \gamma_j^{n-1} x_j^{n-1} + \xi^n \tag{5}$$

where γ_j^{n-1} and ξ^n are to be determined. Substituting into Eq. (3), then,

$$\sum_j \gamma_j^{n-1} x_j^{n-1} + \xi^n = \min_{u^n} \left[\sum_{i,j} \gamma_i^n A_{ij}^n x_j^{n-1} + \sum_i \gamma_i^n b_i^n(u^n) + \xi^{n+1} \right] \tag{6}$$

or, since only one term on the right depends upon u^n,

$$\sum_j \left(\gamma_j^{n-1} - \sum_i \gamma_i^n A_{ij}^n \right) x_j^{n-1} = \min_{u^n} \left[\sum_i \gamma_i^n b_i^n(u^n) \right] + \xi^{n+1} - \xi^n \tag{7}$$

The left-hand side depends upon $\mathbf{x}^{n-1}$ and the right does not, so that the solution must satisfy the difference equation

$$\gamma_j^{n-1} = \sum_i \gamma_i^n A_{ij}^n \tag{8}$$

with the optimal u^n chosen by

$$\min_{u^n} \sum_i \gamma_i^n b_i^n(u^n) \tag{9}$$

The variable ξ^n is computed from the recursive relation

$$\xi^n = \xi^{n+1} + \min_{u^n} \sum_i \gamma_i^n b_i^n(u^n) \tag{10}$$

Comparison of Eqs. (4) and (5) provides the boundary conditions

$$\gamma_i^N = c_i \tag{11}$$

$$\xi^{N+1} = 0 \tag{12}$$

Equations (8), (9), and (11) are, of course, the strong form of the minimum principle, which we established in Sec. 7.8 as both necessary and sufficient for the linear separable system with linear objective. We have not required here the differentiability of $\mathbf{b}^n$ assumed in the earlier proof. This special situation is of interest in that the optimal policy is completely independent of the state as a consequence of the uncoupling of multiplier and state equations, so that Eqs. (8), (9), and (11) may be solved once and for all for any value of $\mathbf{x}^n$ and an optimum defined only in terms of the number of stages remaining.

12.6 THE CONTINUOUS HAMILTON–JACOBI–BELLMAN EQUATION

The analytical aspects of dynamic programming and the resulting Hamilton-Jacobi theory for continuous processes described by ordinary differential equations are most easily obtained by applying a careful limiting process to the definitions of Sec. 12.4. We shall take the system to be described by the equations

$$\dot{x}_i = f_i(\mathbf{x}, u) \qquad 0 < t \leq \theta \tag{1}$$

and the objective as $\mathcal{E}[\mathbf{x}(\theta)]$. If we divide θ into N increments of length Δt and denote $\mathbf{x}(n \, \Delta t)$ as $\mathbf{x}^n$ and $u(n \, \Delta t)$ as u^n, then a first-order approximation to Eq. (1) is

$$x_i{}^n = x_i{}^{n-1} + f_i(\mathbf{x}^{n-1}, u^n) \, \Delta t + o(\Delta t) \tag{2}$$

Then Eq. (4) of Sec. 12.4, the discrete Hamilton-Jacobi-Bellman equation, becomes

$$S^n(\mathbf{x}^n) = \min_{u^n} S^{n+1}[\mathbf{x}^{n-1} + \mathbf{f}(\mathbf{x}^{n-1}, u^n) \, \Delta t + o(\Delta t)] \tag{3}$$

or, writing S as an explicit function of the time variable $n \, \Delta t$

$$S^n(\mathbf{x}^{n-1}) \equiv S(\mathbf{x}^{n-1}, n \, \Delta t)$$

we have

$$S(\mathbf{x}^{n-1}, n \, \Delta t) = \min_{u^n} S[\mathbf{x}^{n-1} + \mathbf{f}(\mathbf{x}^{n-1}, u^n) \, \Delta t + o(\Delta t), n \, \Delta t + \Delta t] \tag{4}$$

For sufficiently small Δt we can expand S in the right-hand side of Eq. (4) about its value at $\mathbf{x}^{n-1}$, $n \, \Delta t$ as follows:

$$S[\mathbf{x}^{n-1} + \mathbf{f} \, \Delta t + o(\Delta t), n \, \Delta t + \Delta t] = S(\mathbf{x}^{n-1}, n \, \Delta t)$$
$$+ \sum_i \frac{\partial S}{\partial x_i{}^{n-1}} f_i \, \Delta t + \frac{\partial S}{\partial (n \, \Delta t)} \Delta t + o(\Delta t) \tag{5}$$

or

$$S(\mathbf{x}^{n-1}, n \, \Delta t) = \min_{u^n} \left[S(\mathbf{x}^{n-1}, n \, \Delta t) + \sum_i \frac{\partial S}{\partial x_i{}^{n-1}} f_i \, \Delta t \right.$$
$$\left. + \frac{\partial S}{\partial (n \, \Delta t)} \Delta t + o(\Delta t) \right] \tag{6}$$

Since $S(\mathbf{x}^{n-1}, n \, \Delta t)$ does not depend upon u^n, it is not involved in the minimization and can be canceled between the two sides of the equation. Thus,

$$0 = \min_{u^n} \left[\sum_i \frac{\partial S}{\partial x_i{}^{n-1}} f_i \, \Delta t + \frac{\partial S}{\partial (n \, \Delta t)} \Delta t + o(\Delta t) \right] \tag{7}$$

As Δt gets very small, the distinction between $\mathbf{x}^{n-1}$ and $\mathbf{x}^n$, u^{n-1} and u^n gets correspondingly small and both $\mathbf{x}^{n-1}$ and $\mathbf{x}^n$ approach $\mathbf{x}(t)$ and similarly for u^n. Again letting $t = n \, \Delta t$, dividing by Δt, and taking the limit as $\Delta t \to 0$, we obtain the Hamilton-Jacobi-Bellman partial differential equation for the optimum objective $S(\mathbf{x}, t)$

$$0 = \min_{u(t)} \left[\sum_i \frac{\partial S}{\partial x_i} f_i(\mathbf{x}, u) + \frac{\partial S}{\partial t} \right] \tag{8}$$

with boundary condition from Eq. (5) of Sec. 12.4

$$S[\mathbf{x}(\theta), \theta] = \mathcal{E}[\mathbf{x}(\theta)] \tag{9}$$

The partial differential equation can be written in classical form as

$$\sum_i \frac{\partial S}{\partial x_i} f_i(\mathbf{x}, \bar{u}) + \frac{\partial S}{\partial t} = 0 \tag{10}$$

where $\bar{u}$ is an explicit function of $\mathbf{x}$ obtained from the operation

$$\min_u \sum_i \frac{\partial S}{\partial x_i} f_i(\mathbf{x}, u) \tag{11}$$

When u is unconstrained or $\bar{u}$ lies inside the bounds, this minimum is found from the solution of the equation

$$\sum_i \frac{\partial S}{\partial x_i} \frac{\partial f_i}{\partial u} = 0 \tag{12}$$

Equations (9) to (11) form the basis of a means of computation of the optimum, and in the next section we shall demonstrate a direct solution of the Hamilton-Jacobi-Bellman equation. Here we shall perform some manipulations analogous to those in Sec. 12.4 to show how the minimum principle can be deduced by means of dynamic programming. The reader familiar with the properties of hyperbolic partial differential equations will recognize that we are following a rather circuitous path to arrive at the characteristic ordinary differential equations for Eq. (10).

First, we take the partial derivative of Eq. (10) with respect to the jth component x_j of $\mathbf{x}$. Since S depends explicitly only upon $\mathbf{x}$ and t and u depends explicitly upon $\mathbf{x}$ through Eq. (11) we obtain

$$\sum_i \frac{\partial^2 S}{\partial x_i \, \partial x_j} f_i + \sum_i \frac{\partial S}{\partial x_i} \frac{\partial f_i}{\partial x_j} + \sum_i \frac{\partial S}{\partial x_i} \frac{\partial f_i}{\partial u} \frac{\partial \bar{u}}{\partial x_j} + \frac{\partial^2 S}{\partial x_j \, \partial t} = 0 \tag{13}$$

The third term vanishes, because of Eq. (12) if $\bar{u}$ is interior or because a small change in $\mathbf{x}$ cannot move u from a constraint when at a bound, in which case $\partial \bar{u} / \partial x_j = 0$. Thus,

$$\sum_i \frac{\partial^2 S}{\partial x_i \, \partial x_j} f_i + \sum_i \frac{\partial S}{\partial x_i} \frac{\partial f_i}{\partial x_j} + \frac{\partial^2 S}{\partial x_j \, \partial t} = 0 \tag{14}$$

Next, we take the total time derivative of $\partial S/\partial x_j$

$$\frac{d}{dt}\frac{\partial S}{\partial x_j} = \sum_i \frac{\partial^2 S}{\partial x_i \, \partial x_j} f_i + \frac{\partial^2 S}{\partial t \, \partial x_j} \tag{15}$$

Assuming that the order of taking partial derivatives can be interchanged, we substitute Eq. (15) into Eq. (14) to obtain

$$\frac{d}{dt}\frac{\partial S}{\partial x_j} = -\sum_i \frac{\partial S}{\partial x_i}\frac{\partial f_i}{\partial x_j} \tag{16}$$

or, defining

$$\gamma_i = \frac{\partial S}{\partial x_i} \tag{17}$$

Eq. (16) may be written

$$\dot{\gamma}_j = -\sum_i \gamma_i \frac{\partial f_i}{\partial x_j} \tag{18}$$

while Eq. (11) becomes

$$\min_u \sum_i \gamma_i f_i \tag{19}$$

If θ is not specified, it is readily established that S is independent of t and Eq. (10) is

$$\sum_i \gamma_i f_i(\mathbf{x},\bar{u}) = 0 \tag{20}$$

Equations (18) to (20) are, of course, the strong minimum principle for this system, and Eq. (17) establishes the Green's function γ as a sensitivity variable.

It is apparent from these results and those of Sec. 12.4 that if we so wished, we could derive the results of the preceding chapters from a dynamic programming point of view. We have not followed this course for several reasons. First, the variational approach which we have generally adopted is closer to the usual engineering experience of successive estimation of solutions. Second, computational procedures are more naturally and easily obtained within the same mathematical framework by variational methods. Finally (and though much emphasized in the literature, of lesser importance), the dynamic programming derivation of the minimum principle is not internally consistent, for it must be assumed in the derivation that the partial derivatives $\partial S/\partial x_i$ are continuous in $\mathbf{x}$ when in fact the solution then obtained for even the elementary minimum-time problem of Sec. 5.3 has derivatives which are discontinuous at the switching surface.

12.7 THE LINEAR REGULATOR PROBLEM

To complete this brief sketch of the dynamic programming approach to process optimization it is helpful to examine a direct solution to the continuous Hamilton-Jacobi-Bellman equation. This can be demonstrated simply for the linear regulator problem, a special case of the control problem studied in Sec. 8.2. The system is described by the linear equations

$$\dot{x}_i = \sum_{j=1}^{s} A_{ij}x_j + b_i u \qquad i = 1, 2, \ldots, s \tag{1}$$

with the quadratic objective

$$\mathcal{E} = \frac{1}{2} \int_0^\theta \left(\sum_{i,j=1}^{s} x_i C_{ij} x_j + u^2 \right) dt \tag{2}$$

or, defining

$$\dot{x}_{s+1} = \frac{1}{2} \sum_{i,j=1}^{s} x_i C_{ij} x_j + \frac{1}{2} u^2 \qquad x_{s+1}(0) = 0 \tag{3}$$

we have

$$\mathcal{E}[\mathbf{x}(\theta)] = x_{s+1}(\theta) \tag{4}$$

The control variable u is assumed to be unconstrained.

The Hamilton-Jacobi-Bellman partial differential equation is

$$0 = \min_u \left(\sum_{i=1}^{s+1} \frac{\partial S}{\partial x_i} f_i + \frac{\partial S}{\partial t} \right) \tag{5}$$

or, from Eqs. (1) and (3),

$$0 = \min_u \left(\sum_{i,j=1}^{s} \frac{\partial S}{\partial x_i} A_{ij}x_j + \sum_{i=1}^{s} \frac{\partial S}{\partial x_i} b_i u + \frac{1}{2} \frac{\partial S}{\partial x_{s+1}} \sum_{i,j=1}^{s} x_i C_{ij}x_j \right.$$
$$\left. + \frac{1}{2} \frac{\partial S}{\partial x_{s+1}} u^2 + \frac{\partial S}{\partial t} \right) \tag{6}$$

From the minimization we obtain

$$u = -\left(\frac{\partial S}{\partial x_{s+1}} \right)^{-1} \sum_{i=1}^{s} \frac{\partial S}{\partial x_i} b_i \tag{7}$$

so that, upon substitution, Eq. (6) becomes a classical but nonlinear partial differential equation

$$\sum_{i,j=1}^{s} \frac{\partial S}{\partial x_i} A_{ij}x_j - \frac{1}{2} \left(\frac{\partial S}{\partial x_{s+1}} \right)^{-1} \left(\sum_{i=1}^{s} \frac{\partial S}{\partial x_i} b_i \right) \left(\sum_{j=1}^{s} \frac{\partial S}{\partial x_j} b_j \right)$$
$$+ \frac{1}{2} \frac{\partial S}{\partial x_{s+1}} \sum_{i,j=1}^{s} x_i C_{ij}x_j + \frac{\partial S}{\partial t} = 0 \tag{8}$$

with boundary condition

$$S[\mathbf{x}(\theta),\theta] = \mathcal{E}[\mathbf{x}(\theta)] = x_{s+1}(\theta) \tag{9}$$

The function $S(\mathbf{x},t)$ satisfying Eqs. (8) and (9) will be of the form

$$S(\mathbf{x},t) = x_{s+1} + \frac{1}{2}\sum_{i,j=1}^{s} x_i M_{ij}(t)x_j \qquad M_{ij}(\theta) = 0 \tag{10}$$

in which case

$$\frac{\partial S}{\partial x_k} = \frac{1}{2}\left(\sum_{i=1}^{s} x_i M_{ik} + \sum_{j=1}^{s} M_{kj}x_j\right) \qquad k = 1, 2, \ldots, s \tag{11}$$

Equation (8) then becomes

$$\frac{1}{2}\sum_{i,j,k=1}^{s} x_i x_j (M_{ik}A_{kj} + A_{ki}M_{kj})$$

$$-\frac{1}{2}\sum_{i,j,k,l=1}^{s} x_i x_j \left[\frac{1}{2}(M_{ik} + M_{ki})b_k\right]\left[\frac{1}{2}b_l(M_{lj} + M_{jl})\right]$$

$$+\frac{1}{2}\sum_{i,j=1}^{s} x_i C_{ij}x_j + \frac{1}{2}\sum_{i,j} x_i \dot{M}_{ij}x_j = 0 \tag{12}$$

or, since it must hold for all $\mathbf{x}$,

$$\sum_{k}(M_{ik}A_{kj} + A_{ki}M_{kj}) - \sum_{k,l}\left[\frac{1}{2}(M_{ik} + M_{ki})b_k\right]\left[\frac{1}{2}b_l(M_{lj} + M_{jl})\right]$$

$$+ C_{ij} + \dot{M}_{ij} = 0 \qquad M_{ij}(\theta) = 0 \tag{13}$$

The solution to Eq. (13) is clearly symmetric, $M_{ij} = M_{ji}$, so that we may write it finally in the form of the Riccati equation (10) of Sec. 8.2

$$\dot{M}_{ij} + \sum_{k}(M_{ik}A_{kj} + M_{kj}A_{ki}) - \left(\sum_{k} M_{ik}b_k\right)\left(\sum_{l} b_l M_{lj}\right)$$

$$+ C_{ij} = 0 \qquad M_{ij}(\theta) = 0 \tag{14}$$

The optimal feedback control is then found directly from Eqs. (7) and (10) as

$$u = -\sum_{i,j=1}^{s} b_i M_{ij}x_j \tag{15}$$

Since this result was obtained in the form of a solution to the Hamilton-Jacobi-Bellman equation, we know that it is sufficient for a minimum, a result also established by other means in Sec. 6.20.

BIBLIOGRAPHICAL NOTES

We shall not attempt a survey of the extensive periodical literature on dynamic programming but content ourselves with citing several texts. Excellent introductions may be found in

R. Aris: "Discrete Dynamic Programming," Blaisdell Publishing Company, Waltham, Mass., 1964

R. E. Bellman and S. E. Dreyfus: "Applied Dynamic Programming," Princeton University Press, Princeton, N.J., 1962

G. Hadley: "Nonlinear and Dynamic Programming," Addison-Wesley Publishing Company, Inc., Reading, Mass., 1964

G. L. Nemhauser: "Introduction to Dynamic Programming," John Wiley & Sons, Inc., New York, 1966

Problems of the type considered in Sec. 12.6 are treated in

S. E. Dreyfus: "Dynamic Programming and the Calculus of Variations," Academic Press, Inc., New York, 1965

and numerous topics of fundamental interest are treated in the original book on the subject:

R. Bellman: "Dynamic Programming," Princeton University Press, Princeton, N.J., 1957

Extensive applications are examined in

R. Aris: "The Optimal Design of Chemical Reactors: A Study in Dynamic Programming," Academic Press, Inc., New York, 1961

R. Bellman: "Adaptive Control Processes: A Guided Tour," Princeton University Press, Princeton, N.J., 1961

S. M. Roberts: "Dynamic Programming in Chemical Engineering and Process Control," Academic Press, Inc., New York, 1964

D. F. Rudd and C. C. Watson: "Strategy of Process Engineering," John Wiley & Sons Inc., New York, 1968

J. Tou: "Optimum Design of Digital Control via Dynamic Programming," Academic Press, Inc., New York, 1963

PROBLEMS

12.1. Develop the Hamilton-Jacobi-Bellman equation for the system

$$\mathbf{x}^n = \mathbf{f}^n(\mathbf{x}^{n-1}, u^n)$$

$$\min \varepsilon = \sum_{n=1}^{N} \mathcal{R}_n(\mathbf{x}^{n-1}, u^n) + F(\mathbf{x}^N)$$

Apply this directly to the linear one-dimensional case

$$x^n = A(u^n)x^{n-1} + \beta[A(u^n) - 1]$$

$$\varepsilon = \alpha(x^N - x^0) + \sum_{n=1}^{N} M(u^n)$$

and establish that the optimal decision u^n is identical at each stage.

12.2. Formulate Prob. 7.2 for direct solution by dynamic programming. Draw a complete logical flow diagram and if a computer is available, solve and compare the effort to previous methods used. Suppose holding times $u_1{}^n$ are restricted to certain discrete values?

12.3. The system

$$\ddot{x} = u \qquad |u| \leq 1$$

is to be taken from initial conditions $x(0) = 1$, $\dot{x}(0) = 1$ to the origin to minimize

$$\varepsilon = \int_0^\theta \mathcal{F}(x,\dot{x},u)\, dt$$

Show how the optimum can be computed by direct application of the dynamic programming approach for discrete systems. (*Hint:* Write

$$\dot{x}_1 = x_2 \Rightarrow x_1(t - \Delta) = x_1(t) - x_2(t)\Delta$$
$$\dot{x}_2 = u \Rightarrow x_2(t - \Delta) = x_2(t) - u(t)\Delta$$

and obtain the Hamilton-Jacobi-Bellman difference equation for recursive calculation.) Assume that u can take on only nine evenly spaced values and obtain the solution for

(*a*) $\mathcal{F} = 1$
(*b*) $\mathcal{F} = \frac{1}{2}(x_2{}^2 + u^2)$
(*c*) $\mathcal{F} = \frac{1}{2}(x^2 + \dot{x}^2)$

Compare the value of the objective with the exact solution for continuously variable $u(t)$ obtained with the minimum principle.

12.4. Using the dynamic programming approach, derive the minimum principle for the distributed system in Sec. 11.4.

Indexes

Name Index

Subject Index

This book was set in Modern by The Maple Press Company, and printed on permanent paper and bound by The Maple Press Company. The designer was J. E. O'Connor; the drawings were done by J. & R. Technical Services, Inc. The editors were B. J. Clark and Maureen McMahon. William P. Weiss supervised the production.